BioMethods
Vol. 6

Series Editors:

Dr. H.P. Saluz
Hans-Knöll-Institut
für Naturstofforschung e.V.
Beutenbergstr. 11
D-07745 Jena

Dr. T. Meier
Department of Physiology
University of Colorado
Health Sciences Center
4200 East Ninth Avenue
Denver, CO 80262
USA

Methods in Comparative Psychoacoustics

edited by
G.M. Klump
R.J. Dooling
R.R. Fay
W.C. Stebbins

Birkhäuser Verlag
Basel · Boston · Berlin

Volume Editors

PD Dr. G.M. Klump
Institut für Zoologie
Technische Universität München
Lichtenbergstr. 4
D-85748 Garching

Prof. R.J. Dooling
Psychology Department
University of Maryland, College Park
Maryland 20742
USA

Prof. R.R. Fay
Department of Psychology
Parmly Hearing Institute
Loyola University of Chicago
Chicago, Illinois 60626
USA

Prof. W.C. Stebbins
Psychology Department
University of Michigan
Ann Arbor
Michigan 48109-1346
USA

A CIP catalogue record for this book is available from the Library of Congress, Washington D.C., USA

Deutsche Bibliothek Cataloging-in-Publication Data
Methods in comparative psychoacoustics / ed. by G. M. Klump ... –
Basel ; Boston ; Berlin : Birkhäuser, 1995
(BioMethods ; Vol. 6)

ISBN 978-3-0348-7465-6 ISBN 978-3-0348-7463-2 (eBook)
DOI 10.1007/978-3-0348-7463-2

NE: Klump, Georg [Hrsg.]; GT

The publisher and editor can give no guarantee for the information on drug dosage and administration contained in this publication. The respective user must check its accuracy by consulting other sources of reference in each individual case.

The use of registered names, trademarks etc. in this publication, even if not identified as such, does not imply that they are exempt from the relevant protective laws and regulations or free for general use.

Softcover reprint of the hardcover 1st edition 1995

Camera-ready copy prepared by the authors
Printed on acid-free paper produced from chlorine-free pulp

9 8 7 6 5 4 3 2 1

We dedicate this book to Bill Stebbins for his pioneering, rigorous, and enduring work in the field of Comparative Psychoacoustics.

Robert Dooling
Richard Fay
Georg Klump

Contents

List of Contributors

Cynthia M. Aleszcyk
Center for Hearing Sciences, Johns Hopkins School of Medicine,
720 Rutland Avenue, Baltimore, MD 21205-2195, U.S.A.

Charles H. Brown
Department of Psychology, University of South Alabama,
326 Life Sciences Classroom Building, Mobile, AL 36688, U.S.A.

Sheryl Coombs
Parmly Hearing Institute, Loyola University of Chicago,
Chicago,IL 60626, U.S.A.

Robert J. Dooling
Psychology Department, University of Maryland,
College Park, MD 20742, U.S.A.

Richard R. Fay
Department of Psychology, Loyola University of Chicago,
6525 North Sheridan Road, Chicago,IL 60626, U.S.A.

H. Carl Gerhardt
Division of Biological Sciences, University of Missouri,
215 Tucker Hall, Columbia, MO 65211, U.S.A.

Henry E. Heffner
Department of Psychology, University of Toledo,
Toledo, OH 43606, U.S.A.

Rickye S. Heffner
Department of Psychology, University of Toledo,
Toledo, OH 43606, U.S.A.

Robert D. Hienz
Center for Hearing Sciences, Johns Hopkins School of Medicine,
720 Rutland Avenue, Baltimore, MD 21205-2195, U.S.A.

Stewart H. Hulse
Department of Psychology, Johns Hopkins University,
139 Ames Hall, Charles and 34th Street, Baltimore, MD 21218, U.S.A.

Aileen Y. Huang
Center for Hearing Sciences, Johns Hopkins School of Medicine,
720 Rutland Avenue, Baltimore, MD 21205-2195, U.S.A.

Georg M. Klump
Institut für Zoologie, Technische Universität München ,
Lichtenbergstraße 4, D-85748 Garching, Germany

Ole Naesbye Larsen
Center for Sound Communication, Odense University,
Campusvej 55, DK-5230 Odense, Denmark

Bradford J. May
Center for Hearing Sciences, Johns Hopkins School of Medicine,
720 Rutland Avenue, Baltimore, MD 21205-2195, U.S.A.

David B. Moody
Kresge Hearing Research Institute, University of Michigan Medical School,
1301 East Ann Street, Ann Arbor, MI 48109-0506, U.S.A.

Cynthia F. Moss
Psychology Department, Harvard University,
William James Hall, 33 Kirkland Street, Cambridge, MA 02138, U.S.A.

Andrew J. Niemiec
Kresge Hearing Research Institute, University of Michigan Medical School,
1301 East Ann Street, Ann Arbor, MI 48109-0506, U.S.A.

Kazuo Okanoya
Cognitive & Information Science, Faculty of Letters, Chiba University,
1-33 Yayoi-cho, Inage-ku,, Chiba, Chiba-city, 263 Japan

Merrilyn J. Penner
Department of Psychology, University of Maryland,
College Park, MD 20742, U.S.A.

Sabine Schmidt
Zoologisches Institut, Universität München,
Luisenstraße 14, D-80333 München, Germany

Andrea Megela Simmons
Department of Psychology and Neuroscience, Brown University ,
Box 1853, Providence, RI 02912, U.S.A.

Joan M. Sinnott
Department of Psychology, University of South Alabama,
University Boulevard, Mobile, AL 36688, U.S.A.

William C. Stebbins
Kresge Hearing Research Institute, University of Michigan Medical School,
1301 East Ann Street, Ann Arbor, MI 48109-0506, U.S.A.

Tim Tucker
Tucker-Davis Technologies Inc.,
4550 N.W. 6th Street, Gainesville, FL 32609, U.S.A.

Steve Varosi
Tucker-Davis Technologies Inc.,
4550 N.W. 6th Street, Gainesville, FL 32609, U.S.A.

Hermann Wagner
Institut für Zoologie, Technische Universität München,
Lichtenbergstraße 4, D-85748 Garching, Germany

Lynne A. Werner
Department of Speech & Hearing Science, University of Washington,
CDMRC, WJ-10, Box 47, Seattle, WA 98195, U.S.A.

Hong Young Yan
School of Biological Sciences, University of Kentucky,
Lexington, KY 40506-0225, U.S.A.

Introduction

Twenty five years ago, Bill Stebbins presented the principles of animal psychophysics in an edited volume (Stebbins, 1970) describing an array of modern, creative methodologies for investigating the range of sensory systems in a variety of vertebrate species. These principles included precise stimulus control, a well defined behavioral response, and a rigorous behavioral procedure appropriate to the organism under study. As a generation of comparative sensory scientists applied these principles, our knowledge of sensory and perceptual function in a wide range of animal species has grown dramatically, especially in the field of hearing.

Comparative psychoacoustics, i.e., the study of the hearing capabilities in animals using behavioral methods, is an area of animal psychophysics that has seen remarkable advances in methodology over the past 25 years. Acoustic stimuli are now routinely generated using digital methods providing the researcher with unprecedented possibilities for stimulus control and experimental design. The strategies and paradigms for data collection and analysis are becoming more refined as well, again due in large part to the widespread use of computers. In this volume, the reader will find a modern array of strategies designed to measure detection and discrimination of both simple and complex acoustic stimuli as well experimental designs to assess how organisms perceive, identify and classify acoustic stimuli. Refinements in modern methodologies now make it possible to compare diverse species tested under similar, if not identical, experimental conditions. Tests on nonverbal human infants described in this volume fit well in this category and have much in common with methodologies used in testing animals.

Comparative psychoacoustics is of critical importance for the field of hearing research because behavioral studies provide the most important reference for anatomical, physiological, and biochemical investigations of auditory function. In addition to the development of animal models of human auditory processes, comparative hearing research has led to significant advances in our understanding of the relation between auditory structures and function, the evolution of hearing, adaptations and specializations in the auditory system, and the role of hearing in complex acoustic communication systems.

The idea for this book, with its focus on methodology, was conceived by one of us (Bob Dooling) during his sabbatical as a fellow of the Humboldt-foundation at the Technical University of Munich. Discussions with numerous colleagues in comparative psychoacoustics reinforced the importance of emphasizing the varieties of methodologies, stimuli, experimental designs, and species studied in comparative psychoacoustics.

The contributors to this volume were selected in part to represent the range of both methodologies and species used in modern comparative psychoacoustics. Because of space limitations, many current applications of these methods to a variety of other species could not be included. It should be clear that these methods are being used successfully by many investigators and creatively modified to fit the requirements of particular animal species. It is hoped that this volume will serve as reference source for graduate students and scientists interested in the comparative aspects of hearing.

October 1994

Robert J. Dooling
Richard R. Fay
Georg M. Klump
William C. Stebbins

Technical Aspects of Animal Psychoacoustics

Introductory Remarks

The advances in the methodology of comparative psychoacoustics over the past 25 years are due, primarily, to the sophistication, speed, and accessibility of laboratory microcomputers. Microcomputers are now routinely used for both stimulus generation and response measurement in all areas of psychoacoustic research. To provide for a better understanding of these techniques, application of computer methods for synthesizing both simple and complex acoustic stimuli and for presenting these stimuli in auditory tests are discussed in the first chapter by Tucker and Varosi. Their contribution describes a major advance in the field of psychoacoustics in general and animal psychoacoustics in particular - the application of digital signal processing methods. Precise stimulus calibration is another topic of immense importance in conducting psychoacoustic experiments. The challenges in achieving this goal are even greater in animal psychoacoustics than in studies in humans, since most experiments are done in the free field rather than with headphones, and the range of stimulus values may fall outside the range of human hearing sensitivity and discriminability. The chapter by Larsen emphasizes the importance of good stimulus calibration and control and demonstrates what should be good practice. Finally, the chapter by Penner describes the advantages and disadvantages of different human psychophysical methods applied to nonhumans. The rigor and precision of these methods serves as a guideline as comparative psychoacousticians must necessarily tailor psychoacoustic paradigms to the specific behavioral adaptations of different animal species.

Methods in Comparative Psychoacoustics
G. M. Klump, R. J. Dooling, R. R. Fay & W. C. Stebbins (eds)

Signal Processing Technology in Animal Psychoacoustics

Tim Tucker and Steve Varosi

Tucker-Davis Technologies Inc., 4550 N.W. 6th Street, Gainesville, FL 32609, USA

Summary. Digital computer systems have revolutionized the field of animal psychoacoustics. Even basic computer systems have the power to perform experiments that were difficult or impossible only 30 years ago. These new abilities have had a profound effect on the field by allowing researchers to generate virtually any stimuli for which they can write a mathematical expression. This chapter discusses the essential theory and practical knowledge needed to generate stimuli using modern Digital Signal Processing hardware.

1 Introduction

This chapter covers the key points involved in using a digital computer to generate and acquire *analog* (real world) signals. Signal generation is the process of going from a discrete-time digital signal in computer memory to a continuous-time, analog signal. Signal acquisition is just the reverse process of going from an analog to a digital signal in memory. Digital systems have revolutionized the types of signals that can be generated and the ways in which they can be presented. Virtually any arbitrary waveform can be generated and controlled precisely, by even a basic computer hardware system. This fact has had a profound effect on the field of psychoacoustics, by enabling any researcher to perform complex experiments with very sophisticated stimuli.

Also addressed is the fundamental method of signal generation and frequency analysis of discrete-time signals: The Discrete Fourier Transform (DFT), which is almost always implemented using the well-known FFT algorithm.

A brief history of some technological milestones pertaining to auditory research is presented first, along with an overview of various modern hardware configurations.

2 History of Auditory Research Technology

Over the past 40 years many technological advances, particularly in digital computer hardware, have profoundly impacted the field of auditory research. The development of the vacuum tube, followed by the transistor, made it possible to produce and control pure tones and noise, the basic experimental stimuli. The first source of high quality tones was made available in 1938: The Hewlett Packard 200-A audio oscillator. This and other types of "analog" equipment were of great aid to auditory science, albeit extremely limited by today's standards. Most notably, it was difficult to produce arbitrary, complex stimuli using these analog instruments.

The advent of the digital computer revolutionized the types and methods of experimental research. The mini-computer was introduced in the 1960's, followed by micro-computers in the 70's and 80's (e.g., PC's). High-speed Digital Signal Processors (DSP's) were introduced in the late 1980's and are now commonly used.

2.1 Mini-computers

The first digital computers appeared on universities in the late 1950's but were mainly data analysis tools. Then in 1963 the mini-computer (namely Digital Equipment Corporation's PDP-8) became the principle research tool used by many hearing-science laboratories. The computer, coupled with the Digital-to-Analog (D/A) converter, became the best method for stimulus generation. The main reason for this, was the ease with which arbitrary, complex wave forms could be produced with great precision and stability in frequency and amplitude. Also the need for constant calibration and adjustment of multiple analog devices was reduced or eliminated. Even today these are still the basic reasons why computers are indispensable in auditory science.

2.2 Micro-computers

The micro-computer, now commonly referred to as the personal computer or PC, was a cheaper, smaller version of the mini-computer. These compact machines quickly became faster than their larger predecessors and rendered them obsolete in the 1980's. Because they were cheaper, it was possible to assign one computer to one task, thus simplifying programming and timing complexities associated with multi-tasking systems. Originally, micro's had slow processors and ran inefficient interpreted programming languages (such as BASIC). Later, the efficiency of programs produced with high-level language compilers made them a more popular alternative. These compiled languages, although more cryptic produced faster, optimized programs.

Although modern PC's have seemingly huge amounts of memory and disk capacity, early PC's had about the same or less storage space as the mini's, therfore production of long, aperiodic signals was still a problem. Also processor and memory speeds limited the rate at which samples could be sent to the D/A, thus limiting stimulus bandwidths. However, even with these limitations, micro's revolutionize the field nearly as much as the introduction of the computer itself, by making it possible for nearly every researcher to afford a good computer-based auditory workstation. The IBM PC's open architecture created numerous second-source suppliers of excellent yet inexpensive hardware and software. The resulting competition drove costs down and performance up at an accelerated pace.

2.3 Digital Signal Processors (DSP's)

The fastest computation power available in integrated circuit form is from DSPs. They are micro-processors optimized for fast numerical calculations on large arrays in memory (e.g. signals or images). Their level of optimization makes DSP's unsuitable for general purpose computing, and early versions were especially difficult to use and program. Highly cryptic machine level coding was required to produce fast optimized programs. More recent advances in specialized software development environments have made the Digital Signal Processor indispensable to those who require real-time or 'on-the-fly' signal generation and/or analysis. Real-time DSP refers to processing signals point-by-point and is generally requireded for long duration or continuous waveforms. On-the-fly is used to describe the generation and/or analysis of descrete length digital waveforms on an 'as-needed' bases. Both of these methods are a departure from the previously required tasks of pre- or post- processing 'off-line.' The speed of the DSP and faster more dense memory devices have allowed researchers, needing ever-higher frequency capability, to benefit from the advantages of computer-generated signals.

3 Modern Hardware Configurations for Psychoacoustics

Choosing the proper hardware setup is critical--it must match the type of work being done, the budget allowed, and the abilities of the operator. For example, if the apparatus is to be used by someone with limited programming experience it would make little sense for them to purchase a hardware system containing a DSP that requires 'low-level' direct device programming. On the other hand, a user-friendly 'turn key' system may not have the speed or may be too inflexible for the tasks at hand.

A typical modern psychoacoustic setup consists of a personal computer interfaced to some combination of Digital to Analog (D/A), Analog to Digital (A/D) and Digital Signal Processing (DSP) hardware. The analog outputs and inputs of the D/A -- A/D hardware are then connected to some number of *analog* signal processing modules. These analog modules simplify overall system operation as well as improve the quality of the generated signal. A typical psychoacoustic setup is shown in Figure 1.

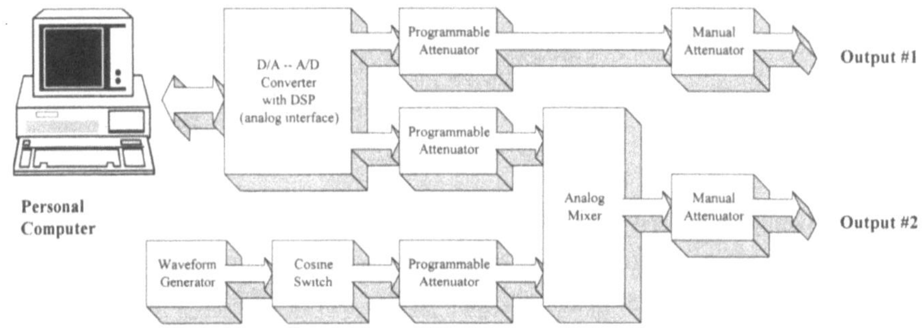

Figure 1. Typical psychoacoustic setup.

The hardware system shown utilizes a two channel D/A capable of generating two independent, but synchronized, wave forms. These signals are routed through programmable and manual attenuators before being transduced into sound via a speaker, headphone or other means. An optional third signal source comes from a 'stand-alone' wave form generator. This signal is piped through a cosine switch (for gating it on and off) and programmable attenuator before being mixed with the computer generated signal as shown. A full treatment of the merits and shortcomings of this hardware setup would go beyond the scope of this chapter. However, as one becomes familiar with the typical signals used in psychoacoustic experimentation, the utility and flexibility of this common configuration will become more clear.

At the heart of any psychoacoustic setup is the personal computer equipped with an 'analog interface'. It is this portion of the overall setup that is responsible for generating signals, recording and storing results, and controlling the overall experimental protocol. Various incarnations of analog interface hardware are available, and posses subtle but important feature differences. These features govern the overall system's power and ease of use.

3.1 Personal Computer with simple D/A - A/D interface

This is a low cost, easy to use solution because the PC's memory is used for signal storage. Typically only one inexpensive D/A - A/D card without memory is required, which plugs easily into the PC's expansion bus. All programming can be done directly on the PC from a high-level language like 'C'. Signals are transferred to and from PC memory via its DMA (Direct Memory Access) controller, which limits the sampling rate for D/A or A/D conversion to audio bandwidths (< 20 kHz). Also the PC microprocessor must do all computations, and may still be too slow to generate stimuli in real-time, or 'on-the-fly.' However, this setup is ideal for low-complexity signals, and speech recording and playback.

Upgrading to a D/A - A/D card with dedicated memory is more costly, but allows signals to be stored out of PC memory which has many advantages. This type of card also has fast, dedicated hardware to sequence the samples from its memory to the D/A and from the A/D into memory, removing the speed limitation of the PC's DMA mentioned previously. The amount of memory on the card still determines the maximum signal duration for aperiodic signals, however, with memory becoming less expensive this is typically not a limiting factor. The PC must still generate and analyze all signal data, but it is free to perform other tasks (like computing the next stimulus) while signals are being played from the D/A or recorded by the A/D. The added speed potential of this configuration makes it attractive to high-frequency researchers working with rodents, bats, and other animals which can often hear well above 50kHz.

3.2 Personal Computer, D/A - A/D and DSP

Integrating a DSP, often directly on the D/A - A/D card, is a very popular alternative to the 'off-line' signal generation approach discussed above. Possibly the greatest advantage of including a DSP is that it allows for 'in-the-signal-path' processing algorithms like digital filtering, and real-time signal generation, e.g. continuous, non-repetitive noise. Very long, arbitrary wave forms can be generated on a point-by-point basis. The PC serves as 'host' merely telling the DSP what to do, and is free to perform other tasks while signal generation and processing is in progress. Depending on the speed of the DSP and the numerical complexity of the signal generation/processing algorithm, audio to ultrasonic bandwidths are attainable.

DSP-based cards remain somewhat difficult to program, and code development time is quite long especially if the user is not familiar with DSP software architecture. There are, however, pre-packaged software routines available for certain general purpose tasks such as continuous digital filtering and noise generation, which may suffice for non-custom applications. This configuration has gained prevalence in hearing aid algorithm research where fast and adaptive signal processing algorithms must run in 'real-time'.

3.3 Personal Computer, D/A - A/D - DSP with Memory

One of the most common DSP-based configurations for auditory research combines the DSP with substantial, dedicated signal memory. This takes advantage of the DSP's ability to process large signal arrays in memory very quickly. As a result, long wave forms can be generated or analyzed very quickly, allowing for 'on-the-fly' stimulus generation and/or processing. This configuration has become the most commonly used for psychoacoustic research where complex discrete signals are generated 'on-the-fly'. When continuous signals are required (i.e. noise maskers) they are often generated on dedicated stand-along devices such as noise generators and programmable filters. These signals are often gated off and on using a cosine gating device and mixed with the computer generated discrete signal using an analog summer (see Figure 1).

4 Fundamentals of Discrete-time Digital Signals

Although mastery of digital signal processing theory is not required to operate modern Digital Signal Processing (DSP) hardware, users should be acquainted with reconstruction and sampling of analog signals, and digital signal representation in computers. A brief review of some basic concepts associated with moving between the analog and digital signal domains are presented in this section. For a more in-depth treatment of these concepts and related signal processing topics, consult an introductory text on DSP (see suggested readings).

4.1 Signal Generation by Computer

The computer generates stimuli directly from mathematical expressions which are numerically evaluated at a discrete set of points (called samples) corresponding to equally spaced intervals of time (the sampling period). For example, suppose we have a mathematical function which defines a desired stimulus. To produce the stimulus for a 10 ms duration using 200 computed samples, we would define that a sample occurs every 10/200 = 0.05ms. This set of samples would then be sent sequentially to the D/A converter, one every 0.05ms, to produce the *analog* stimulus. The purpose of various signal generation parameters and the criteria governing their proper selection are presented in the following paragraphs.

4.2 Sample Period and Signal Reconstruction

As stated above, an analog signal can be reconstructed from discrete time samples stored in a computer's memory. These could be digitally recorded samples of an analog signal, or numerically computed values using some mathematical formula or algorithm. To produce a signal, digital numbers are converted to analog voltage levels by a D/A converter at discrete time intervals. The analog level of each sample is held constant for the sampling period until the next sample, resulting in a staircase analog output as shown below:

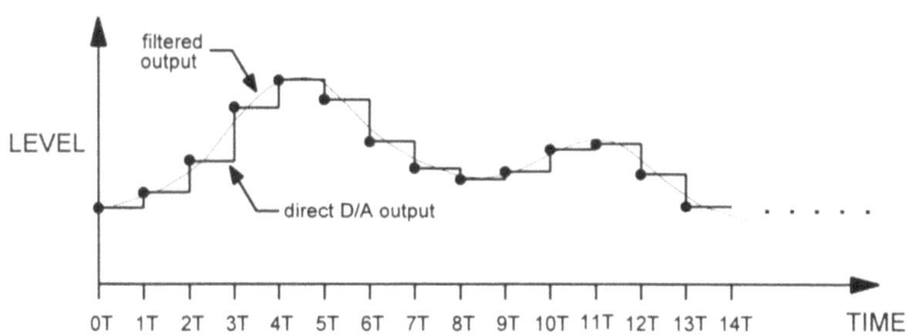

Figure 2. Typical staircase output generated by D/A converter.

Here, the 14 samples (0-13) of a signal are used to *reconstruct* the analog signal for 14T seconds. A low-pass anti-imaging filter on the D/A output is used to "smooth" out the staircase into a continuous wave form, shown super-imposed (note the slight delay which results from filtering).

The time interval between samples is T called the sampling period. Similarly the sampling rate or sampling frequency (number of samples per second), is simply the reciprocal of the sampling period: sampling frequency = $f_s = 1/T$. The total number of samples is related to the duration of signal generation in the following manner:

$$\text{signal duration} = \text{number of samples} \cdot \text{sampling period} = \frac{\text{number of samples}}{\text{sample frequency}}$$

In theory, to faithfully reconstruct (generate) a signal whose maximum frequency is bandlimited to some maximum frequency f_{max}, the sampling frequency must be at least twice the maximum frequency: $f_s \geq 2 f_{max}$. This is known as the Shannon-Nyquist sampling theory. The minimum sampling frequency is often called the Nyquist frequency. In addition, the low-pass (anti-imaging) filter must cutoff all frequencies above f_{max}.

Although in theory the sampling frequency need only be twice the maximum frequency, in practice the sampling rate should be 2.5 to 3 times f_{max} with the low-pass filter cutoff frequency at 0.4 to 0.5 times f_s. This added spectral spacing will allow for practical limitations in filter roll-off and D/A hardware. The low-pass filter removes high-frequency "images" of the signal spectrum caused by the staircase jumps in reconstruction. In audio applications, these images might be perceptible by the human ear. However, if the sampling frequency is high enough, the audio equipment or the human ear may filter the output sufficiently. Consider the reconstruction of a three-component tone wave form. The spectrum of the direct D/A output will include a spectral image as shown below.

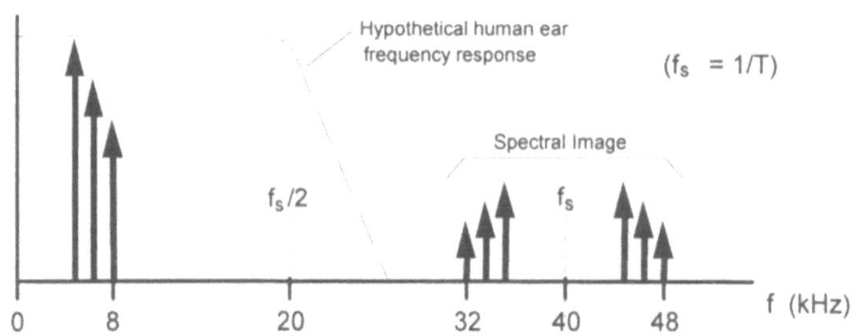

Figure 3. Spectrum of unfiltered D/A output.

In this case, the chosen sampling frequency was high enough to place the lowest frequency component of the imaged spectrum (32kHz) well above the cut-off frequency of the human ear, therefore, filtering is not required.

4.3 Signal Sampling - Acquisition

The process of signal acquisition (sampling) using an A/D converter is very similar to the D/A signal generation process. The terms sampling period and sampling frequency hold the same meaning for signal sampling as they did for signal generation. In the analog domain (i.e., the physical world) time is continuous. In all digital data acquisition systems, samples of an analog signal are taken at discrete time intervals. The level of each analog sample is converted to a corresponding digital number (A/D conversion), and stored in a memory location. The total number of samples is related to the duration of signal acquisition just as with signal generation. Figure 4 illustrates the sampling process and the associated time parameters.

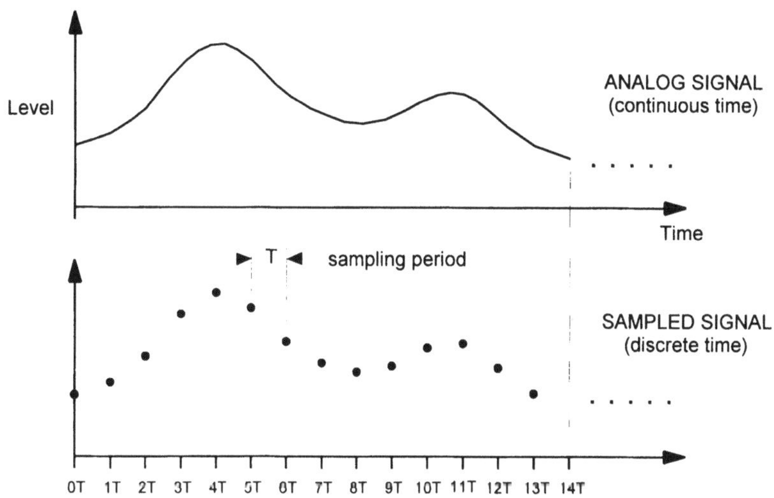

Figure 4. Typical analog signal and discrete time equivalent.

The analog signal is sampled every T seconds for 14 samples (0-13) representing a total of $14T$ seconds of the analog signal. As the time between samples (the sampling period) decreases, the samples form a more accurate representation of the original signal (e.g., if you were to plot the samples on a computer screen). Phrased another way, because the samples are separated in time, some information about the signal between samples is lost in the sampling process. The amount of information lost depends on how much the signal changes during the time between samples, which is directly related to the signal's frequency content. In theory, if a signal's frequency spectrum is bandlimited to some maximum frequency f_{max}, an exact representation of the signal can be acquired if the sampling frequency is at least twice the maximum frequency: $f_s \geq 2 f_{max}$. This is the other part of the Shannon-Nyquist sampling theory from Section 4.2.

Intuitively, sampling a signal faster will retain more information content at higher frequencies. In fact, a sampling frequency faster than the theoretical minimum is often needed to faithfully capture transient signal behavior encountered in practice. However, as the sampling frequency is increased, more samples are needed for the same total signal duration and memory requirements increase accordingly. Also, the sampling frequency is limited by the speed of the A/D converter hardware and/or the system's maximum data throughput capability.

Failure to adequately bandlimited an analog signal for the chosen sampling frequency, i.e., $f_{max} > f_s /2$, will result in a highly undesirable phenomenon known as *aliasing*. Suppose one wished to digitize speech waveforms (without an anti-aliasing filter) from a typical microphone having a published 3dB bandwidth extending to 10kHz. Cosidering the rules associated with sampling discussed above one might choose a 30kHz sampling rate. Figure 5 illustrates the potential problem associated with digitizing speech in this way.

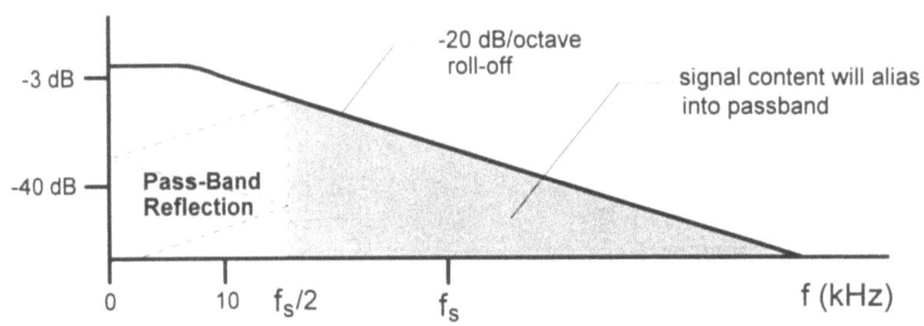

Figure 5. Aliased broad-band spectrum.

The frequency response of the microphone begins to roll off gradually at 10kHz. Although most audible information is below about 10kHz, high-frequency room noise can extend the signal spectrum appreciably. With f_s = 30kHz, the shaded area of the spectrum is "folded over" about 15kHz and added to the spectrum of the *sampled* signal as indicated by the pass-band reflection area in Figure 5.

To minimize the effects of aliasing and/or reduce the required sampling frequency, a low-pass anti-aliasing filter is used to limit the spectral content of the *analog* signal before sampling and A/D conversion. It should have low magnitude and phase distortion in its passband and a very steep roll-off characteristic. The spectrum of Figure 6 shows the result of filtering the broad-band microphone spectrum with a typical anti-aliasing filter.

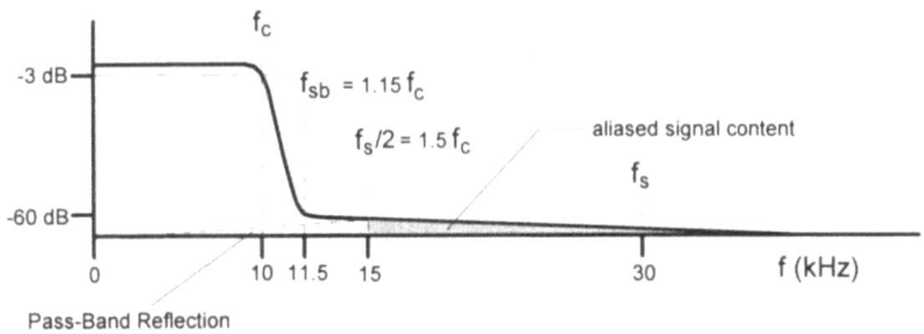

Figure 6. Broad-band spectrum with anti-aliasing filter.

The frequency f_c at which the filter begins to limit the spectrum is the corner frequency. A typical 9-pole elliptical filter has a stop-band frequency f_{sb} of $1.15 f_c$, beyond which a signal attenuation of 60dB is guaranteed. The sampling frequency should be at least twice the stop-band frequency: $f_c \geq 2 f_{sb}$. Comparing the filtered and non-filtered spectral plots, it is evident that filtering reduces the amount of aliasing which corrupts the discrete-time waveform. Incidentally, there is no difference between anti-aliasing or anti-imaging filters; the terms refer to the filter's function in the system.

4.4 Digital Signal Representation

The preceding discussions of signal reconstruction and sampling make reference to D/A and A/D conversion and digital numbers representing analog signal levels. A D/A (Digital-to-Analog) converter produces an analog signal level corresponding to a digital binary number while an A/D (Analog-to-Digital) converter produces a digital binary number corresponding to an analog signal level. A major concern in moving between digital and analog signals is that an analog signal is continuous in level as well as time, but a digital signal can have only a finite number of levels/values. This means that an analog signal level can only be represented digitally with a finite precision depending on the number of bits in the binary sample value (each bit being 0 or 1). For example, a three-bit binary number can represent $2^3 = 8$ different values: 000, 001, 010, 011, 100, 101, 110, 111. Similarly, 8 bits can represent $2^8 = 256$ different values, and 16 bits can have $2^{16} = 65536$ different values. In user software programs, these values are usually interpreted as signed integers (e.g., from -32768 to +32767 for 16-bits) when dealing with D/A and A/D processes. Signals represented by finite precision binary numbers are called digital signals. The number of bits used to represent an analog level will depend on the particular D/A or A/D device employed and is called the conversion resolution.

Typically, 32-bit floating point values (essentially infinite precision for most purposes) are used for mathematical computation of signals, which are then rounded to finite precision integers before D/A conversion. Likewise, when a signal sample is acquired by an A/D converter it's level is rounded to an integer digital value. The rounding effect is known as quantization error, and is clearly less of a concern as the number of bits in the conversion increases. For most D/A and A/D devices, "analog level" refers to signal and output and input voltages. Most modern devices are designed for voltage signals in the range of ±5 or ±10 Volts, and thus amplifiers or attenuators are needed to handle very small or very large signals. In general, for a converter with total voltage swing V and number of bits n:

$$\text{analog voltage} = \frac{V \cdot \text{integer value}}{2^n} \quad (\text{for D / A})$$

$$\text{integer value} = \text{int}\left[\frac{2^n \cdot \text{analog voltage}}{V}\right] \quad (\text{for A / D})$$

From the first expression, it can be seen that the difference between adjacent analog voltage levels (the quantization level) is $q = V/2^n$. The dynamic performance of a digital signal processing system is directly related to the size of q. The theoretical maximum dynamic range (in dB) of an n-bit A/D or D/A converter is given by:

$$20\log_{10}\left[\frac{\text{overall voltage swing}}{\text{quantization step level}} = \frac{V}{q} = 2^n\right] = 6.02n$$

The error introduced by the quantization process can be modeled as additive noise. The model assumes the error between the actual or intended signal level and the quantized (rounded) signal level is treated as a random variable (noise) between $\pm\ q/2$. Through derivations ommited here, it can be shown that the RMS voltage of this equivalent noise is $q/\sqrt{12}$. From a spectral analysis viewpoint, this signal shows up as broad-band quantization noise from 0 to $f_s/2$ with total energy down 96dB relative to the total spectral energy. On a per-Hz basis, the noise spectrum is down below $96+10\log_{10}(f_s/2)$.

For example, consider the reconstruction of a 1000Hz tone with a 16-bit D/A converter sampling at 40kHz. The sinusoid is computed with floating point precision, multiplied by 32767 for full scale D/A output (maximum dynamic range), and then quantized by rounding to the nearest integer. The difference between the *intended* sinusoidal output and the *true* D/A output will be seen as quantization noise down $96 + 10\log_{10}(20000) = 139$dB across the spectrum. In reality, this theoretical maximum dynamic range is never achieved because of distortions and non-linearities in the D/A converter. Therefore, a common specification for D/A converters is THD (Total Harmonic Distortion) + noise, which includes effects of both theoretical quantization error and physical limitations. A 16-bit D/A converter will have a broadband THD + noise of about 80 to 90 dB. This means that if a signal is being generated at 0dB and utilizes the entire voltage range of the D/A, the *broadband* noise and distortion (measured with an RMS voltmeter for instance) will be -80 to -90 dB. A typical D/A spectral output generating a sinusoid of frequency f_0 is shown in Figure 7. The noise level at any particular frequency (excluding peaks) is about -127dB, but the broadband noise sum over all frequencies from 0 to $f_s/2$ is about $-127 + 10\log_{10}(20000) = -84$dB.

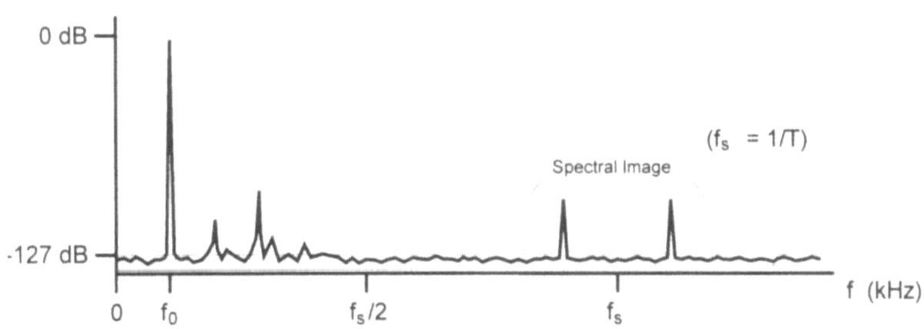

Figure 7. Typical D/A output showing distortions, images, and noise floor.

To fully utilize the conversion resolution (dynamic range) of a D/A converter the digital signal should be scaled (in software) so that its maximum and minimum integer values are as close as possible to +32767 and -32768, which will resulting in an analog output having the maximum voltage swing. If the output signal level needs to be reduced, it should be done with an analog attenuator to maintain maximum signal resolution (dynamic range). The level can be adjusted by scaling the digital signal, but as the digital signal is lowered, analog voltage resolution is lost. For example, suppose a digital signal in memory is scaled by 1/4 to reduce the analog output level by 1/4. Four quantization levels are then lost due to rounding: e.g., 998/4, 999/4, 1000/4, and 1001/4 will all be rounded to 250. The conversion resolution has been effectively reduced by two bits, resulting in a drop of 6.02·2 = 12dB in dynamic range. However, scaling in software is convenient, and because there are 65536 levels to begin with, the loss in resolution may not be critical in many instances.

To fully utilize the conversion resolution of an A/D converter, the analog input signal's minimum and maximum levels should be amplified to the specified input voltage the A/D being used. If the signal level goes outside the allow input voltage, a highly undesirable form of distortion known as <u>clipping</u> will result. Conversely, if the signal level is confined to a small voltage range, the digital signal will have unacceptable resolution. For example, the output level of a microphone can be as low as ±1.2mV and if it were connected directly to the input of an A/D module, the resulting digital signal's integer values would be between only ±4 (~18dB of dynamic range) which is an extremely coarse representation of the original signal. The subtleties of the analog signal will have been rounded off and lost.

5 Fourier Transforms of Digital Signals

It has been assumed that the reader is familiar with the idea of the frequency spectrum of an analog signal. Conceptually frequency is still a continuous variable for discrete-time signals, however, from a practical standpoint frequency-domain analysis is usually applied to a finite number of signal samples using a discrete algorithm which returns spectral information at a finite number of frequencies. One such algorithm is the DFT (Discrete Fourier Transform) which transforms a discrete-time signal into a discrete frequency spectrum revealing certain characteristics not evident in the original signal. The *inverse* DFT (IDFT) transforms a discrete frequency spectrum into a discrete-time signal and is used for generating an arbitrary signal from its spectral specifications. The DFT and IDFT are algorithmically identical. Before we discuss generating signals by this method, some brief theory on the DFT will be presented.

5.1 DFT Basics

Consider a digital signal of N *real* values, indexed from 0 to $N-1$. The DFT of this signal will consist of N *complex* values also indexed from 0 to $N-1$, yet, only the first 0 to $N/2$ values are necessary (the values from $N/2+1$ to $N-1$ are redundant conjugates). Now suppose that the digital signal values are samples of an analog signal taken with a sampling period of T seconds so that each signal value index k corresponds to an analog time of:

$$t = kT, \quad k = 0, 1, 2, \dots, N-1.$$

Then the index n of each DFT value corresponds to an analog frequency component of :

$$f_n = \frac{n}{NT}, \quad n = 0, 1, 2, \dots, \frac{N}{2}$$

which are just multiples or harmonics of the fundamental frequency $1/NT$. A sinusoid at this frequency will complete one full cycle in exactly NT seconds. The DFT value at n=0 corresponds to the zero frequency which is just the average value of the signal (DC component). In this way, the DFT can provide information about an analog signal's spectral content at a discrete set of frequencies separated by:

$$\Delta f = \frac{1}{NT} = \text{binwidth}$$

This frequency resolution is known as the frequency binwidth (which is equal to the fundamental frequency). The DFT array index or bin number for a particular frequency component f is simply:

$$n = \text{round}\left[f \cdot N \cdot T = \frac{f}{\text{binwidth}} \right] = \text{bin number}$$

which is just the index n of the DFT harmonic f_n nearest to f in frequency (because of rounding). The binwidth therefore determines the resolution of frequency placement and analysis. This is one of the primary limitations in using the IDFT to generate stimuli. Figure 8 shows a sampled signal and its discrete frequency spectrum resulting from a DFT. The DFT values are complex numbers so only their magnitudes are shown. The IDFT simply transforms a discrete frequency spectrum back into a sampled signal which will have the frequency and phase components specified in each frequency bin.

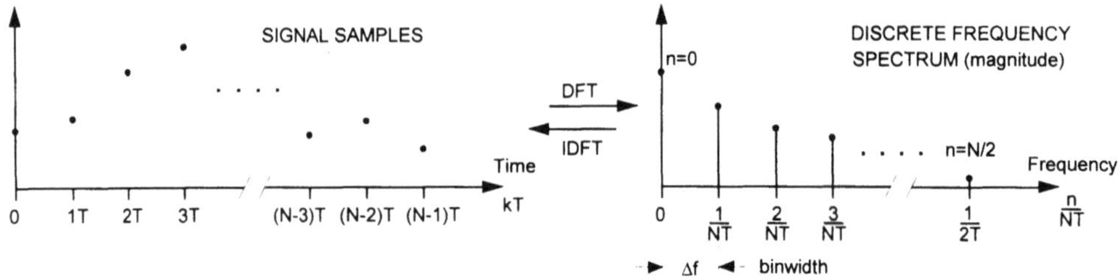

Figure 8. Parameters associated with time and frequency representations of a typical discrete signal.

As the total number of samples N increases, the binwidth Δf becomes smaller and the frequency resolution *increases* over the same total frequency range. As T becomes shorter (f_s increases), the binwidth becomes larger; the frequency resolution *decreases*, but the frequency range, determined by the highest harmonic $f_{N/2} = 1/2T$, increases. It is important to note that the frequency *range* is independent of N. With these factors in mind, N and T can be adjusted for the requirements of specific applications. The Fast Fourier Transform (FFT) is almost always used for computation of the DFT and IDFT and thus N must be an integer power of 2 (Radix-2) which imposes some constraints. Typically T is varied to set a binwidth such that the DFT frequency bins will be as close as possible to specific frequencies of interest. Then, if more frequency resolution is desired, N can be increased (by factors of two) in effect subdividing the existing frequency bins.

When using the DFT or IDFT, the signal's frequency components at and near f_{max} should be very close to zero in magnitude compared to the rest of the spectrum. If this is not the case, then aliasing and/or unwanted distortion will occur.

5.2 Computational Considerations and the FFT

Most DFT and IDFT software algorithms require and return complex values in *rectangular* format. For interpretation and plotting, these complex values are most often represented in *polar* form where each frequency bin has an associated magnitude and phase component. The polar to rectangular transformation is trivial and usually included as an option with FFT functions. Polar form is also more convenient for specifying spectral components for an IDFT since the magnitude and phase are just the amplitude and phase of the sinusoid at the particular frequency.

For computational efficiency, the DFT/IDFT is almost always computed using the FFT/IFFT. For a signal of N samples, the DFT requires on the order of N^2 multiplications, while the FFT requires only $N\log_2(N)$ multiplications, drastically reducing computation time. One constraint is that the number of points on which the FFT operates must be radix-2 (an integer power of 2, e.g., $2^{10} = 1024$, $2^{11} = 2048$, etc.). If the length N of a digital signal is not radix-2, it must either be truncated or "padded" with zeros to the nearest radix-2 number. This constraint is unfortunate and causes some adverse effects (Section 5.3), but note that for $N = 2^{13} = 8192$ an ordinary DFT would require 8192/13 = 630 times more multiplications than the FFT!

5.3 Signal Generation Using the IFFT

The IFFT simply transforms a discrete frequency spectrum, specifying magnitudes and phases at desired frequencies, back into time domain signal samples. The resulting signal is *exactly* what would have been obtained by computing individual sinusoids of the desired frequency, magnitudes and phases, and adding them together. If only a few different frequencies are needed, computing sinusoids may be slightly faster, however the IFFT method takes far less time as the number of different frequencies increases.

A disadvantage of the IFFT method is limited frequency placement resolution described by Δf in Section 5.1. This discrete resolution makes the IFFT inappropriate for generating signals requiring very exact frequency control. For example, a reasonable size IFFT (8192 points) and typical sampling period (25.0 μs) will yield a frequency resolution of: $1/(8192 \cdot 25 \cdot 10^{-6}) \approx 4.9$ Hz. This frequency placement resolution would be inadequate for generating signals to be used in a frequency discrimination task.

One advantage of using the IFFT is that the resulting signal will *always* be periodic, so it may be looped or duplicated without discontinuities when placed end-to-end to extend the duration of stimuli generated by this method.

5.4 Signal Analysis Using the FFT

An analog signal acquired using an A/D can be analyzed for frequency content using the DFT/FFT. When analyzing a periodic waveform, it is ideal to have a sample length N which contains an exact integer number of complete periods (i.e. it ends at the same phase of the cycle as it started). One might think that this is a simple matter of adjusting N accordingly, but we cannot use the FFT unless N is radix-2 (which is not likely in general). Non-radix-2 DFT's take so long to compute they are impractical for most purposes. So keeping in mind a radix-2 constraint on N, it is possible (but not always practical) to adjust the acquisition sampling period T such that the waveform's fundamental frequency will fall exactly on an FFT frequency bin f_n. Otherwise, some amount of "spectral smearing" will result. If T cannot be adjusted, spectral smearing can be reduced by extending waveform acquisition time (i.e., increasing N) to collect as many periods as possible, and by windowing the digital signal data (before the FFT) to smooth the beginning and ending transitions (e.g. using a Hanning or Hamming window). Figure 9 compares spectra obtained with and without windowing from a 2048 point FFT of a complex modulated tone generated by the expression:

$$x_k = [1 + \sin 2\pi 300kT]\sin(2\pi 1000kT), \ k = 0,1,\ldots,N-1$$

where T is 20ms and N = 2048. The parameters are such that the tone does not contain an integer number of complete periods.

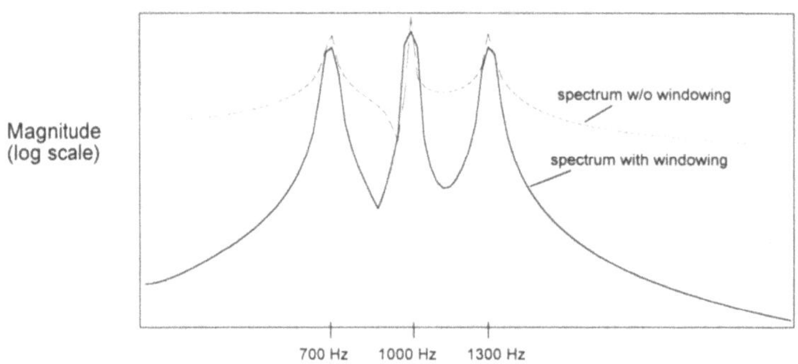

Figure 9. Comparison of frequency spectra with and without windowing of signal data.

The solid plot of Figure 9 is the magnitude spectrum after imposing a Hanning window on the time data, while the dashed plot is the spectrum *without* windowing. If the time-data array had contained an integer number of periods, or if we had used an $N = 2000$ point DFT, the spectrum (without windowing) would show just three peaks at the three frequencies and appear to be nearly zero at all other frequencies. As can be seen, the Hanning window does a good job of reducing spectral smearing and capturing the desired spectral peak information in a non-ideal situation. Windowing is often applied to non-periodic and time-limited transient type signals for the same reasons.

6 Practical Signal Generation

When generating complex acoustic stimuli it is important to choose an appropriate method of generation. Attempting to generate a signal via an unsuitable technique often leads to frustration and ultimate failure. Choosing the best method depends not only on the signal to be generated, but also the hardware available to do the job and the abilities of the operator. Often times what seems to be the appropriate path later results in failure due to a previously unnoticed limitation in the chosen method. Unfortunately, expounding on the many aspects and nuances of choosing an appropriate method for generating the signals needed in the seemingly endless array of psychoacoustic protocols goes well beyond the scope of this chapter. The best road to proficiency in this area will come through long hours of hands-on experience.

6.1 Generating a Tone

Many DSP hardware manufacturers have written high-level software function libraries which greatly simplify programming for signal generation. Suppose we wish to generate a 100 ms, 1000 Hz sinusoidal tone blip using a typical DSP card. This could be done in time domain (using the sine function) or in the frequency domain (using the IFFT). A time-domain program example might look like:

```
f = 1000;       /* frequency of tone blip: 1000 Hz          */
DUR = 100e-3;/* duration of blip 100ms                   */
T = 20e-6;      /* sampling period of 20 microseconds ( fs = 50kHz)      */
N = DUR/T;      /* # of samples needed for duration at the sample period   */
dpush(N);       /* create empty array of appropriate length          */
tone(f, T);     /* compute signal samples corresponding to tone frequency  */
                /* and sampling period specified             */
qwind(2.0, T);/* apply a 2ms rise/fall window to eliminate 'clicks'   */
scale(32000);/* scale for playing from 16-bit D/A converter      */
```

The number of samples N required is determined by duration and sampling period. The frequency placement resolution will be very precise with this method, but note that unless the tone produced has an integer number of periods by ensuring the condition $f\,NT = m$, where m is a positive integer, it will have audible discontinuities if it is looped. However, this is not an issue if the tone is windowed (as above) and used for a single presentation.

To accomplish a similar task in the frequency domain, we would specify only a single frequency bin (corresponding to the tone frequency) in a spectrum array, then apply the IFFT. The program might look like:

```
N = 4096;     /* number of samples/points for IFFT (must be Radix-2)    */
T = 20e-6;    /* sampling period of 20 microseconds ( fs = 50kHz)    */
f = 1000      /* frequency of tone blip: 1000 Hz              */
DUR = N*T;      /* duration of blip = 81.92 ms                */
n=(int)(f*N*T);  /* integer bin number for the specific frequency, f     */
dpush(N/2); /* create empty array for magnitude vs frequency spectrum,   */
            /* (length only N/2 since other half is assumed conjugate)   */
value(0.0); /* initialize magnitude of all frequency bins to zero   */
make(n, 1.0)   /* specify amplitude for desired frequency         */
dpush(N/2); /* create empty array for phase vs frequency spectrum, */
value(0.0); /* initialize all phase bins to zero          */
rect();       /* convert from polar to rectangular format        */
rift();       /* apply real inverse fast Fourier transform (IFFT) to */
              /* produce samples of tone blip at the specified frequency   */
qwind(2.0, T); /* apply a 2ms rise/fall window to eliminate 'clicks'   */
              /* note: omit qwind if the signal is to be looped.      */
scale(32000); /* scale for playing from 16-bit D/A converter     */
```

In the time-domain program N computes to be 5000, while the value N = 4096 shown in the program, was used because it is the closest Radix-2 value to 5000. Thus the duration of the resulting signal was lowered slightly to 81.92 ms. However since the IFFT was used (Section 5.2), this duration can easily be extended by duplicating or looping the result, without any special considerations for avoiding end-to-beginning discontinuities. Therefore the signal may be played indefinitely or truncated to the any sample length for a specific duration.

Rounding in the statement n = (int)(f*N*T) introduces frequency placement *error*: n = (int)(1000*4096*20e-6) = (int)(81.92) = 82, and this bin number corresponds to the frequency, $f_n = n/NT$ = 82/(4096*20e-6) = 1000.977 Hz. Thus, our actual tone frequency will be off by almost 1 Hz, which may be unacceptable for some applications. Frequency placement *resolution* is $\Delta f = 1/NT = 12.2$ Hz, e.g. the next nearest possible frequencies are 1000.977 ± 12.2 Hz. Increasing N to the next highest Radix-2 value of 8192 and/or increasing the sample period T will improve the placement resolution Δf, however, for some applications the IFFT method may still prove inappropriate.

For a single frequency tone there may be no advantage to using the IFFT method, but this example illustrates the basic procedure. A multi-component tone is easily achieved with additional make() statements to specify magnitudes/phases at other frequency bins.

6.2 Generating Noise Signals

Basically, noise consists of a random number for every time sample. The spectral distribution of the noise signal is of primary importance. It can be modified via the probability distribution of the random number generator (uniform, Gaussian, etc.), or by digital filtering of the noise samples. It is also possible to specify a desired spectral shape and use the IFFT to generate random samples accordingly.

The simplest way to generate noise is with an algorithm which fills an array with random numbers. This type of function is supplied with most signal generation software. Suppose we wish to generate a 100 ms, noise burst using a typical DSP card. The program example might look like:

```
DUR = 100e-3;  /* duration of burst 100ms                    */
T = 20e-6;    /* sampling period of 20 microseconds ( fs = 50kHz)    */
N = DUR/T;    /* # of samples needed for duration at the sample period */
dpush(N);     /* create empty array of appropriate length       */
gauss();      /* compute random samples with Gaussian distribution     */
qwind(2.0, T); /* apply a 2ms rise/fall window to eliminate 'clicks'  */
scale(32000);  /* scale for playing from 16-bit D/A converter     */
```

Using the IFFT, a noise signal with very specific spectral shaping can be generated by filling appropriate bin ranges within the magnitude spectrum and randomizing the phase spectrum. This method is ideal for arbitrary spectral shapes with very steep cutoffs. For example, to generate a noise burst with spectral content from 500 Hz to 2000 Hz a typical set of DSP function calls would look like the sequence shown on the next page:

```
N = 4096;     /* number of samples/points for IFFT (must be Radix-2)    */
T = 20e-6;    /* sampling period of 20 microseconds ( fs = 50kHz)    */
DUR = N*T;      /* duration of busrt = 81.92 ms                 */
n1 = 500*N*T; /* bin number for start frequency           */
n2 = 2000*N*T;    /* bin number for end frequency               */
dpush(N/2); /* create empty array for magnitude vs frequency spectrum,  */
        /* (length only N/2 since other half is assumed conjugate)   */
value(0.0); /* initialize magnitude of all frequency bins to zero   */
block(n1,n2);  /* block off portion of array between bins n1 & n2     */
value(1.0)  /* specify uniform amplitude over blocked frequency band  */
dpush(N/2); /* create empty array for phase vs frequency spectrum, */
rand();     /* randomize all phase bins              */
rect();     /* convert from polar to rectangular format       */
rift();     /* apply real inverse fast Fourier transform (IFFT) to */
        /* produce samples of noise burst             */
qwind(2.0, T); /* apply a 2ms rise/fall window to eliminate 'clicks'   */
        /* note: omit qwind if the noise is to be looped.   */
scale(32000); /* scale for playing from 16-bit D/A converter    */
```

6.3 Generating Long-Duration Stimuli

The methods of Sections 6.1 and 6.2 are good for short duration or repetitive stimuli, but for generating signals such as continuous non-repetitive noise other methods must be used. The best way to generate continuous, long-duration stimuli is to use a dedicated DSP to compute samples of the signal in real time. This is commonly a stand-alone waveform generator peripheral whose sole task is to generate a continuous tone or noise. If an experimental protocol requires continuous diotic noise for example, such a stand-alone device should be utilized to allow the PC - D/A system to perform other tasks.

If spectral shaping of a continuous signal is needed, a digital filtering device should be employed. Like the waveform generator, digital filters are DSP based devices capable of filtering signals with complex and arbitrary filter functions.

7 Suggested Readings

Oppenheim, A.V., Schafer, R.W. (1989) *Introduction to Discrete-time Systems*. New Jersey: Prentice Hall.

Proakis, J.G., Manolakis, D.G. (1988) *Introduction to Digital Signal Processing*. New York: Macmillan.

Varosi, S.M., Tucker, T. J. (1991) *Digital Signal Processing Applications*. Tucker-Davis Technologies.

8 References

Oppenheim, A.V. & Schafer, R.W. (1989) *Introduction to Discrete-time Systems.* New Jersey: Prentice Hall.
Proakis, J.G. & Manolakis, D.G. (1988) *Introduction to Digital Signal Processing.* New York: Macmillan.
Varosi, S.M. & Tucker, T. J. (1991) *Digital Signal Processing Applications.* Tucker-Davis Technologies.
Wightman, F.L. Chap 7, The Minicomputer and Research in Hearing. In: Mayzner, M. & Doland, T. (1978) *Minicomputers in Sensory and Information-Processes Research.* Erlbaum.

Methods in Comparative Psychoacoustics
G. M. Klump, R. J. Dooling, R. R. Fay & W. C. Stebbins (eds)

Acoustic Equipment and Sound Field Calibration

Ole Naesbye Larsen

*Centre for Sound Communication at Odense University,
Campusvej 55, DK-5230 Odense M, Denmark*

Summary. High-quality microphones make it possible to select loudspeakers with appropriate specifications and to test the sound field. Shortcomings in the loudspeakers' frequency responses may be overcome by using digital signal processing techniques. Only under free-field conditions do sounds reach the experimental subject from the loudspeaker alone. Free-field conditions rarely exist even in anechoic rooms, since echoes from the setup create an inhomogeneous sound field, but sound reflecting objects can be located by measuring the acoustic impulse response of the setup, and their disturbance of the field can be minimized by reducing their dimensions or covering them with sound absorbing material. Outside the sonic range special problems prevail, especially at very low frequencies.

1 Introduction

Equipment considerations and sound field calibration are boring exercises, which invite to cutting corners in order to get on with the exciting psychoacoustic experiments. Such experiments are designed to monitor the responses of the experimental subjects to relevant sound stimuli. The trivial assumptions are that the experimenter can generate the relevant stimuli and that these are the actual signals to which the experimental subjects, animals or humans, respond. In the real world, however, the fulfilment of these assumptions is a non-trivial matter.

When generating sound signals the experimenter faces the problem that all links in the chain of equipment are limited in their performance and that most of them distort signals to some extent. For instance, they may not reproduce high frequencies, but introduce frequency components, which were not present in the input signal. In a well designed setup the limitations take effect outside the range of the intended experiment and the distortions are smaller than the thresholds of the experimental subjects.

In open field experiments the radiation characteristic of the loudspeaker changes with frequency; the experimental setup, normally placed in a sound attenuated booth or room, creates sound shadows, adds echoes, and absorbs parts of the spectrum of the sound stimulus to generate an inhomogeneous sound field in the space normally occupied by the head of the experimental subject (for definitions of sound fields see section 4). In addition, the subject may move its head and body during the experiment. Consequently, it is difficult to know the exact nature of the sound stimulus to which the experimental subject responds, unless the sound field has been made as homogeneous as possible. These problems may be avoided using sealed acoustic systems (earphones) - and hoping for cooperative experimental subjects! While such conditions of presentation provide much better stimulus control and a much more manageable experimental environment than the open field, it is often difficult to generalize the results obtained to the natural listening conditions of the real world. Closed field conditions are not considered here. Interested readers are referred to the excellent account by Pfeiffer (1974), to Kleindienst *et al* (1981), and to Zwicker and Fastl (1990, pp. 5-8 and 317-318) for more recent references.

Below I point out some problems encountered when designing a psychoacoustic experiment and suggest how to solve them, but it does not mean that one should run through all the tests for every type of experiment or necessarily always work in a free field. Use common sense.

2 Sound Receivers

A calibrated, high-quality sound measuring setup consisting of microphone and amplifier and preferably connected to a frequency analyzer (or a fast computer with signal analysis software) makes it is possible to improve the quality of the sound field (section 4.1-4.2), to judge the quality of loudspeakers in the actual setup, and to select the best one for a given experiment (section 3.2-3.5). The microphone should serve as a standard, against which all acoustic aspects of the setup are compared.

2.1 Microphone types

Ideally the microphone should measure the acoustic stimulus without distortion, with a high sensitivity over a suitable range of intensities and frequencies, and its presence should cause negligible disturbance of the sound field. No single real microphone meets all these requirements, since the construction of a microphone is the compromise between mutually exclusive demands. A microphone with a large membrane, for instance, is very sensitive, but can measure only low frequencies and distorts the sound field by its presence. A small microphone, on the other hand, responds to a very wide frequency band and causes little distortion of the sound field, but is less sensitive. It is highly recommended to invest in high-quality microphones such as the condenser microphones manufactured by Brüel & Kjær (B&K), although the price may seem forbidding. Over limited frequency ranges such microphones come very close to the ideal. They are very reliable and their inevitable distortions and less than perfect responses are very well documented.

An ideal microphone of zero size in a free progressive plane wave experiences a uniform pressure applied over the entire membrane. However, when a real microphone is introduced into the sound field, reflection and diffraction (see section 4) occur around the microphone. At progressively higher frequencies, where the dimensions of the microphone are comparable to and eventually become much larger than the wavelength, the sound pressure acting on the microphone membrane increases by about 6 dB relative to that in the undisturbed field, if it is pointing directly towards the source, and its frequency response becomes strongly dependent on the angle of sound incidence. In a 'free-field condenser microphone' the membrane is damped to compensate for pressure increases at high frequencies. This design compensates for the disturbance caused by the microphone's own presence in the sound field, provided that the direction of propagation of sound is perpendicular to its membrane. In contrast, a 'pressure microphone' has a uniform frequency response to the sound field as it exists, including its own disturbance of the field, i.e. it produces a signal proportional to the sound pressure actually existing at its diaphragm. The sound pressure in the undisturbed sound-field can be measured with a pressure microphone, but it must be positioned so that the membrane is parallel to the direction to the sound source. If no reflections from objects in the field are present, the pressure microphone may give a correct reading, but it may severely overestimate the true reading (by 8-10 dB), if reflections dominate the field (see section 4)! Pressure microphones should mainly be used in sealed enclosures, while free-field microphones should be selected for measurement in open fields with well defined positions of the sound sources.

2.2 Microphone Calibration

High-quality condenser microphones are very reliable and keep their specifications for years. For many types of experiment it therefore suffices to calibrate the measuring set-up, from microphone to read-out, with a known acoustical reference at a single frequency (a pistonphone or a sound level calibrator) and adjust the amplifier sensitivity until the readout corresponds to the very precise sound pressure input. The frequency response is then given by the calibration chart from the factory. The method works well, provided that good sealing is maintained and that the correct coupler volume is used.

For critical measurements it may be important to know the phase response of the microphone also. Before and during experiments, we use the substitution method to calibrate the measuring setup (Michelsen *et al* 1994). The frequency response of the microphone under test is compared with that of a standard microphone at anechoic conditions. An eighth-inch microphone serves as the standard, since its frequency response is very flat (<< ±1 dB deviation) and its output voltage lags the input sound by up to only 10o in the frequency range below 20 kHz. The frequency response of the loudspeaker is recorded with the eighth-inch microphone and stored on disk (the loudspeaker response should be relatively flat in the frequency range of interest). Then the procedure is repeated with the microphone under test (often a probe microphone) in the same position. Finally, the frequency response of the microphone under test is calculated by dividing its spectrum (both amplitude and phase) by that of the eighth-inch microphone. The resulting transfer function can then be used for correcting measurements obtained with the tested microphone.

Once every second year or so we use the electrostatic actuator method to check the pressure and phase response of our microphones. A specially designed, isolated metal grid is placed in immediate proximity to the microphone membrane and both a DC polarization voltage and a smaller AC voltage are applied to the grid. This produces a sinusoidal force on the diaphragm simulating a uniform acoustic pressure, and makes possible the measurement of the pressure response both in amplitude and phase up to high frequencies. The pressure response can then be compared with the calibration chart from the factory, where deviations resulting from damage (or aging) of the microphone will immediately show up.

If your funds are limited and you don't have access to an advanced spectrum analyzer, buy two high-quality microphones, e.g. a free-field quarter-inch and a probe microphone. Measure their frequency responses immediately on receipt by placing them in the setup such that they experience the same sound pressure (the quarter-inch pointing directly towards the loudspeaker and the probe touching its grid from the side). The amplitudes and phase lags at a number of discrete frequencies measured on an oscilloscope will do. Repeat the procedure once every three months. If you spot changes in the frequency responses, seek expert help from the factory's agent. This strategy will spare you much annoyance!

3 Sound Stimulus Sources

Sound stimuli are produced by a chain of equipment in principle consisting of a signal generator, an attenuator, a power amplifier, and a loudspeaker.

In modern setups the signal generator virtually always is a fast digital computer connected to a suitable digital-to-analog converter and a smoothing low-pass filter, and often equipped with a DSP (digital signal processor) expansion card for creating complex stimuli. The digital-to-analog converter should be (at least) 16 bit corresponding to a theoretical dynamic range of

96 dB and with a typical signal-to-noise ratio of 85 dB. In principle, its clock frequency should be twice the highest stimulus frequency of interest (the Nyquist criterium), but in practice much higher, which in the sonic range means well above 40 kHz. (Experiments involving mammals other than man often require much higher sampling frequencies as their high-frequency cut-off is often surprisingly high. Consult Fay's databook (Fay 1988)!). The low-pass filter, which is sometimes referred to as a reconstruction or anti-aliasing filter, should cut-off at the highest stimulus frequency of interest with a steep roll-off (9-pole or higher order filters). If the sampling frequency is several times higher than the highest frequency of interest (oversampling), a less steep low-pass filter may suffice. Oversampling should be preferred to very steep filters as they introduce phase distortions (see e.g. Pohlmann (1990), pp. 115-123, and Stoddard (1990)).

Manual analog attenuators (passive resistive divider networks) are still used, but typically they have a smallest step size of 1 dB and are not suited for fast attenuation change. Consequently, programmable attenuators are gradually replacing the manual ones as they have a resolution of up to 0.1 dB and are much faster (state-of-the-art equipment such as that from Tucker-Davies Technologies has a settling time of typically 5 ms).

Though the really weak link in the sound generation setup is the loudspeaker, it pays to buy an expensive power amplifier (which is still far cheaper than most of the other equipment used in the typical psychoacoustics setup), since high-quality power amplifiers (such as those from DENON and NAD) outperform most ears as they have a frequency range from a few Hz to more than 100 kHz. Their harmonic distortion is claimed to be less than 0.002%, their signal-to-noise ratio is 100-120 dB, and their output power 100-450 W.

3.1 The Ideal Loudspeaker

The ideal all-round loudspeaker for psychoacoustic experiments has a perfectly flat frequency response both in amplitude and phase ranging from 1 Hz to above 100 kHz. In this frequency range it generates sound pressures up to 100 dB SPL (measured at a distance of 1 m) without distortions neither in amplitude nor phase. It radiates sound energy from a single point in space, and it is small and lightweight. Such a loudspeaker does not exist, since physical laws limit the performance of real loudspeakers. A detailed discussion of loudspeaker construction, performance, and limitations is outside the scope of this chapter, but interested readers are referred to Zwicker and Zollner (1987, pp. 123-148 and 165-195).

3.2 The Frequency Response of Real Loudspeakers

Even a high-quality loudspeaker can efficiently produce sound in only a limited frequency range; exceptionally good loudspeakers perhaps from 20 Hz to 20 kHz with variations in the amplitude of ±2 dB. The loudspeaker then consists of two or more units, 'woofers' and 'tweeters', which cover low and high frequencies, respectively. Electronic filters, known as cross-over networks, divide the signal to the loudspeaker into low- and high-frequency bands separated at the cross-over frequency. Though the total amplitude spectrum may be rather flat, in such a construction the phase-spectrum rarely is, because rapid changes often occur close to the cross-over frequency. In some types of experiments this could be a problem, since a waveform is only reproduced accurately, if all the frequency components are reproduced not only at the correct relative amplitudes, but also at the correct relative phases. Phase excursions produce variations in the delays of the different frequency components. Humans - and perhaps some animal species also - are sensitive to such delays, especially in short duration sounds (Blauert and Laws 1978, Moore and Glasberg 1989, Patterson and Green 1970). If phase distortions could possibly ruin an experiment, perhaps by providing listeners with an extra cue, one should select the loudspeaker only after measuring the phase response, since this aspect of loudspeaker performance seldomly (if ever) is specified by the manufacturer. The same type of problem is introduced, if a graphic equalizer is used to flatten the frequency response, since it consists of a series of filters, which nearly always distort the waveshapes of complex sound signals. Using a single unit loudspeaker and correcting the response by means of digital signal processing techniques is a much better way to improve its frequency response both in amplitude and in phase.

3.3 Dynamic Range and Nonlinear Distortion

The loudspeaker should operate linearly, i.e. a change in input amplitude should be directly proportional with that of the output, in the intensity and frequency range of interest. This can easily be verified by measuring the frequency response at successive 10 dB input attenuations (Fig. 1A). At low stimulus levels the dynamic range will be limited by system noise, sometimes with a predominant low-frequency hum, but the latter can almost always be abolished by a proper grounding of the setup (see Pfeiffer 1974). With a high-quality amplifier in the setup the upper limit of the dynamic range is reached only when the speaker blows due to overheating, but this never happens at physiological sound levels below 80 dB SPL!

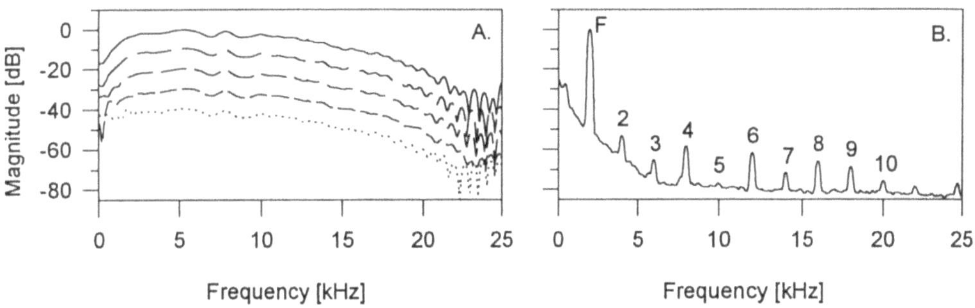

Figure 1. Some loudspeaker properties. A. Linearity. The graph shows the frequency response (amplitude only) of a loudspeaker for 5 attenuator settings in successive 10 dB steps. The speaker does not deviate appreciably from linearity in the range 1-20 kHz. B. Harmonic distortion. A sine wave signal of 1.984 kHz is fed into the loudspeaker. The transmitted sound is recorded with a 1" B&K microphone and frequency analyzed on a dynamic spectrum analyzer. 'F' indicates the fundamental, while numbers 2-10 indicate higher harmonics. Total harmonic distortion was computed by the analyzer to be 0.27%.

A less than perfectly flat frequency response will change the temporal structure of the signals, but it will never introduce new spectral components in the output signal. Such distortion components are introduced by the action of non-linear spring and damping elements in the suspension of the loudspeaker diaphragm. When a pure tone of frequency *f* is fed into the loudspeaker, smaller spectral components are observed at n*f*, where n = 2, 3, 4, 5.... (cf. Fig. 1B). These extra components change the input waveshape to a periodic waveform, which will deviate more and more from the sinusoid, the bigger the extra components are. The sum of these spectral components is called the 'harmonic distortion', and its magnitude is expressed as the percentage of their amplitude relative to the amplitude of the input frequency, the fundamental. If for instance the amplitude of 2*f*, the second harmonic, is 40 dB lower than that of the fundamental, then the second harmonic produces 1% harmonic distortion. The tolerable amount of harmonic distortion depends on the experimental design, but the distortion of higher harmonics should normally be kept below 0.1%. With the sound measuring setup and a spectrum analyzer this can be checked in a matter of minutes for the frequencies of interest (cf. Fig. 1B). Another type of distortion is the so-called 'intermodulation distortion'. When two tones, *f1* and *f2* are played simultaneously, they will produce not only their respective harmonic distortion, but in addition some extra frequency components by the non-linear interaction between them at *f1-f2*, *f1+f2*, *2f1-f2*, *2f1+f2*, etc. In the future, the suspension non-linearity of loudspeakers may be compensated for by digital signal processing and emerging techniques such as time-delay neural networks. Brave readers are referred to Chang *et al.* (1994) for an introduction and references.

3.4 Perfecting the Frequency Response by Digital Signal Processing

In our lab we use digital signal processing techniques for perfecting the frequency response of loudspeakers. Once the linearity of the loudspeaker has been checked, its frequency response (Fig. 2A) is measured by emitting a short frequency sweep and recording the transmitted sound in the free field. The frequency response is then used to correct the digitally generated stimulus signal by division in the frequency domain. The result of this operation (which is equivalent to deconvolution in the time domain) is that signals, be it pure tone or complex vocalization, will seem to be emitted from a loudspeaker with a very flat frequency response (Fig. 2B). With this technique it is possible to generate a frequency response flat within ±1 dB, which is not detectably different from a perfectly flat response. The method requires a DSP expansion card to be run on-line and a sufficiently large dynamic range both of the converters and of the loudspeaker. The digital-to-analog and analog-to-digital converters should be 16 bit, and the loudspeaker should produce sufficiently high sound pressures in the frequency range of interest without any dramatic dips, or unwanted distortion in the corrected signals will occur.

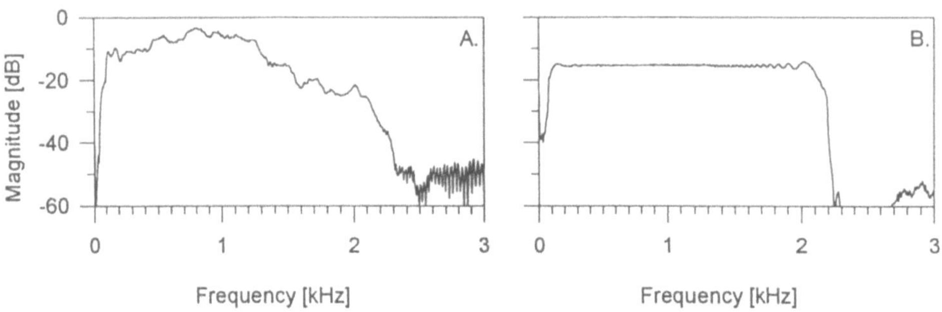

Figure 2. The frequency response (only the amplitude part of the transfer function) of a loudspeaker, used for the study of directional hearing in frogs, before A. and after B. correcting it by means of digital signal processing. Such a correction is possible only in frequency ranges without substantial dips in the loudspeaker's frequency response.

This approach was used in a study of directional hearing in frogs, where a moveable loudspeaker is traditionally used to provide free-field stimuli from different directions. In such a setup it takes some time to move the loudspeaker to a new position, which may be a problem in time-critical experiments. Alternatively, an array of loudspeakers may be used, but their individual frequency responses differ and they may create different sound fields due to differences in the geometry. Using the method described above it was possible to produce sound signals, which when emitted from the eight different loudspeakers in the circular array differed by less than ±0.5 dB and ±5o at the position of the frog (Jørgensen and Christensen-Dalsgaard in prep.).

3.5 Directional Characteristics

As a very coarse rule of thumb, household sized loudspeakers in enclosures emit sound almost equally in all directions at frequencies below 400 Hz, but become increasingly directional for successively higher frequencies as they 'beam' the sound mainly in the direction in front of the diaphragm. Too narrow a beam makes the position of the experimental subject very critical. Too wide a beam is also undesirable, unless the speaker is placed in a truly anechoic room. In a room with lots of reflecting surfaces both the location of such a loudspeaker and the clarity of its sound may be blurred, since the perceptual suppression of echoes by the precedence effect may not work. If directionality is important for the experiment, the frequency response of the loudspeaker should be measured at a number of angles in an anechoic room and polar diagrams produced for relevant frequencies. For our studies of directional hearing we use single unit Dynaudio loudspeakers (Michelsen *et al* 1994) or KEF loudspeakers with a coincidently mounted woofer and tweeter, which share not only the same axis, but have acoustic centres in the same plane and similar directivities in the crossover region (Klump and Larsen 1992).

4 Sound Fields

Close to a sound source there is no simple relation between sound pressure and other physical parameters such as sound intensity, particle displacement, and distance to the source. This region of space is called the acoustic *near-field*. At sufficiently great distances from a sound source, the wave fronts become plane. The sound waves propagate away from the source and the sound intensity obeys the inverse square law (geometrical attenuation). This means that the sound pressure is halved, i.e. is reduced by 6 dB, when the distance is doubled (dd) between the source and a reference point (for the mathematical derivation see e.g. Crocker and Price 1975, pp. 16-25). This region of space farther away from the source is called the acoustic *far-field*. The extent of the near-field depends in a complicated way on the geometry and vibrating parts of the source. There is a gradual transition between near-field and far-field, but as a coarse rule of thumb the near-field component has almost disappeared about one wavelength away from the source, provided that the dimensions of the source are small relative to the wavelength. Since the sound pressure in the near-field is ill defined, experimental subjects should be positioned in the far-field. In principle, the far-field can be located by measuring the sound pressure at different distances from the loudspeaker and controlling that it is reduced by 6 dB/dd, but the procedure is very time consuming and difficult (Ingerslev *et al* 1968). If the experiments are performed in a room (or e.g. in a forest) one will soon discover that the inverse square law holds true only over a limited range of distances from the source. Close to the walls or other obstacles the sound pressure decreases with distance in a complicated way by less than 6 dB/dd. Accordingly, the acoustic far-field is said to consist of two parts, the *free field* closest to the source and the *reverberant field* farther away. In the free field, as the term suggests, sound propagates as if in open air, without reflecting surfaces to interfere with its propagation. In the reverberant field the sound pressure is dominated by *reflections* from the walls and other large reflecting surfaces. To complicate the matter further, objects with dimensions of the same order of magnitude as the wavelength of the stimulus sound will not cause reflections, but will redirect the sound in all directions in a complicated - and sometimes counter-intuitive - manner. This redirection is called *diffraction*.

4.1 Locating Sound Reflecting Objects

So, although free-field conditions existed in the large anechoic room, before the psychoacoustic setup was established there, the sound field now is a mixture of the direct sound from the loudspeaker and reflected and diffracted sound from objects in the setup.

Consequently, the two ears of the experimental subject may receive very different sound pressures and delays reflecting the geometry of the setup, not the physiology of the animal. This could ruin for instance a study of directional hearing. Consequently, sound reflecting objects should be identified and their influence reduced as much as possible. It may be difficult to locate reflecting objects and to judge their relative influence, but very short intense sound emitted by the loudspeaker will reach the microphone first, while reflections arrive later

(Fig. 3). From the relative delays it is often possible to locate the reflecting objects (Fig. 3B-C). Sounds of short duration are difficult to generate (Schiolten *et al* 1981), but the acoustic impulse response of the setup may also be computed by recording the frequency response and subjecting it to an inverse Fourier transform, which is how the signals in Fig. 3 were obtained.

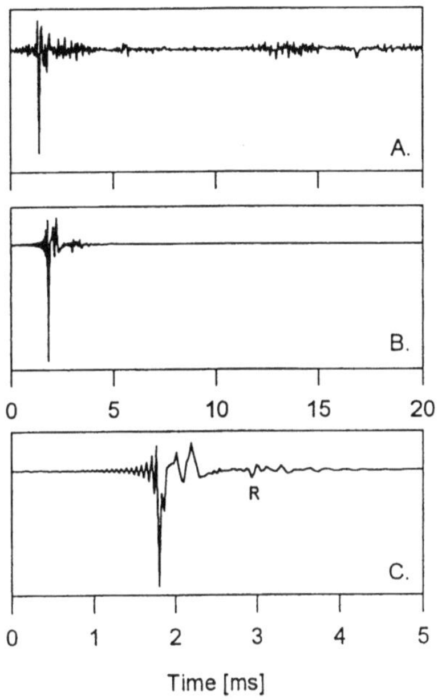

Fig.3. A. The acoustic impulse response in a normal laboratory environment. The large downward deflection is the direct wave from the loudspeaker, while the smaller deflections are echoes from reflecting surfaces arriving later. B. Illustrates the impulse response when free-field conditions are approached. Only a single echo is observed after the direct wave. This is shown on a bigger scale in C. and indicated with 'R'. The echo arrives 1.28 ms later than the direct wave and is caused by reflection from a vertical rod 2 cm in diameter, which was placed on a tripod about 20 cm behind the microphone (344 m/s · 1.28 ms = 44 cm).

4.2 Improving the Sound Field

Once the reflecting objects and surfaces have been located, the sound field may be improved by reducing the echoes and/or increasing the direct sound. As a crude rule of thumb an object will start to diffract the sound such that the pressure measured by a microphone (or ear) deviates measurably from that in the free field, when the dimensions of the object exceed 10% of the wavelength of the sound (but measurable phase deviations are found for even smaller dimensions!). To control the improvement you need a suitable microphone, but beware! The microphone dimensions should be so small relative to the wavelength of the sound that it does not change the sound field by its presence. Trying to evaluate the sound field close to the head of a small bird with a one inch microphone is simply meaningless. In our lab we prefer B&K probe microphones (type 4182) with tip diameters of 1 mm for such exercises. Vertical rods are often included in physiological and psychoacoustic setups. If the highest frequency of interest is 10 kHz (wavelength = 34 mm), a rod diameter of < 3 mm will not cause appreciable diffraction and reflection in the frequency range of interest, provided that the rod is not very close to the microphone. Zollner (1982) has shown that a maximum deviation from the free field of 2 dB somewhere in the frequency range 0-20 kHz will be caused by a 5 mm diameter rod 4 cm from the microphone, while a 30 mm diameter rod must be placed 25 cm away to cause the same effect. Closer to the microphone the effect will be much larger. One way to improve the sound field therefore is to reduce the dimensions of objects in the field to the smallest practical size. Some objects can not be reduced in size, but reflections may be substantially reduced by covering them with a few cm thick layer of soft mineral wool. Bringing the loudspeaker close to the microphone (or ear) may also improve conditions, since the direct wave will be much more intense than the reflected waves, which are attenuated geometrically by the extra distance travelled.

The three methods, reducing the object dimensions, covering reflecting surfaces with sound absorbing material, and bringing the loudspeaker closer, may improve the sound field such that the sound pressure measured with a small microphone anywhere in the space normally occupied by the experimental subject varies by about ±1 dB over a narrow frequency range - if you are very lucky! In many psychoacoustic setups, however, one often has to be content with variations of up to ±3 dB, especially over wider frequency ranges.

5 Leaving the Sonic Range

5.1 Special Considerations Regarding Low Frequency Sound

Perception of low frequency sound, i.e. below about 100 Hz, in the open field has been little studied so far - and for good reasons. First, special precautions must be taken to reliably produce low frequency sound at sufficiently high sound pressures. Secondly, intense low frequency sounds invariably bring into vibration parts of the experimental setup, and reactions of the experimental animal may be caused by other sensory pathways than the acoustic ones (Christensen-Dalsgaard and Narins 1993). Thirdly, anechoic rooms seldomly (and small Sonex foam covered single-walled booths definitely never) obey the inverse square law for frequencies below 100-400 Hz, since the low frequency cut-off is inversely proportional to the size (and prize!) of the room. A cut-off at 50 Hz requires that the smallest dimension of the free space between wedge tips on opposite walls is >6 m! Finally, the level of ambient noise in the frequency range below 50-100 Hz is virtually always high both in the biotope and in the "sound proof" room in the basement of the typical university department. Consequently, a high sound output of the loudspeaker is required to overcome the masking effect of the ambient low frequency noise, which from our experience is virtually impossible to attenuate, if special precautions were not taken in constructing the anechoic room (Ingerslev *et al* 1968).

Commercial sub-woofers seldomly produce a sufficient sound pressure below about 35 Hz, while professional ones push the range to about 17 Hz, but are very bulky, heavy, and expensive. Decreasing sub-woofer size, while increasing its sound output may be solved by using new designs such as microcomputer controlled, servo-motor driven pistons (e.g. the "Contrabass" manufactured by Intersonics). The inevitable sound-induced (and substrate-borne) vibrations of the experimental subject at low frequencies may be cancelled by actively vibrating the setup in anti-phase (Christensen-Dalsgaard and Jørgensen in prep.). This procedure works with frogs, but may be impossible to realize for larger and heavier animals. As an alternative to a large and expensive anechoic room, psychoacoustic experiments in the low frequency range could be performed in open space, where free-field conditions may be closely approximated using a setup similar to that of Stevens and Newman (1936) placed on top of a tall building. But such a setup will of course be very vulnerable to meteorological conditions and to ambient noise.

5.2 Special Considerations Regarding Ultrasound

Ultrasonic loudspeakers with a flat frequency response between 15 and 130 kHz (±1.5 dB) can easily be build (Machmerth *et al* 1975). Small high-quality condenser microphones such as the eighth-inch B&K microphone with appropriate amplifiers reliably measure sounds at frequencies up to at least 150 kHz. Ambient noise levels are very low, as high frequencies are attenuated in air. Under standard conditions a 100 kHz sound is attenuated by about 4 dB/m (in addition to the geometric attenuation of 6 dB/dd). This means that its sound pressure has been reduced by 60 dB after propagating 10 m relative to the pressure 1 m from the source (e.g. Kinsler and Frey 1962, pp. 227-234). Echoes from the walls of the room therefore are negligible, provided it is large. Echoes from the floor will also be small, since an ultrasonic loudspeaker 'beams' the sounds almost like a flashlight beams the light (the angle of the beam typically varies between 40o and 10o in the range 20-100 kHz (Machmerth *et al* 1975)). But echoes from parts of the setup close to the experimental subject will predominate, since the small wavelength of the sound (at 100 kHz it is 3.4 mm) calls for extraordinarily small dimensions of objects in the setup. However, echoes may be reduced by 30 dB relative to the direct sound signals by covering the surfaces of the setup with loose felt or mineral wool and moving the loudspeaker close to the subject (Surlykke and Miller 1985).

6 Acknowledgements

I am grateful to Morten Buhl Jørgensen for the data for Fig. 2, and to him, Jakob Christensen-Dalsgaard, Axel Michelsen, and Simon Boel Pedersen for critical and constructive comments on the manuscript.

7 Suggested Reading

Crocker, M.J. & Price, A.J. (1975). *Noise and Noise Control. Vol. I.* Pp. 18-29, 100-126, and 173-202. Cleveland: CRC-Press.

Michelsen, A. (1983). Biophysical basis of sound communication. In: *Bioacoustics. A comparative approach.* (B. Lewis, ed.), pp. 3-38. London: Academic Press.

Moore, B.C.J. (1989). *An Introduction to the Psychology of Hearing.* 3rd ed., pp. 291-304. London: Academic Press.

Pfeiffer, R.R. (1974). Consideration of the acoustic stimulus. In: *Handbook of Sensory Physiology. Vol. V/1.* (W.D. Keidel & W.D. Neff, eds.), pp. 9-38. Berlin: Springer-Verlag.

Zwicker, E. & Zollner, M. (1987). *Elektroakustik.* Pp. 133-208. Berlin: Springer-Verlag.

8 References

Blauert, J. & Laws, P. (1978). Group delay distortions in electroacoustical systems. *Journal of the Acoustical Society of America 63(5)*, 1478-1483.

Chang, P.-R., Lin, C.G. & Yeh, B.-F. (1994). Inverse filtering of a loudspeaker and room acoustics using time-delay neural networks. *Journal of the Acoustical Society of America, 95*(6), 3400-3408.

Christensen-Dalsgaard, J. & Jørgensen, M.B. (In prep.). Sound and vibration sensitivity of VIIIth nerve fibers in the grassfrog, *Rana temporaria*.

Christensen-Dalsgaard, J. & Narins, P.M. (1993). Sound and vibration sensitivity of VIIIth nerve fibers in the frog *Leptodactylus albilabris* and *Rana pipiens pipiens*. *Journal of Comparative Physiology A 172*, 653-662.

Fay, R.R. (1988). *Hearing in Vertebrates: A Psychophysics Databook.* Winnetka: Hill-Fay Associates.

Ingerslev, F., Pedersen, O.J., Møller, P.K. & Kristensen, J. (1968). The anechoic chambers at the Technical University of Denmark. *Brüel & Kjær Technical Review 2*, 3-29.

Jørgensen, M.B. & Christensen-Dalsgaard, J. (In prep.). Directionality of auditory nerve fiber responses in the grassfrog, *Rana temporaria*. I. Tonal stimuli.

Kinsler, L.E. & Frey, A.R. (1962). *Fundamentals of Acoustics.* 2nd ed. New York: John Wiley & Sons.

Kleindienst, H.-U., Koch, U.T. & Wohlers, D.W. (1981). Analysis of the cricket auditory system by acoustic stimulation using a closed sound field. *Journal of Comparative Physiology 141*, 283-296.

Klump, G.M. & Larsen, O.N. (1992). Azimuthal sound localization in the European starling (*Sturnus vulgaris*): I. Physical binaural cues. *Journal of Comparative Physiology A, 170*, 243-251.

Machmerth, v.H., Theiss, D. & Schnitzler, H.-U. (1975): Konstruktion eines Luftultraschallgebers mit konstantem Frequenzgang im Bereich von 15 kHz bis 130 kHz. *Acustica 34*, 81-85.

Michelsen, A., Heller, K.-G., Stumpner, A. & Rohrseitz, K. (1994). A new biophysical method to determine the gain of the acoustic trachea in bushcrickets. *Journal of Comparative Physiology A 175*, 145-151.

Moore, B.C.J. & Glasberg, B.R. (1989). Difference limens for phase in normal and hearing impaired subjects. *Journal of the Acoustical Society of America 86(4)*, 1351-1365.

Patterson, J.H. & Green, D.M. (1970). Discrimination of transient signals having identical energy spectra. *Journal of the Acoustical Society of America 48(4)*, 894-905.

Pohlmann, K.C. (1990). *Principles of Digital Audio.* 2nd ed. Carmel: SAMS.

Schiolten, P., Larsen, O.N. & Michelsen, A. (1981). Mechanical time resolution in some insect ears. I. Impulse responses and time constants. *Journal of Comparative Physiology 143*, 289-295.

Stevens, S.S. & Newman, E.B. (1936). The localization of actual sources of sound. *American Journal of Psychology, 48*, 297-306.

Stoddard, P.K. (1990). Audio computers - Theory of operation and guidelines for selection of systems and components. *Bioacoustics 2(3)*, 217-239.

Surlykke, A. & Miller, L.A. (1985). The influence of archtiid moth clicks on bat echo location; jamming or warning? *Journal of Comparative Physiology A, 156*, 831-843.

Zollner, M. (1982). Einfluss von Stativen und Halterungen auf den Mikrofonfrequenzgang. *Acustica 51*, 268-272.

Zwicker, E. & Fastl, H. (1990). *Psychoacoustics. Facts and models.* Berlin: Springer-Verlag.

Methods in Comparative Psychoacoustics
G. M. Klump, R. J. Dooling, R. R. Fay & W. C. Stebbins (eds)

Psychophysical Methods

Merrilyn J. Penner

University of Maryland, Psychology Department
College Park, Maryland, 20742 USA

Summary. A brief description of classical and modern psychophysical techniques and the theory motivating these techniques is presented. Generally, it is desirable to employ those psychophysical techniques which permit separation of sensory sensitivity from response proclivity. Extant data obtained using methods which do not specifically control for response criteria are compared with those that do. Implications of these data for psychophysical results from animals are discussed.

1 Introduction

In sensory psychology, a variety of questions concerning the stimulus may be addressed. Some experiments are concerned with the problem of detection (whether a stimulus was presented), some with the problem of discrimination (whether two stimuli were the same), and some with the problem of recognition (what the stimulus was). In this chapter, techniques to determine detection thresholds in humans are discussed. The techniques may be easily generalized to include tests of discrimination or recognition by humans. It is hoped that briefly reviewing the techniques which have been developed for rigorous psychophysical testing in humans will, as has historically been the case, provide both a guideline for the evolution of rigorous psychophysical testing methods in animals and a standard to which current methods may be compared.

To begin, some sounds are so faint that they are never heard. Some sounds are so intense that they are always heard. In between these extremes is a region in which sounds are sometimes heard and sometimes not heard. In this region of uncertainty, responses can only be described probabilistically. Investigators keep track of the relative frequency, or probability, of a detection response as a function of the stimulus level.

The form of this function, called a *psychometric function*, is often depicted as a cumulative normal distribution. Its shape is a matter of interest. Smooth curves can appear when the data are averaged over many trials, even though on any one trial there might be a single point at which detection probability changes from 0 to 1 (as long as the position of the point changes from one trial to the next). There is a debate as to whether detection is discrete (Larkin & Norman, 1964; Stevens et al., 1941) or continuous (Green & Swets, 1966). Whatever the form of the psychometric function, the absolute threshold is defined as the stimulus value for which the signal is detected half the time and it is the determination of the corresponding stimulus value which concerns us here. All psychophysical methods employed to measure threshold assume that there is a psychometric function.

The stimulus magnitude at threshold often serves as the dependent variable and changes in it are studied as a function of other stimulus parameters which are represented on the abscissa. For example, in one classic study demonstrating that only the noise in a nearby frequency region masked a pure tone (Fletcher, 1940), the stimulus magnitude which was just detected was plotted on the ordinate and the bandwidth of the noise was displayed on the abscissa. Changes in stimulus magnitude presented on the ordinate served as a basis for inferences concerning the operation of the sensory system. It follows that the stimulus level required for constant performance (detection or discrimination) must be carefully determined. This chapter deals with psychophysical methods used in the pursuit of these levels which form the foundation of the attempt to make sense of the senses.

2 Response Bias in Yes-No Paradigms

Experimenters are concerned with determining the subject's sensitivity, independent of response proclivities (Green & Swets, 1966; Robinson & Watson, 1972). That is, interest should be centered about what is actually audible, not what is reported to be audible. The listener's response criteria influence the value of the measured threshold. For example, two subjects may be equally sensitive, but because one is more reluctant than the other to report the presence of a stimulus, his threshold may be judged as higher.

Such concerns are more than academic: older observers are thought to adopt stricter criteria than younger observers in detection tasks (Craik, 1969; Potash & Jones, 1977; Rees & Botwink, 1971). Another difficult potential scenario is that a subject may change response criteria from one condition to another so that comparisons between stimulus magnitudes in various conditions become meaningless. Additional examples of the effects of criteria on psychophysical methods are presented after describing the procedures used to control criteria.

Signal detection theory (SDT) offers a framework for separating response bias and sensitivity (Green & Swets, 1966). SDT is now described in the context of a yes-no psychophysical task. For ease of presentation, it may help to consider a yes-no detection paradigm in which a signal is presented on some trials but not on others. The events of a single trial define a *paradigm* and the technique which tracks threshold is called a *method*.

In a yes-no paradigm, the subject is asked whether a signal was contemporaneous with one clearly marked temporal interval. A binary decision, signal present or signal absent, is required. The SDT analysis of this situation is illustrated in Figure 1.

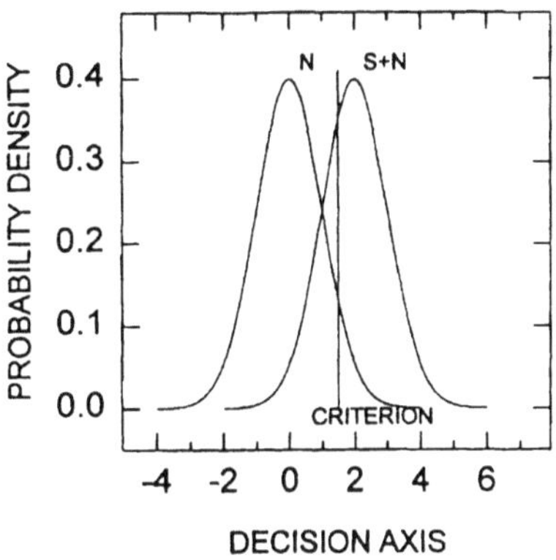

Figure 1

As seen in Figure 1, repeated samples from the signal absent condition (labeled N for noise) result in a normal distribution of sensations shown on the left in Figure 1. A distribution of sensations arises because of variations in internal noise or, possibly, due to cognitive factors. The external signal may sometimes generate a smaller and sometimes a larger internal signal. A tonal signal is assumed to add a fixed increment (determined by its magnitude) to the noise distribution without changing the variability, resulting in the normal distribution on the right (labeled S+N for signal plus noise).

The subject should respond "no" when a sensation is so far to the left in Figure 1 that it could never have resulted from the signal. Conversely, the subject should respond "yes" when the sensation is so far to the right that it could only have arisen from the signal. Subjects who are hesitant to report a signal generally require larger sensations (i.e., have a stricter criterion) than subjects who are not (i.e., have a laxer criterion). The criterion depicted in Figure 1 may be viewed as a strict criterion in the sense that a sensation at the criterion is more likely to arise from the S+N distribution than from the N distribution alone.

To the extent that the two curves overlap, the subject cannot always be correct. If there is overlap, then the subject is forced to make a binary decision when in fact the sensations arising from the signal present and signal absent conditions are not mutually exclusive. Of course, if the signal magnitude increased markedly, then the signal distribution would move farther to the right and so the two distributions might not overlap. In these circumstance, there would be no region of ambiguity, and a higher proportion of correct responses.

In the yes-no paradigm, there are four distinct outcomes on each trial. These outcomes are delineated in Table 1 below. Given the signal, the observer must respond either yes or no, and so the sum of the probabilities in the top row of the matrix must equal one. Likewise, if the signal were not presented, the observer must respond either yes or no, and so the sum of the probabilities in the second row of the table must also equal one. There are then only two independent probabilities in the table.

Table 1. Response probabilities in a yes-no task

	Response: Yes	Response: No
Signal present	Hit	Miss
Noise Present	False-alarm	Correct rejection

Typically, the two probabilities used to describe the results are the hit and the false-alarm rates. These probabilities are graphed on the ordinate and abscissa of what is called a receiver operating characteristic (ROC), as seen in Figure 2.

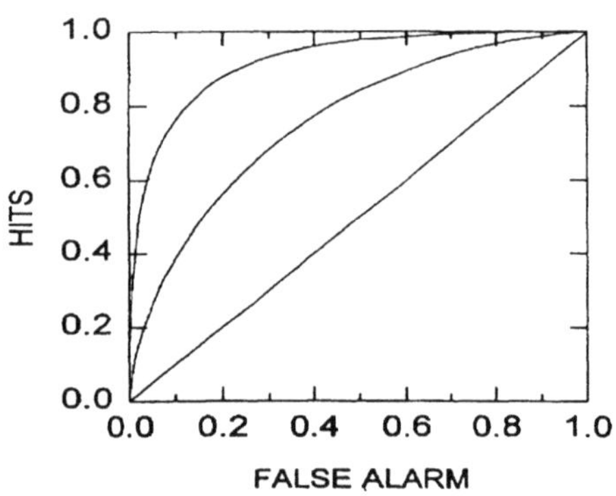

Figure 2

It is easy to intuit the form of the ROC curve. Suppose first that the magnitude of the signal is so low that nothing is actually heard. The subjects might respond "no" 100% of the time when the signal is present and "no" 100% of the time when the signal is absent, giving both a hit rate and a false-alarm rate of 0. This produces a point in the lower left corner of the ROC curve. However, if a subject had a response bias leading him to consider false alarms unimportant and hits very important, then the subject should always respond "yes" (even though hearing nothing). This produces a point in the upper right-hand corner of the ROC. For intermediate biases, points could be produced interior to the square, but so long as nothing is heard, the subject's performance will always lie on the diagonal line connecting the lower left corner to the upper right corner.

As the signal intensity rises so that it may be heard some of the time, then the observed data points should lie somewhere in the upper left quadrant of the square. For any given pair of signal and noise distributions, the area under the noise alone curve to the right of the criterion is the false-alarm rate and the area under the signal curve to the right of the criterion is the hit rate. ROC curves for several different signal levels are presented in Figure 2.

Note that different ROC curves tend to cluster tightly together for small and large false-alarm rates. This means that a single point with a low or high false-alarm rate cannot be used to distinguish between different ROC curves.

The ROC curve reveals two important concepts. First, as the false-alarm rate increases, the hit rate also increases. Second, all points on one ROC curve represent the same sensory response. Different points on one ROC curve represent different response criteria. As the subject's criterion moves, successive points on the same ROC curve are traced. Sensory sensitivity is reflected in the position of the entire ROC curve: the closer the ROC curve is to the upper left-hand corner, the greater the sensitivity.

Because interest focuses on the sensory capability and not on the subject's criterion, the area under the ROC curve provides a useful measure of detectability because it ranges from 0.5 to 1.0. The distance between the N and S+N curves divided by the standard deviation of the curves is defined as d' and is also a measure of detectability. Finally the distance, d_e, from the positive diagonal to the ROC curve (measured along the negative diagonal) provides another measure of sensitivity. The interesting point about the area measure is that it can be shown to be the percentage correct in a two-interval forced-choice task (2IFC) paradigm, utilizing the same stimulus values (Green & Swets, 1966).

Before ending the discussion of the influence of response bias in a yes-no paradigm, it is interesting to note that the logic in SDT is similar to the logic in statistical testing. SDT conceptualizes a yes-no detection paradigm as a situation in which the observer chooses between two mutually exclusive alternatives, signal or noise. Likewise, in statistics, an experimenter chooses between two mutually exclusive alternatives, the null hypothesis (H_o) and the experimental hypothesis (H_e). Statistical errors arise if H_e (or signal) is chosen when H_o (or noise) was correct. This error, called a false alarm in SDT, is called a Type I error in statistics and is generally fixed at 0.05 or 0.01. Statistical errors may also arise if H_o is chosen when H_e is true. This error, called an incorrect rejection in SDT, is called a Type II error in statistics and its probability is unknown.

3 Forced-Choice Paradigm

In the simplest of the forced-choice tasks, the observer is asked to indicate which of two clearly marked temporal intervals is contemporaneous with a signal. The signal is presented with equal probability in either the first or the second interval. If the signal magnitude was so low that it was never heard, then the subject has no reason to select one interval or the other and if the choice is random, the correct interval will be chosen on half the trials. If the signal magnitude is so large that it is always heard, then the subject should always be correct. Therefore, the 2IFC task has a psychometric function which ranges from 0.5 to 1.

Green and Swets (1966) showed that the area under the yes-no ROC curve is the percent correct in a 2IFC task. Recall that the ROC curve represented all possible outcomes for all possible decision criteria. The probability of a correct response in a 2IFC task provides the same information as the entire ROC curve. Thus, the probability of being correct in a 2IFC task is a criterion-free measure of detectability.

The threshold in a 2IFC task requires 100 trials or so to determine, whereas determination of an ROC curve requires many thousands of trial. Because of the speed with which the FC tasks may determine threshold, use of the FC paradigm is widespread in human psychophysics.

4 Psychophysical Methods

Detailed discussions of what have become known as classical method for tracking threshold date back to Fechner (1860). The classical techniques are the method of limits, the method of constant stimuli, and the method of adjustment. Descriptions of these techniques are beyond the scope of this chapter and summaries of them may be found in almost all books involving an introduction to hearing (eg., Christman, 1971; Penner, 1978; Woodworth & Schlossberg, 1938; Yost & Nielsen, 1977).

Briefly, though, the method of limits (MOL) alternates between ascending and descending series. In the descending MOL, the experimenter begins with a signal which is clearly audible and decreases its strength until it is no longer heard. In the ascending MOL, on the other hand, the signal strength is systematically increased. The change in signal strength, called the step size, must be carefully chosen so that a single change in stimulus level does not exceed the extent of the psychometric function. The method of constant stimuli (MOCS) is similar to the MOL except that the stimuli spanning the psychometric function are presented in random order. In the MOL and the MOCS, data spanning an entire psychometric function are collected. In the final method, the method of adjustment (MOA), the listener adjusts the strength of a standard stimulus until it just becomes inaudible (descending) or audible (ascending). In the MOA task, only the thresholds are recorded and so the psychometric function is not obtained. The MOA is generally efficient; the time required for a threshold estimate may be minimal.

The different psychophysical methods themselves may produce divergent thresholds (Herrick, 1967; Pollack, 1968). Furthermore, the dependent variable in many psychophysical studies corresponds to one point on the psychometric function. The MOL and the MOCS provide more information than this: they trace the entire psychometric function. Unless the psychometric function itself is of interest, it is not necessary to determine the entire function in order to find a single point on it. This observation may have led researchers (Raab et al., 1963; Levitt, 1971) to propose adaptive psychophysical methods. The Levitt (1971) technique has become almost ubiquitous for measures of human psychophysical responses.

An adaptive psychophysical procedure is one in which the stimulus value on any one trial is determined by the preceding stimuli and the subject's responses. In the most widely used version of the Levitt (1971) technique, the following strategy is adopted. Whenever the subject is wrong in a 2IFC task, the stimulus value increases; if the subject is right twice in a row for the same stimulus value, the value decreases; otherwise, the value is unchanged. This procedure converges to a stimulus value for which the probability of bcing right twice in a row, p^2, equals the probability of not being right twice in a row, $1-p^2$. That is, the procedure converges to that stimulus level for which $p = 0.707$. Levitt (1971) also presents a variety of other up-down rules tracking stimulus levels corresponding to other points on the psychometric functions.

There are many discussions of the optimal response probability at which to estimate the physical stimulus value (e.g., Penner, 1978). Green (1990) has used a computer simulation to demonstrate that the optimal stimulus placement level produces between 84% and 94% correct responses in a 2IFC task.

The step size in FC tasks must also be appropriate. Wetherill (1963) has suggested that the step size equal 1/6 the range of the psychometric function (i.e., a step corresponding to an 8% change in detectability), close to the 7.5% suggested in Green et al.'s (1989) paper.

Finally, most threshold estimates are based on a specific number of reversals. It is also possible to base estimates on the reversals in some fixed total number of trials reasoning that the observer should not be provided with the option of shortening the total number of trials simply be missing a few easy stimuli (Green et al., 1989).

5 Psychophysical Methods and "Classical" Data: Humans

Prior to the 1960s, no psychophysical data were collected in a forced-choice task and a non-zero false-alarm rate was considered the hallmark of a poor subject. In contrast, a good subject is now defined as one who makes as many false alarms as misses (i.e., puts his criterion in Figure 1 at the point at which the two curves have equal ordinates). Do data prior to 1960 in which response criteria were not controlled also reflect sensory sensitivity?

Although there is no single answer to that question, in broad terms, current advances have not negated classical results. Even modern textbooks on hearing (Gulick et al., 1989; Moore,

1989) display data from Wegel and Lane (1924), Fletcher (1940), and Egan and Hake (1950). These data would not be presented today if the criterion-free data were, in fact, markedly different. However, in many of the older studies, the experimenters were also subjects, raising the question of whether replicable constant-criteria data might be obtained from other subjects.

Comparisons of thresholds of relatively naive subjects using a Bekesey tracking task in which criteria may influence the data, and a forced-choice task have been published (Marshall & Jesteadt, 1986). Bekesey tracking may be viewed as a type of adjustment task. The 2IFC thresholds obtained by Marshall & Jesteadt (1986) at 0.5 and 4.0 kHz were 6.5 dB lower than those obtained using Bekesey tracking. Thus, data from tasks in which the criteria may influence the results may be somewhat different from forced-choice results.

6 Psychophysical Methods and Animal Psychophysics

As in human psychophysics, the focus in animal psychophysics is almost always on the organism's sensory capability rather than its response criterion. The task most commonly`used in animal psychophysics is the Go/No-Go task which is generally applied in such a way that it is functionally equivalent to the Yes-No paradigm in human psychophysics. Not surprisingly, the critical issue with this method in animal psychophysics is identical to that in the Yes-No paradigm with human subjects - how to deal with the problem of response bias in the assessment of an organism's sensitivity.

One method that has sometimes been used is derived from the high threshold theory of classical psychophysics and involves the application of statistical procedures to the data such as a measure (pc*) which is the percent correct adjusted for guessing (Hienz et al., 1977; Okanoya & Dooling 1990a,b; Sinnott et al., 1980). The problem with this method for minimizing the effects of variation in false alarms is that high-threshold theory predicts linear ROC curves which are not in fact found, at least for humans. Thus, high-threshold models are generally rejected in human psychophysics (Green & Swets, 1966; Gescheider, 1985). Whether the same holds true in animal psychophysics remains to be seen. The practical effect of applying this statistical correction is always a more conservative estimate of an animal's threshold. Unfortunately, however, the threshold obtained with the high-threshold correction is not necessarily the threshold which would be obtained with a constant behavioral false-alarm rate.

One common strategy for controlling response bias and minimizing the effect of variation in false alarm rate in a Go/No-Go task is to conduct the experiment in a way that the false-alarm rate is constrained. For example, it is a common practice to exclude the small fraction of test sessions (usually less than 10 %) in which false alarm rates rise above 20% (see, for example,

Langemann and Klump, 1992; Okanoya & Dooling, 1990a,b). For that matter, it is also common practice to exclude sessions in which an animal's miss rate for the most detectable or discriminable stimuli in the test set increases above 20% (e.g. see Langemann and Klump, 1992). The rationale here is that the animal is, temporarily at least, no longer under stimulus control.

The danger in this approach is that sensitivity is not perfectly reflected in comparisons of hit rates across conditions unless the false-alarm rate is actually constant, not approximately constant. For humans, large changes in d' accompany small changes in false-alarm rates. For example, if the false-alarm rate varies from 0% to 20% and if threshold is defined as 50% correct detection (typical of a Go/No-Go task) and if the signal alone and signal plus noise distributions are normal, then d' changes from 2.32 to 0.84 as the false-alarm rate changes from 0% to 20%. Thus, in practical terms, it would seem essential for the experimenter to conduct a sufficient number of sham (noise alone) trials to ensure that the false alarm rates across stimulus conditions are not different. Often the proportion of sham trials in a session is only 20-30 out of 100 total trials (see, for example, Langemann and Klump, 1992; Okanoya and Dooling, 1990a,b). This means that data from several sessions should be used in establishing a reliable value for the false alarm rate and for a threshold estimate for an animal. To the extent that any one animal exhibits consistent thresholds over time or a group of subjects exhibits similar thresholds, one can infer that the data are not markedly affected by shifts in criterion.

As indicated earlier, for theoretical reasons, the best approach to the separation of sensitivity and response proclivity in a Yes/No or Go/No-Go task would be to explore an entire ROC curve. There have been a number of ROCS reported for animals over the last 30 years (see, for example, Blough, 1967; Gray, 1992). The main reason that such procedures are not widely used in animal psychophysics is that the determination of a complete ROC curve or family of ROC curves requires collection of so much data that the procedure is simply too time-consuming. Furthermore, in contrast to human subjects, trained animal subjects induced to produce high false alarm rates (necessary for determining the entire ROC-curve) can rarely be retrained to behave as more conservative listeners.

The FC task is clearly the paradigm of choice for humans and it has been successfully applied in comparative psychoacoustics (Hulse, this volume; Klump et al. 1986, Klump, this volume; Schmidt, this volume). But this method is not yet widely used because it is not without problems. First, it is generally more difficult to train an animal to perform an FC task than a Go/No-Go task. Second, even well-trained animals may develop response interval or response position biases that, at best, complicate the interpretation of the data (Klump, this volume; Owren, 1990). Third, it is difficult to keep the animals under stimulus control because chance responding in a FC task results in food reinforcement 50% of the time - appealing odds for an organism accustomed to foraging for food under natural conditions. Nevertheless, if animals could be efficiently trained to perform FC tasks reliably, then data could be gathered in fewer test trials and potential problems associated with response proclivity avoided.

In sum, there is evidence that some psychophysical results from humans (audiometric thresholds) are not markedly dependent on psychophysical method. Further, even perceptual judgments of naive observers (Stevens, 1955) were considered sufficiently valid to warrant the International Standards Organization (1975) rule of 10 dB per doubling of loudness. Nonetheless, the specter of criterion-dependent data, means that psychophysical methods separating sensitivity and proclivity should be employed. If they are not, caveat emptor.

7 Suggested Readings

Green, D.M. & Swets, J.A. (1966). *Signal detection theory and psychophysics*. NY: Wiley.

Gulick, W.L., Gescheider, G.A. & Frisina, R.D. (1989). *Hearing*. Oxford University Press: Oxford, England.

Penner, M. J. (1978). Psychophysical methods and the minicomputer. In M. S. Mayzner & T. R. Dolan (Eds), *Minicomputer in sensory and information processing*. Lawrence Earlbaum Associates: New Jersey.

8 References

Christman,, R.J. (1971). *Sensory Experience*. London: International Textbook Company.

Craik, F.I.M. (1969). Application of signal detection theory to studies of aging. *Interdisciplinary Topics in Gerontology*, 4, 147-157.

Egan, J. P. & Hake, H. W. (1950). On the masking pattern of a simple auditory stimulus. *Journal of the Acoustical Society of America*, 22, 622-630.

Fechner, G.T. (1860). *Elemente der Psychophysik*. Leipzig: Breitkopf & Hartel.

Fletcher, H. (1940). Auditory patterns. *Review of Modern Physics*, 12, 47-65.

Gray, L. (1992). An auditory psychometric function from newborn chicks. *Journal of the Acoustical Society of America*, 91, 1608-1615.

Green, D. M. (1990). Stimulus selection in adaptive psychophysical procedures. *Journal of the Acoustical Society of America*, 87, 2662-2674.

Green, D. M., Richards, V. M. & Forrest, T.G. (1989). Stimulus step size and heterogeneous stimulus conditions in adaptive psychophysics. *Journal of the Acoustical Society of America*, 86, 629-636.

Green, D.M. & Swets, J.A. (1966). *Signal detection theory and psychophysics*. New York: Wiley.

Gulick, W.L., Gescheider, G.A. & Frisina, R.D. (1989). *Hearing*. Oxford University Press: Oxford, England.

Hienz, R.D., Sinnott, J. M. & Sachs, M.B. (1977). Auditory sensitivity of the redwing blackbird (*Agelaius phoeniceus*) and brown-headed cowbird (*Molothrus ater*). *Journal of Comparative Physiology and Psychology*, 91, 1365-1376.

Herrick, R.M. (1967). Psychophysical methodology: Comparison of thresholds of the method of limits and of the method of constant stimuli. *Perceptual and motor skills, 24, 915-922.*

International Standards Organization (1975). Recommendation 532. New York: International Organization for Standardization.

Langemann, U. and Klump, G.M. (1992). Frequency discrimination in the European starling (*Sturnus vulgaris*): a comparison of different measures. *Hearing Research*, 63, 43-51.

Larkin, W.D. & Norman, D.A. (1964). An extension and experimental analysis of neural quantum theory. In R.C. Atkinson (Ed.), *Studies in mathematical psychology*. Stanford: Stanford University Press.

Levitt, H. (1971) Transformed up-down methods in psychoacoustics. *Journal of the Acoustical Society of America*, 49, 467-477.

Marshall, L. & Jesteadt, W. (1986). Comparison of pure-tones\ audibility thresholds obtained with audiological and two-interval forced-choice procedures. *Journal of Speech and Hearing Research*, 29, 82-91.

Moore, B.C.J. (1989). *An Introduction to the Psychology of Hearing*.Academic Press : San Diego, California.

Oakanoya, K. & Dooling, R.J. (1990a). Detection of auditory sinusoids of fixed and uncertain frequency by budgerigars (*Melopsittacus undulatus*) and zebra finches (*Poephila guttata*). *Hearing Research*, 50, 175-184.

Okanoya, K. & Dooling, R.J. (1990b). Detection of gaps in noise by budgerigars (*Melopsittacus undulatus*) and zebra finches (*Poephila guttata*). *Hearing Research*, 50,185-192.

Owren, M. J. (1990). Acoustic classification of alarm calls by vervet monkeys (*Cercopithecus aethiops*) and humans (*homo sapiens*): II. Synthetic calls. *Journal of the Acoustical Society of America*, 104, 29-40.

Penner, M. J. (1978). Psychophysical methods and the minicomputer. In M. S. Mayzner & T. R. Dolan (Eds), *Minicomputer in sensory and information processing*. Lawrence Earlbaum Associates: New Jersey.

Pollack, I. (1968). Computer simulation of threshold observations by the method of limits. *Perceptual and Motor Skills*, 26, 583-586.

Potash, M. & Jones, B. (1971). Aging and decision criteria for the detection of tones in noise. *Journal of Gerontology*, 32, 436-440.

Raab, D.H., Osman, E. & Rich, E. The effects of waveform correlation and signal duration on detection of noise bursts in continuous noise. *Journal of the Acoustical Society of America*, 35, 1942.

Rees, J.N. & Botwick, J. (1971). Detection and decision factors in auditory behavior of the elderly. *Journal of Gerontology*, 26, 133-136.

Robinson, D.E. & Watson, C.S. (1972).Psychophysical methods in modern psychophysics. In J.V. Tobias (Ed.), *Foundations of Modern Auditory Theory*, Volume 2.

Sinnott, J.M., Sachs, M.B. & Hienz, R.D. (1980). Aspects of frequency discrimination in passerine birds and pigeons. *Journal of Comparative Physiology and Psychology*, 94, 401-415.

Stevens, S.S. (1955). The measurement of loudness. *Journal of the Acoustical Society of America*, 9, 11-23.

Stevens, S.S., Morgan, C.E. & Volkmann, J. (1941). Theory of neural quantum in the discrimination of loudness and pitch. *American Journal of Psychology*, 53, 315-355.

Wegel, R.L. & Lane, C.E. (1924). The auditory masking of one sound by another and its probable relation to the dynamics of the inner ear. *Physics Review*, 23, 266-285.

Wetherill, G.B. (1963). Sequential estimates of quantal response curves. *Journal of the Royal Statistics Society*, 25, 1-48.

Woodworth, R. S. & Schlossberg, H. (1938). *Experimental Psychology*. New York: Holt & Company.

Yost, W.A. & Nielsen, D.W. (1977). *Fundamentals of Hearing*. New York: Holt, Rinehart and Winston.

The Study of Basic Hearing Mechanisms

Introductory Remarks

There is a long history of a reciprocal generation of hypotheses and theories between psychoacoustics on the one hand and physiological and anatomical studies of hearing on the other. Thus, the search for basic mechanisms underlying hearing and perception occurs at the nexus of psychoacoustics and physiology. Comparative psychoacoustics makes a unique contribution here since the exploration of animal models allows detailed investigations into the physiological and anatomical foundations of hearing in ways not possible with human subjects. Moreover, the comparison of hearing capabilities and mechanisms among different species provides a kind of natural experiment where one can gain insight into the adaptation and evolution of hearing mechanisms and the role that environmental selective pressures may have played. Comparing analogous and homologous auditory structures across vertebrates provides a much richer understanding of the forms, functions, and anatomical bases of hearing in any one species, including the human. None of this would be possible without the precise measurement of behavioral detection and discrimination thresholds. The modern technologies presented in this section illustrate the rigor necessary for cross-species comparisons of absolute thresholds and discrimination limens.

Mammals

The study of hearing in other mammals has special relevance to our understanding of human hearing. The structural similarities in critical features of the auditory pathway make comparisons among the mammals especially likely to provide insights into basic auditory processes in humans. Patterns of similarities and differences among mammals have lead to the development of some species as animal models for human disease processes. On the other hand, as the field of comparative psychoacoustics has revealed, there is great diversity among the mammals in hearing capabilities. Thus, within the mammals, there are ample opportunities to study the auditory specializations that have presumably evolved to meet the demands of particular environmental niches. The study of these specializations provides insight into the evolution of hearing mechanisms.

The first chapter in this section by Niemiec and Moody compares several psychophysical procedures for measuring absolute auditory thresholds in the macaque monkey. The remarkable similarity of results obtained using these procedures attests to the validity of current comparative psychoacoustic methods when carefully applied. Heffner and Heffner describe a method of conditioned avoidance that has been used with a considerable range of mammalian species with great success. The chapter by May and his colleagues describes an application of the method of constant stimuli to the measurement of hearing in cats - one of the most popular animal models for human auditory processes. Brown describes a set of well-developed methods, derived from classical psychophysics, for measuring directional hearing in animals with specific application to primates. Schmidt presents an application of the two-alternative forced-choice procedure in the study of hearing in bats - an auditory specialist whose acoustic world is largely outside the frequency range of human hearing. Finally, the study of hearing in nonverbal human infants has much in common with the study of hearing in animals - in both cases, creative strategies tailored to the subjects' response proclivities must be developed to induce the organism to reveal what it hears. Werner describes a new and successful procedure for use with human infants called observer-based approaches to human infant psychoacoustics. This procedure may have great potential for use with animal species that are normally intractable in psychoacoustic research.

Methods in Comparative Psychoacoustics
G. M. Klump, R. J. Dooling, R. R. Fay & W. C. Stebbins (eds)

Constant Stimulus and Tracking Procedures for Measuring Sensitivity

Andrew J. Niemiec and David B. Moody

Kresge Hearing Research Institute, The University of Michigan Medical School
Ann Arbor, Michigan 48109-0506, USA

Summary. The methods of constant stimuli, adaptive tracking, and transformed tracking are fundamental tools for the study of sensory function of humans and animals. This chapter describes the adaptation of these procedures to the study of auditory sensitivity in the macaque monkey, outlines the primary advantages and disadvantages of each of these techniques, and describes situations in which one of these techniques might be more appropriate than another.

1 Introduction

The most basic questions in the study of sensation and perception involve determining the relationship between the physical parameters of a stimulus and the sensation it produces. Of these questions, perhaps the most fundamental involves determining the relationship between the magnitude of a stimulus and the sensory system's ability to detect it. With the publication of his *Elements of Psychophysics* in 1860, Gustav Fechner developed the framework for the scientific study of these types of questions in human subjects. Almost one hundred years later, following an integration of technology and behavior analysis, the field of animal psychophysics was born, adapting many of Fechner's methods to the study of animal sensory systems (Blough, 1966; Stebbins, 1970, 1990). This paper describes the application of Fechner's method of constant stimuli and variations on his method of limits, i.e. adaptive tracking and transformed tracking, to the measurement of auditory sensitivity in macaque monkeys.

2 Training and Methods

Because we cannot use a common language to ask animal subjects to report the detection of an auditory stimulus, we must find alternative ways to determine whether or not our subjects can detect the stimuli presented to them. One relatively straightforward way to make these determinations is to use operant conditioning techniques to train the animal subject to make an "observing response" in the absence of a stimulus and a "reporting response" upon detection of the stimulus. We find a hold-release procedure to be effective. In this procedure the observing response consists of initiating and maintaining key contact and the reporting response consists of releasing the key. The observing response is essentially an indication that the animal is ready to work on the procedure and is listening for a stimulus presentation. An additional benefit of the observing response, particularly with small animals being tested in a free field, is that the observing response ensures that the animal is in a relatively fixed position so that the sound level at the animal's ear does not differ greatly from trial to trial. The reporting response requires the animal to perform a distinct task to indicate detection of the stimulus. Training a subject to produce stable behavior usually takes several weeks and can require subtle adjustments to the procedure to bring behavior under appropriate control of the auditory stimuli. A brief description of the training process is presented here and a more detailed description is available in Moody, et al. (1976).

For training and testing, the monkey is placed into a primate-restraining chair that allows sitting in a comfortable position during the testing sessions. The training takes place in a sound attenuating chamber equipped with a cue light, a response manipulandum (a metal cylinder), headphones for stimulus presentation, and a food pellet dispenser for rewarding the animal. Contact with the metal cylinder serves as the behavioral response and is sensed by a circuit that detects a minute current flowing from the cylinder through the monkey to the grounded chair. Once the animal is placed into the testing chamber, headphones are mounted onto the primate chair such that they comfortably cover both of the animal's ears and, therefore, allow monaural testing. For threshold determinations, stimuli are pure tones ranging from 125 Hz to 32 kHz in one octave intervals. Stimulus presentation and food reward can be controlled manually and by a computer.

Training the observing response involves getting the monkey to contact the response cylinder and to maintain that contact until a stimulus is detected. The trainer initially shapes this response by rewarding the animal whenever its hand approaches the response cylinder. Using the technique of successive approximation, in which reward is withheld until the animal more closely approximates the desired behavior, the monkey eventually learns to use its hand to make contact with the cylinder, at which time the computer presents a clearly audible tone. Upon release of contact with the cylinder (the "reporting response"), the computer dispenses a banana-flavored food pellet that serves as a reinforcer. When the monkey is reliably touching and releasing the response manipulandum, a time window is placed on the release of the response manipulandum such that the monkey must release within three seconds of the onset of the tone to receive the banana-pellet reward. Simultaneously, the interval between the onset of the observing response and the tone presentation is gradually increased. This interval is determined by the computer on a trial-by-trial basis and, in the final procedure, varies randomly from one to eight seconds.

Once this level of performance is reached, catch trials are inserted into the testing procedure. Catch trials, which serve to monitor and control the probability that the animal will make reporting responses in the absence of the test stimuli, are identical to tone trials except that the tone is not presented. The correct behavior is to maintain the observing response throughout the duration of the catch trial. If the animal releases the response manipulandum during a catch trial, reporting that it heard a tone when none was presented, it is put into a time-out which serves as a mild punisher that is effective in reducing the occurrence of inappropriate reporting responses. During a time-out, the cue light is turned off and responses are ineffective in producing test tones. We have found time-out durations of five to eight seconds to be most effective, but occasionally adjust this parameter to longer values. Over several sessions, the proportion of catch trials is gradually increased to 20%. Once the animal reliably reports supra-threshold tones and correctly withholds responses on at least 80% of the catch trials, one of the three psychophysical methods can be employed to manipulate the level of the stimulus and measure the animal's threshold for the tone.

3 The Method of Constant Stimuli

In the method of constant stimuli, the experimenter selects several stimulus levels that bracket the subject's estimated threshold and presents each level several times. The subject's response to each presentation of a particular stimulus level is tabulated to yield the proportion of correct detections for each level. This procedure yields a psychometric function that plots the percentage of correct detections as a function of stimulus level. The absolute threshold for the stimulus is then estimated from an arbitrary point on the psychometric function. Typically, threshold is defined as the level of the stimulus that can be detected correctly on 50% of the presentations. Another procedure that is sometimes used to estimate threshold and control for the effects of bias is to take the percent correct midway between the catch trial response rate and 100% correct as the threshold percent correct and to determine the threshold stimulus level based on this number. Thus, if a subject responded on 20% of all catch trials, threshold would be estimated at 60%, halfway between 20% and 100%.

The main advantages of the method of constant stimuli are: (1) the stimulus presentation levels used cover a wide range, allowing the experimenter to estimate many points on the psychometric function (Levitt, 1970) and (2) the majority of the stimuli are presented above the 50% correct point, making the overall task of detecting the stimuli relatively easy for the subject (Moody, et al., 1976). This point is particularly useful when an animal is required to make very difficult detections or discriminations. Because most trials are placed above threshold, the animal is not forced to make a difficult detection or discrimination on each trial, therefore, the relatively few trials below the animal's threshold should not adversely affect stimulus control of the behavior. The main disadvantages of this method are: (1) a preliminary estimate of threshold is necessary in order for the experimenter to choose the appropriate range of stimulus levels and (2) it requires many test trials so that if the experimenter is interested in estimating only one point on the psychometric function, it is relatively inefficient since many of the trials presented are not used in the threshold determination (Levitt, 1970). However, more of the data available from the psychometric function can be incorporated into threshold estimation by implementing a procedure such as curve fitting or computing a linear regression over the straight portion of the psychometric function.

Our implementation of the method of constant stimuli begins by determining a rough estimate of threshold for a randomly chosen test frequency. This estimate is determined by presenting the monkey with a supra-threshold stimulus at the test frequency. Each time the monkey correctly detects the stimulus, the level is reduced by 10 dB. This process continues until the monkey misses one of the stimulus presentations. This procedure produces a rough estimate of the animal's threshold that is used to select the range of stimulus levels to be presented to the subject to measure a psychometric function. Six stimulus levels in 10-dB increments are chosen over a 50-dB range on the basis of this rough threshold estimate such that one stimulus is at the missed level and the remaining stimuli are above the missed level. Ideally, this method of choosing the stimulus levels results in five stimuli above the 50% correct point on the psychometric function and one stimulus below 50% correct. If the animal begins to detect stimuli at the lowest presentation level, the range of stimuli is adjusted downward by the computer.

Once chosen, the six stimulus levels are presented to the animal in a random order and the animal's response to each stimulus level is recorded. Each stimulus level is presented several times. After presenting the complete set of stimulus levels at a given test frequency, the computer randomly picks a new test frequency, determines a rough threshold estimate for the new frequency, and calculates the range of stimulus presentation levels for that frequency. The computer then randomly presents the levels for the new test frequency to determine threshold sensitivity. This procedure is repeated until psychometric functions have been measured for all nine test frequencies. This typically takes about one hour, allowing measurement of a complete sensitivity function in each testing session. Figure 1 shows examples of psychometric functions measured for seven of the nine test frequencies using the implementation of the method of constant stimuli described above.

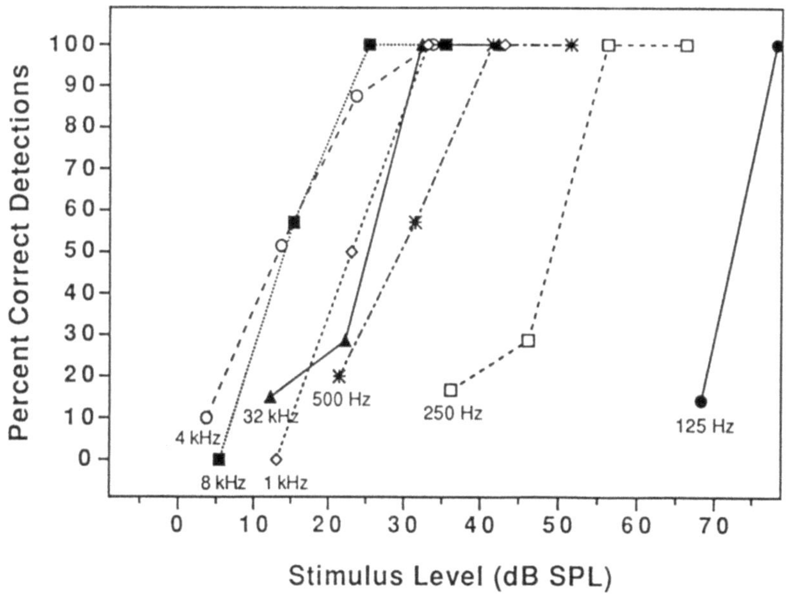

Figure 1. Examples of psychometric functions obtained using the method of constant stimuli. Functions for seven of the nine test frequencies are presented. The lowest four points of each function are plotted for all but the 125 Hz function. For the 125 Hz function, only the lowest two points are plotted.

Daily thresholds for each of the nine test frequencies are estimated by interpolating the stimulus level that yields 50% correct detection on the psychometric function. Data from testing sessions that have an overall catch trial response rate over 20% are discarded. Final threshold estimates for each test frequency are determined when the subject meets a stability criterion that requires: (1) four out of five daily threshold estimates must fall within ±5 dB of the median daily threshold estimate and (2) there is no apparent trend toward an increase or decrease in threshold. When the stability criterion is met for a given test frequency, the median of the five stable threshold estimates is taken as the final threshold at that test frequency. Testing continues until stable estimates of threshold are measured at all test frequencies. Figure 2 shows the absolute thresholds for all nine test frequencies measured using the method of constant stimuli. The monkey's best sensitivity is in the 1-8 kHz range, with thresholds at higher and lower frequencies showing decreasing sensitivity.

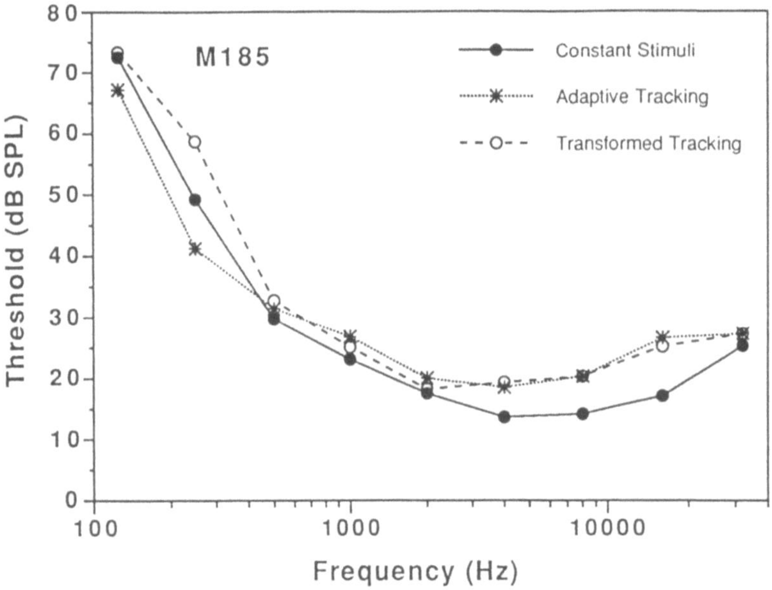

Figure 2. Absolute thresholds for a young male Rhesus monkey. The three sensitivity functions were measured at nine test frequencies using the method of constant stimuli, adaptive tracking, and transformed tracking.

4 Adaptive Tracking

The one-down/one-up adaptive tracking procedure is essentially a variation of Fechner's method of limits. In the method of limits, the stimulus intensity is increased until the subject reports detection (ascending series) or decreased until the subject fails to report detection (descending series). Typically, several sets of ascending and descending series of stimulus levels are presented and for each series the experimenter determines the level at which the subject switches from detecting the stimulus to non-detection or vice-versa. Thresholds are then determined by averaging estimates from a number of ascending and descending series. In adaptive tracking, the ending point of one series is used as the starting point of the next series.

Like the method of constant stimuli, the adaptive tracking procedure uses two response categories: detection (releasing key contact) or non-detection (maintaining contact through the stimulus presentation). The adaptive tracking procedure uses a one-down/one-up tracking rule: after each correct detection, the stimulus level is reduced and after each miss, the stimulus level is increased. Adaptive tracking differs from transformed tracking, described below, in that the latter uses more complex rules for determining what stimulus will be presented. Tracking is begun by initially presenting the stimulus at a supra-threshold level. From this point on, the stimulus level is decreased by 10 dB after each correct detection and increased by 10 dB after each miss. The adaptive tracking procedure incorporates both ascending and descending series of stimulus levels because once a stimulus is missed or detected, the direction in which the stimulus level changes is reversed. In our procedure, reversing the direction of stimulus level presentation following each detection or miss is repeated until at least six level reversals take place. After the required reversals are obtained, the first two are discarded and the remaining reversals are averaged to derive an estimate of threshold. We discard the first two reversals on the assumption that they will be the most variable and that the later reversals will yield the most stable thresholds. Figure 3 shows a typical adaptive tracking run at a 1-kHz test frequency for one subject.

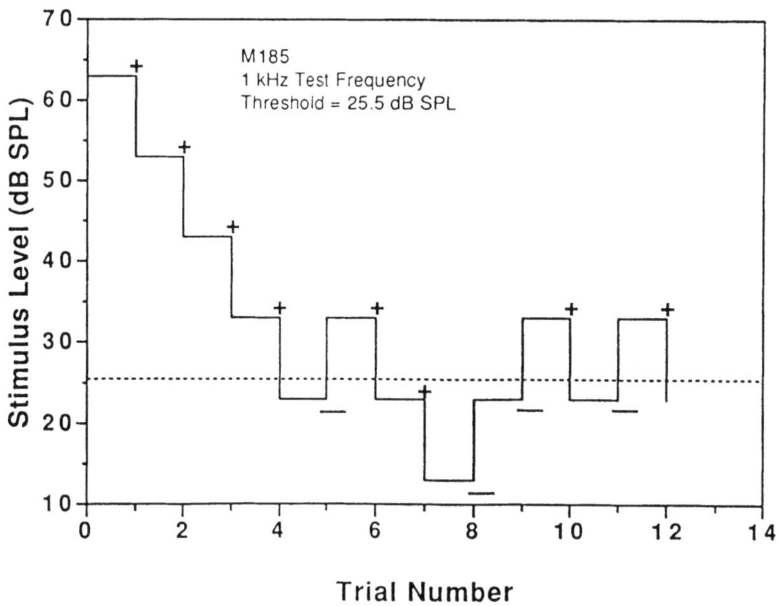

Figure 3. A graphical representation of a single run of the one-down/one-up adaptive tracking procedure. Each correct detection (+) resulted in a 10 dB decrease in the stimulus level while each miss (-) resulted in a 10 dB increase. Six reversals are shown in this figure. The first two reversals were discarded and the remaining four reversals were averaged to estimate threshold.

Once a threshold is determined at a given test frequency, the computer randomly selects the next test frequency and begins tracking the subject's threshold. This procedure is repeated until threshold estimates have been measured for all test frequencies. In a typical one-hour testing session, this procedure allows the measurement of three threshold estimates for each of nine test frequencies. These three estimates are averaged together for each frequency and this average is used as the daily threshold estimate, subject to the stability criterion described previously. Testing continues until stable estimates of threshold are measured at all test frequencies. The absolute thresholds for all nine test frequencies measured using this procedure are shown in Figure 2. Again, the monkey's best sensitivity is in the 1-8 kHz range, with the thresholds measured using the one-down/one-up adaptive tracking procedure falling very close to the thresholds measured using the method of constant stimuli.

The primary advantage of the one-down/one-up adaptive tracking procedure is the increase in efficiency over the method of constant stimuli and the method of limits since only those trials necessary to determine threshold are presented (Moody, et al., 1976). We can see the increased efficiency in data collection by the fact that switching from the method of constant stimuli to the adaptive tracking procedure in the present example roughly tripled the number of threshold estimates measured per testing session. Also related to this issue are the tradeoffs involved in deciding on the number of stimulus presentations to be used. Under our implementation of the method of constant stimuli, increasing the number of stimulus presentations from six to seven would require six additional trials per threshold determination, one at each test stimulus level, whereas one additional reversal under our implementation of the adaptive tracking procedure could require as few as two additional trials per threshold determination. Clearly the adaptive tracking procedure is more efficient. Another possible advantage of the adaptive tracking procedure may be an increase in the reliability of the threshold estimates. Because most observations are placed near threshold, the threshold estimates may be less variable than those derived from the method of constant stimuli which relies on points that may be far above and far below threshold to interpolate the threshold estimates.

One of the primary disadvantages of this procedure is that it is difficult for the animal because, except for the initial descending series, the stimulus is always presented at near-threshold levels. Another disadvantage results from the sequential nature of the tracking rule. That is, it may be possible for the animal to anticipate the next stimulus level and to adjust its responses so that the stimulus level is easy to detect. One final disadvantage is that guessing can sometimes drive the level of the stimulus far below threshold, causing a further increase in guessing behavior due to the inability of the animal to detect the sub-threshold stimulus (Moody, et al., 1976). While the first disadvantage makes this procedure unsuitable for use with difficult animals, the second disadvantage has rarely been observed in our laboratory. The third problem, however, is often seen in animals that do not show good stimulus control over the behavior prior to the onset of testing. There are a number of strategies that can be employed if this problem develops. One possibility is to increase the proportion of catch trials in order to give the animal more practice with catch trials. Another option would be to increase the duration of the time-out for catch trial responding. While this is an effective technique, it should be used judiciously since long time-outs reduce reinforcement density and may reduce the animal's motivation to perform the task. One final option is to reinforce the animal for holding through catch trials (correct rejections). This results in the animal being reinforced for making correct decisions regardless of whether the decision is a correct detection or a correct rejection.

5 Transformed Tracking

Like the one-down/one-up adaptive tracking procedure described above, the transformed tracking procedure is essentially another variation of Fechner's method of limits. The transformed tracking procedure uses a two-down/one-up tracking rule so that the stimulus level is reduced following two correct detections in a row and increased following each miss.

As in adaptive tracking, transformed tracking is begun by initially presenting the stimulus at a supra-threshold level. From this point on, the stimulus level is decreased by 10 dB after two correct detections in a row and increased by 10 dB after every miss. After at least six reversals have occurred, the first two are discarded and the remaining reversals are averaged to derive an estimate of threshold. Figure 4 shows a typical transformed tracking run at a 1-kHz test frequency for our subject.

As before, once a threshold estimate for a given frequency is determined, the computer randomly chooses another frequency and begins tracking the subject's threshold. The procedure is repeated until threshold estimates have been measured at all nine test frequencies. Typically, this procedure allows the measurement of two threshold estimates at each test frequency during the one-hour testing session. These two estimates are averaged together for each frequency and the average is used as the daily threshold estimate that is subjected to the stability criterion. As with the psychophysical methods described previously, testing continues until stable estimates of threshold are measured at all test frequencies. Figure 2 shows the absolute thresholds for all nine test frequencies measured using transformed tracking. Thresholds measured using this technique are very similar to the thresholds measured using the method of constant stimuli and the adaptive tracking technique.

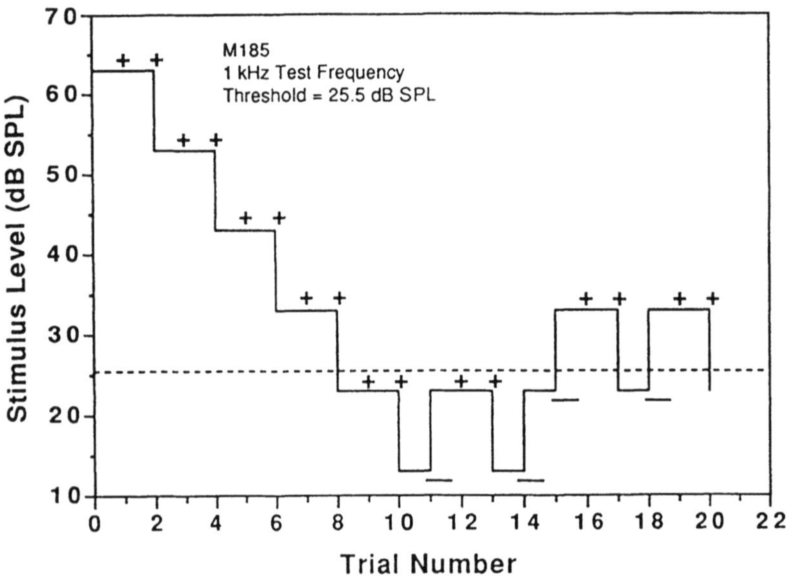

Figure 4. A graphical representation of a single run of the two-down/one-up transformed tracking procedure. Stimulus level was decreased by 10 dB after two correct detections (+) in a row and increased by 10 dB following each miss (-). Six reversals are shown in this figure. The first two reversals were discarded and the remaining four reversals were averaged to estimate threshold.

Although the transformed tracking procedure is more efficient than the method of constant stimuli and the method of limits, it is less efficient than the one-down/one-up adaptive tracking procedure. This is due to the increase in the number of trials required to get two correct detections in a row. The decreased efficiency in data collection can be seen by comparing the adaptive and transformed tracking procedures with the method of constant stimuli. Adaptive tracking tripled the number of threshold estimates measured per testing session whereas transformed tracking only doubled the number of threshold estimates measured per testing session.

Transformed tracking does have advantages over adaptive tracking, however. First, in adaptive tracking, the stimulus is presented above and below threshold an equal number of times, on average. This task is difficult for the animal because it is forced to work at near-threshold stimulus levels. Transformed tracking is not as difficult for the animal because on average, the stimulus is presented twice as often above threshold as below threshold. This feature of the procedure helps reduce the amount of guessing during a tracking run so that the stimulus level is less likely to be driven far below threshold. The result is better stimulus control over the animal's behavior and a procedure that is more suitable for problem animals than the adaptive tracking procedure. In addition, the transformed tracking procedure is less susceptible to random guessing than the adaptive tracking procedure. To illustrate, Figure 5 compares the effects of a random guess on both adaptive and transformed tracking procedures. In the figure we used the tracking runs presented in Figures 3 and 4 and inserted a correct detection of a stimulus that was, in fact, below the subject's true threshold. Since this trial is below threshold, the animal fails to detect the stimulus on the very next trial which must also be below threshold. From this point on, the animal continues to perform as in Figures 3 and 4. From Figure 5 we can see that the random guess had no effect on the threshold estimate measured using the transformed tracking procedure. However, threshold decreases by 2.5 dB in the adaptive tracking procedure.

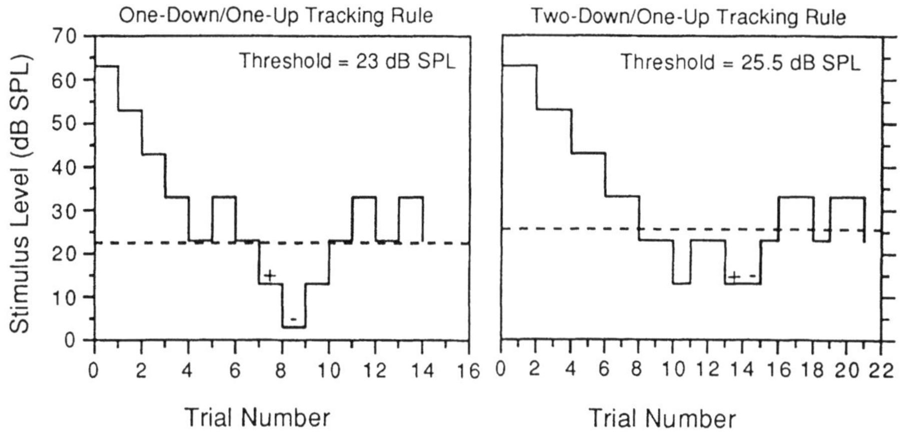

Figure 5. The effect of a random guess that results in a correct detection. The chance correct detection (+) occurs on Trial 7 in the one-down/one-up adaptive tracking procedure and on Trial 13 in the two-down/one-up transformed tracking procedure. In both procedures, the animal misses the very next trial (-). In the adaptive tracking procedure, threshold decreases by 2.5 dB whereas in the transformed tracking procedure, threshold remains unchanged.

Another possible advantage of the transformed tracking procedure, which could result from the more complex nature of the two-down/one-up tracking rule, is that it would be more difficult for the animal to anticipate the next stimulus level and to adjust its responses so that the stimulus level is easy to detect.

6 Discussion

This chapter compares the method of constant stimuli, a one-down/one-up adaptive tracking procedure, and a two-down/one-up transformed tracking procedure for use in measuring auditory sensitivity in a macaque monkey. All three procedures produce similar threshold estimates at all test frequencies, however, each procedure differs in terms of difficulty for the animal and in its efficiency. The method of constant stimuli is easiest for the subject, primarily due to the large number of trials that occur above threshold, however, it is also the least efficient method for estimating threshold due to the large number of trials required to generate the psychometric functions. The adaptive tracking procedure is most efficient at estimating threshold, however, because most of the stimuli are presented near threshold, it is also the most difficult procedure from the animal's perspective. Transformed tracking offers a good compromise for both difficulty and efficiency, falling somewhere between the method of constant stimuli and adaptive tracking. Transformed tracking requires more trials than adaptive tracking, but fewer than the method of constant stimuli. In addition, these trials are presented at levels that are generally above threshold, making it easier to maintain stimulus control over the animal's behavior.

7 Acknowledgements

The authors would like to thank Catherine Thompson and Colleen Garbe for technical assistance. The preparation of this chapter was supported by program project grant P01-DC00078 from NIDCD-NIH.

8 Suggested Reading

Berkley, M.A. and Stebbins, W.C. (1990) *Comparative Perception: Basic Mechanisms* (Vol. 1). John Wiley & Sons: New York.

Fay, R.R. (1988). *Hearing in Vertebrates: A Psychophysics Databook.* Hill-Fay Associates: Winnetka, IL.

Gescheider, G.A. (1985). *Psychophysics: Method, Theory, and Application.* Lawrence Erlbaum Associates: Hillsdale, NJ.

Levitt, H. (1970). Transformed up-down methods in psychoacoustics. *Journal of the Acoustical Society of America* 49(2): 467-477.

Moody, D.B., Beecher, M.D., and Stebbins, W.C. (1976). Behavioral methods in auditory research. In C.A. Smith and J.A. Vernon (eds) *Handbook of Auditory and Vestibular Research Methods.* Charles C. Thomas: Springfield, IL, pp.439-497.

Stebbins, W.C. and Berkley, M.A. (1990) *Comparative Perception: Complex Signals* (Vol. 2). John Wiley & Sons: New York.

9 References

Blough, D.S. (1966). The study of animal sensory processes by operant methods. In W.K. Honig (ed) *Operant Behavior: Areas of Research and Application.* Appleton-Century-Crofts: New York, pp.345-379.

Fechner, G. T. (1860). *Elemente der Psychophysik.* Breitkopf und Härtel: Leipzig. English translation of Vol. 1 by H.E. Adler. (D.H. Howes and E.G. Boring (eds)) Holt, Rinehart and Winston: New York, 1966.

Levitt, H. (1970). Transformed up-down methods in psychoacoustics. *Journal of the Acoustical Society of America* 49: 467-477.

Moody, D.B., Beecher, M.D., and Stebbins, W.C. (1976). Behavioral methods in auditory research. In C.A. Smith and J.A. Vernon (eds) *Handbook of Auditory and Vestibular Research Methods.* Charles C. Thomas: Springfield, IL, pp.439-497.

Stebbins, W.C. (1970). *Animal Psychophysics: The Design and Conduct of Sensory Experiments.* Appleton-Century-Crofts: New York.

Stebbins, W.C. (1990). Perception in animal behavior. In M.A. Berkley and W.C. Stebbins (eds) *Comparative Perception: Basic Mechanisms (Vol. 1).* John Wiley & Sons: New York, pp. 1-26.

Methods in Comparative Psychoacoustics
G. M. Klump, R. J. Dooling, R. R. Fay & W. C. Stebbins (eds)

Conditioned Avoidance

Henry E. Heffner and Rickye S. Heffner

Laboratory of Comparative Hearing and the Department of Psychology, University of Toledo, Toledo, Ohio 43606, USA

Summary. The procedure described here involves training an animal to make steady contact with a reward spout in order to receive food or water and then pairing a stimulus with mild electric shock delivered through the spout. The animal quickly learns to avoid the shock by breaking contact with the spout whenever it detects the stimulus. The breaking of contact with the spout is then used to indicate that the animal detected the stimulus. This procedure can be used to assess sensory and perceptual abilities in a wide variety of animals.

1 Introduction

In devising a psychophysical procedure for use with animals, it is important to choose a task which utilizes an animal's natural responses and is therefore easily learned. One response common to many animals is to freeze or suppress ongoing behavior when a stimulus that signals danger is detected. This suppression of behavior was experimentally investigated by Estes and Skinner in 1941, and has been used extensively by James C. Smith of Florida State University as a psychophysical procedure (Smith, 1970; Thaw & Smith, 1992). Referred to as *conditioned suppression*, it involves training an animal to make a response, such as pressing a lever to obtain food, and then presenting a stimulus followed by an electric shock. After a few stimulus-shock pairings an animal will cease responding when the stimulus is presented; this cessation of responding is then used to indicate that the animal detected the stimulus. It should be noted that this procedure is a *two-choice* task in which a subject makes one response in the presence of one stimulus and a different response when that stimulus is absent or a different stimulus is present.

Over the years, we have gradually modified the procedure developed by Smith in order to simplify training and accelerate testing. The procedure we now use represents a significant departure from the original conditioned suppression procedure in that it allows an animal either to avoid or escape the shock. Like conditioned suppression, this *conditioned avoidance* procedure has proved useful in testing the sensory and perceptual abilities of a wide variety of animals.

2 Conditioned Avoidance Procedure

The following is a description of the conditioned-avoidance procedure which has been developed in the course of the comparative study of mammalian hearing. Although most of the examples in this chapter have been drawn from that field, this procedure can be applied, *mutatis mutandis*, to any two-choice discrimination involving animals.

2.1 Overview of the Procedure

A hungry or thirsty animal is placed in a test cage and allowed to consume a steady trickle of food or water which is dispensed through a "reward" spout as long as the animal is in contact with the spout. Next, a suprathreshold stimulus is presented at random intervals and followed by a mild electric shock delivered through the reward spout. The animal soon learns to associate the stimulus with the shock and breaks contact with the spout whenever it detects the stimulus thereby avoiding the shock. The presentation of the stimulus constitutes a *warning* trial and breaking contact with the reward spout during a warning trial is taken as an indication that the animal detected the stimulus.

The response of an animal on each warning trial is recorded by a computer which determines whether or not the animal was in contact with the spout immediately before the shock was delivered. In signal detection terminology, breaking contact during a warning trial is referred to as a *hit*, while failure to do so is a *miss*. Because an animal occasionally breaks contact in the absence of a warning stimulus, its *false alarm* rate is obtained by determining its response rate during *safe* trials, that is, intervals when a stimulus could have been, but was not, presented.

A detection threshold is determined by reducing the intensity of the stimulus in successive blocks of trials until the animal no longer responds to the stimulus above the level expected by chance—in other words, the response rate during the warning trials no longer differs statistically from that during the safe trials. Similarly, a difference threshold is determined by reducing the difference between two stimuli until performance falls to chance. Threshold is defined as the stimulus level (or difference) resulting in a performance level of 50%.

2.2 The Test Cage

The design of the test cage is determined by the requirements of the stimulus as well as the species being tested. In auditory research where an animal is placed within a sound field, the cage is constructed of a sound-transparent material, such as wire mesh, and obstructions to sound are minimized (Fig. 1). An important feature of the test cage is the reward spout.

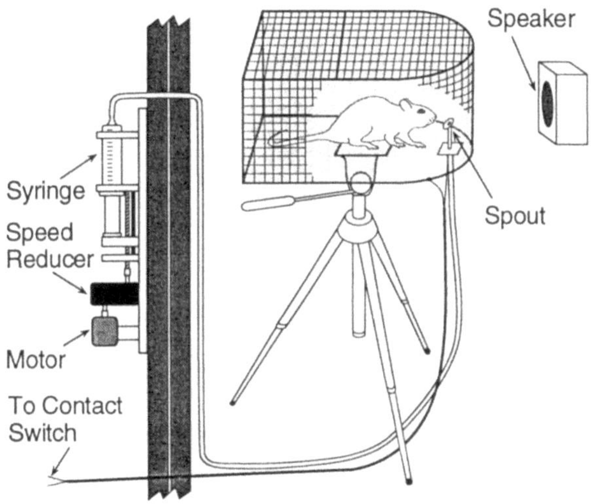

Figure 1. Semi-schematic drawing of a test cage and syringe pump.

Because the animal maintains contact with the spout, it can be used to position the animal precisely within the cage. In auditory testing, a reward spout which comes up through the bottom of the cage is preferred because it minimizes obstructions between the loudspeaker and the animal's ears. The spout can be made of copper or stainless-steel tubing with a small lick plate mounted on the top at an angle of approximately a 45°. The exact configuration of the spout depends on the species being tested—the goal is to construct a spout that requires an animal to hold its head in the desired position when making contact with the spout. In some cases, an animal may try to turn sideways while licking the spout, as when attending to sounds coming from one side. One way to prevent this is by placing shoulder-high wire mesh barriers within the cage to require the animal to face directly forward while licking the spout.

An animal's contact with the reward spout is detected with a contact switch connected between the spout and the cage floor (Fig. 1). Some animals, such as rabbits and least weasels, have fur on their feet which prevents them from making good electrical contact with the floor. This problem can be solved by wetting their feet or placing a damp sponge on the cage floor.

Larger animals, such as horses and other hoofed mammals, can be tested in a stall using a stainless steel bowl as a reward spout (Heffner and Heffner, 1984a). Contact with the reward bowl is detected by a contact switch connected between the bowl and a metal plate on the stall floor or an electrode taped to the animal's flank.

Primates are often tested in primate chairs, in which case the reward spout is mounted on the chair in front of the animal. One configuration consists of two drink tubes mounted parallel and close enough (1 cm apart) so that a monkey can comfortably place its mouth on both spouts. The spouts are electrically isolated from each other so that a contact switch can be used to detect when the animal places its mouth on them and the shock can be delivered between them. A reward, such as water, is delivered through either one or both of the drink tubes and auditory stimuli may be presented via insertion earphones or loudspeakers (Heffner and Heffner, 1990a).

The use of a reward spout to fix an animal's head may be helpful in testing other modalities, such as vision and olfaction, where placement of the head is important, as well as in somatosensory testing of the face or vibrissae (e.g., Hutson and Masterton, 1986; Smith, 1970). The range of tests depends primarily on the ingenuity of the experimenter: taste can be tested by injecting flavors into a water reward; somatosensory tests of a foot pad can be done by requiring an animal to place its foot on a stimulator in order to turn on the reward.

2.3 The Reward

The purpose of the appetitive reward is to keep an animal in continuous contact with the reward spout, breaking contact only when a warning stimulus is presented. To do this, it is necessary to use a reward for which an animal will reliably work and which can be delivered continuously or in many small amounts. For most mammals, the ideal reward is water, although in some cases a food puree or paste is preferable. The issues here are the type of reward, how to deliver it, and how to deprive an animal.

a) Water Reward

Water is an ideal reward for this procedure because most mammals readily work for it and, unlike food pellets, it can be continuously dispensed. An inexpensive way of delivering water is to use a constant-pressure water reservoir that is connected via an electrically operated water valve to the reward spout. The water reservoir can be a graduated cylinder with an outlet at the bottom. The cylinder is capped with a rubber stopper with an air inlet tube passing through the stopper to below the water level (see Heffner et al, 1994). The water pressure remains constant as long as the bottom of the air inlet tube is submerged; the water height is measured from the bottom of the air inlet tube. The water flow rate is controlled by first adjusting the height of the reservoir and then operating the water valve with a train of electrical pulses (e.g., 50 msec duration) that can be continuously varied (e.g., 2 to 8 pulses/sec) to provide fine control.

A drawback of the water reservoir/electric valve delivery system is that it can be difficult to dispense small amounts accurately, especially when an animal consumes 5 ml or less per session. In addition, the height of the water reservoir in relation to the reward spout must be kept constant and the reservoir height must be readjusted if the cage height is changed. A solution to this problem is to use a syringe pump and adjust the flow rate by varying its speed (Fig. 1). Although commercially-available syringe pumps are relatively expensive, it is possible to construct a satisfactory syringe pump in a modestly equipped shop (Thompson et al., 1990).

b) Food Reward

There are some animals for which food is the preferred reward. In general, these are animals that normally obtain most or all of their water from their food. They include desert rodents, such kangaroos rats and gerbils, which obtain metabolic water from dry food (Schmidt-Nielsen, 1979), and underground rodents, such as gophers and mole rats, which obtain water from the roots they consume. Because these animals cannot easily be deprived of water without also depriving them of food, a solution is to use a food paste or puree which can be continuously dispensed. Examples for rodents are strained vegetable or fruit baby food, and applesauce mixed with peanut butter. These diets can then be supplemented as needed with dry food (Heffner and Heffner, 1992, 1993). Animals whose diets consist primarily of insects may also work better for food. An example is the big brown bat (*Eptesicus fuscus*), which is typically maintained in the laboratory on a diet of mealworms. In this case, a food paste can be made of blended and strained mealworms, with cottage cheese added to obtain a uniform consistency. Finally, although water can be used to reward domestic cats, provided they are maintained on dry cat food (e.g., Masterton et al., 1994), cats are highly motivated by meat and often work better for meat paste. Although commercial baby food has been used (Berkley et al., 1971; Thompson et al., 1990), a more economical reward is canned cat food blended with water or milk and baby cereal to achieve the desired consistency.

Food pastes can be dispensed with a syringe pump (Thompson et al., 1990). The food should be carefully blended to eliminate lumps and clogging and the pump should be located directly beneath the test cage to minimize tubing length. For auditory testing, this necessitates the selection of a relatively quiet pump motor so as not to mask the auditory stimulus. Alternatively, a hydraulic system may be constructed in which the drive sits outside the test room and powers a piston which depresses the plunger of a food syringe located below the test cage.

c) Deprivation

In order to train an animal using an appetitive reward, it is necessary to remove the animal's food or water from its home cage and have it obtain its daily ration in the test cage. Although some animals may be trained to work for special treats, their performance breaks down when the discrimination becomes difficult (e.g., around threshold). The same may hold true for animals routinely given free access to the reward following a test session. Thus, data obtained under such situations may be suspect on the grounds that the animals were insufficiently motivated.

The usual procedure is to place an animal on deprivation and begin training the following day. The animal's body weight is recorded daily prior to testing and serves as a useful indication of its deprivational state. The animal is placed in the test cage and accustomed to maintaining steady contact with the spout. The reward rate is adjusted so that the animal works long enough to allow sufficient data to be collected and receives adequate reward to maintain a stable body weight. In rare instances, a species may not maintain its weight in a single daily feeding and can either be tested twice daily or else given supplements. Most animals can be trained to work steadily for about an hour, although some small animals that consume little may work for less time. Avoidance training is begun as soon as an animal is reliably maintaining steady contact, usually within one to three sessions.

The body weight at which an animal works well is usually between 80 and 90% of ad lib weight, although this depends on the species. For some species, an individual's weight must be reduced to well below 80% before it is sufficiently motivated (Heffner and Heffner, 1992), while others will work at or near 100% ad lib weight once they have adapted to the testing regimen (e.g., chinchillas). The goal is to keep an animal's weight as high as possible while maintaining sufficient motivation. An animal which is too hungry or thirsty may fail to respond to the warning stimulus until it has consumed enough to reduce its hunger or thirst. With experience, one can determine both an animal's optimal working weight and the amount of reward it needs to maintain that weight.

There are two important effects of deprivation on the health and well-being of an animal. First, animals living in the wild rarely have continuous access to food and water and by the standards applied to laboratory animals would be considered deprived. For example, wild pigeons brought into the laboratory and placed on ad lib feeding gained 9 to 30% body weight even though they had been trapped amid abundant food supplies (Poling et al., 1990). Furthermore, young guinea pigs placed on food deprivation for a behavioral study showed the same growth curves as guinea pigs living in the wild (Petersen et al., 1977). Thus, animals whose food or water intake is restricted in order to motivate them to perform in behavioral experiments appear to be operating at deprivation levels to which they are naturally adapted.

Second, there is a large literature documenting the fact that reducing the caloric intake of laboratory animals by 30 to 70% of ad lib feeding results in animals that are significantly healthier and longer lived than those on free feed. Specifically, dietary restriction greatly decreases the incidence and severity of degenerative diseases, retards the onset of tumors and reduces their incidence, and increases both lifespan and life expectancy (Bucci, 1992). Thus, restricting the food or water intake of animals not only reduces their weights to those of normal wild animals, but results in healthier animals.

Finally, it should be noted that although one may encounter the belief that water deprivation is more stressful than food deprivation (Orlans, 1991), there is little evidence to support this contention. Those wishing to study this issue should consult the article by Desimone et al. (1992).

2.4 Electric Shock

The purpose of the electric shock is to make the animal break contact with the reward spout whenever it detects a warning stimulus. Unlike conditioned suppression, the shock is avoidable, a feature which increases the number of warning trials that can be given in a session. Because the shock is avoidable, it is presented simultaneously with a signal, such as a light or buzzer, which indicates that the shock is on and provides feedback for successful avoidance.

The shock is adjusted to the lowest level that produces reliable avoidance. Too low a level results in a low hit rate and underestimates an animal's ability; too high a level results in a high false alarm rate which may make the data unusable. Ideally, the shock level is adjusted to give a false alarm rate of 1 to 10%, although false alarm rates as high as 20% can give usable data if the proper correction is applied (see below). The shock level is initially adjusted for each animal by presenting warning trials with the shock level set near or at zero voltage and rapidly increasing the level until the animal breaks contact with the spout when it senses the shock. The level can be gradually increased further until it is sufficiently aversive to cause the animal to break contact when it detects the warning stimulus. The shock level should be occasionally increased or decreased during testing to insure that it is at optimal level.

An important factor which allows the use a relatively low level of shock is the fact that an animal is required to break contact with the spout for a very brief interval (e.g., the last 200 msec of the trial, Fig. 2A). In contrast, the original conditioned suppression procedure required an animal to stop responding for 10 or more seconds (Ray, 1970; Smith, 1970). Because the cost to the animal of making a response is the temporary loss of access to the reinforcer, the shorter the required response time, the lower the level of shock needed. Moreover, the response cost to the animal can be compensated for by momentarily increasing the reward rate following a successful avoidance to make up for the small loss of reward (i.e., rewarding hits).

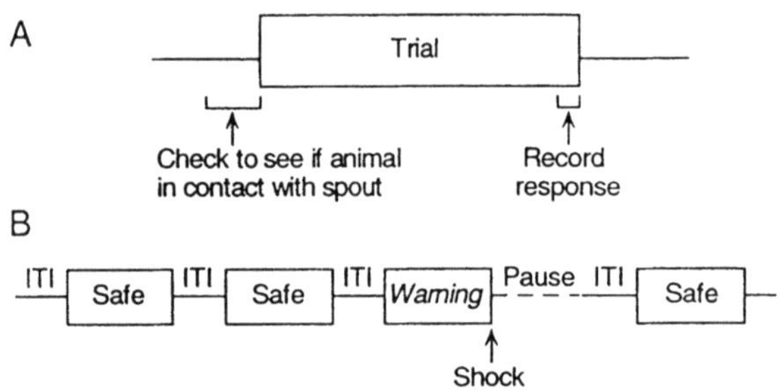

Figure 2. A: Schematic representation of a trial. B: Schematic representation of a trial sequence with the warning signal occurring on trial 3. Trial presentation is paused following a warning trial until the animal returns to the reward spout. ITI, inter-trial-interval.

The electric shock is a high voltage, low current stimulus, the level of which depends on the species and the degree of contact which the animal makes between the reward spout and cage floor. As a result, it is rarely helpful to specify the voltage and current settings; the preferred method is to specify the shock in terms of its behavioral effect on the animal. In general, small animals require lower levels of shock than larger animals. The shock duration is usually set between 100 and 300 msec. However, unlike conditioned suppression, the shock is escapable and the duration an animal experiences the shock is dependent on its reaction time and is generally well under 100 msec. Occasionally, an animal breaks contact when the warning stimulus is presented but returns to the spout before the trial has ended; this behavior can be eliminated by temporarily increasing the duration of the shock to 1 sec or longer.

There are several advantages to shocking the animals through the spout. Not only does it makes it easier for them to learn to break contact, but the sensitivity of the lips and tongue make it possible to use much lower levels than if they were shocked between their feet. In addition, the construction of the test cage is simplified because a grid floor is not needed. In the case of monkeys tested in a primate chair, the shock is delivered between the two water spouts.

The electric shock can be produced by a generator designed for behavioral research. Alternatively, a satisfactory shocker can be devised by using an inexpensive AC fence charger and controlling the shock level by adjusting the input voltage with a variable transformer.

With regard to the subjective sensation of the shock, it may be noted that electric shock is not a natural stimulus and while it can be quite aversive, its sensation is not adequately described as painful. For illustrative purposes, a helpful comparison is the neuromuscular stimulator commonly used on humans for physiotherapy. Tests in our laboratory with squirrels have demonstrated that such stimulators can serve as an adequate, if expensive, shock source (e.g., Medtronic, model 3128). Thus, the shock levels used with animals are typically within the range of those experienced by humans in therapeutic situations. Moreover, because the animals do not develop a fear of the reward spout and readily return to it after receiving a shock, the shock level is properly described as mild.

2.5 Trial Presentation

The test procedure consists of presenting a series of trials which ends either with the presentation of a warning trial (Fig. 2B) or after a fixed number of safe trials has been presented, i.e., a "sham" trial sequence. The duration of a trial, the inter-trial-interval (ITI), and the maximum number of trials in a sequence can be varied to suit the requirements of the particular test. For example, a trial can be 3-sec long, with an ITI of 2 sec, and the warning trial occurring from 1 to 7 trials after the last warning trial. In addition, a pause can be inserted after a warning trial to give an animal time to return to the spout or, alternatively, the testing sequence can be halted until the animal has resumed contact with the spout. Once an animal has resumed contact, a warning trial may be presented within 2 to 30 sec—or longer as sham sequences are occasionally inserted to prevent an animal from automatically responding after 30 sec.

It is important to distribute the warning trials so that each position in a sequence (i.e., positions 1 through 7 of the preceding example) has the same probability of containing a warning signal. If one randomly presents the same number of warning trials in each trial position, the probability of a warning trial will increase with position number. Thus, it is necessary to construct a "look-up" table in which the number of times a warning trial can occur in a particular position in the sequence is adjusted so that each position has approximately the same probability of containing a warning trial (Table I). This table can be used to construct a fixed sequence of safe and warning trials which is then repeated, rather than selecting the trials randomly. This is done to prevent an animal from receiving too many identical sequences in a row, especially sham sequences and sequences in which the first trial is a warning trial; sequences containing more than 80 warning trials are unlikely to be memorized by an animal. Typically, a sequence does not contain more than 3 warning trials in a row.

Table I. Look-up table for sequences from 1 to 7 trials long with overall probability of a warning trial equal to .221.

Position of Warning Trial in the Sequence	Number of Warning Trials in that Position	Number of Safe Trials in that Position	Probability of a Warning Trial
1st	10	36	.217
2nd	8	28	.222
3rd	6	22	.214
4th	5	17	.227
5th	4	13	.235
6th	3	10	.231
7th	2	8	.200
Sham*	8		

*A sequence of 7 safe trials not followed by a warning trial.

Our procedure for determining the ability of an animal to detect or discriminate a stimulus consists of presenting a particular stimulus value (e.g., a specific intensity) in blocks of 6 or more warning trials (for a titration procedure, see Masterton et al., 1994). Thresholds are initially estimated by gradually reducing the level of the stimulus until performance falls to chance. Next, detailed testing is conducted by presenting trials at levels just above, at, and below the estimated threshold. Typically, a block of trials involving a difficult discrimination is followed by a block of easier trials to ensure that an animal is still under control of the stimulus. However, it is occasionally necessary to continue a difficult discrimination in order to train an animal to "attend," as is the case when an animal must learn to listen for sounds near threshold.

The response of an animal is typically determined by measuring spout contact during the 200 msec preceding the shock and recording a response if the animal breaks contact for at least half of that 200-msec interval (Fig. 2A). The hit and false alarm rates are recorded separately for each block of trials as the false alarm rate often varies, increasing when the discrimination becomes more difficult and decreasing when it is easy. Because an animal may temporarily cease responding for other reasons (e.g., to groom), the results of a trial are automatically discarded if the animal is not in contact with the spout immediately preceding a trial (e.g., during the preceding .5 sec, Fig. 2A). Because this criterion is applied equally to safe and warning trials, it does not bias the results.

2.6 Data Analysis

The performance of an animal for a particular stimulus value is calculated by correcting the hit rate for the false alarm rate. The classic method for this is the formula: Performance = (Hit rate - False Alarm Rate)/(1-False Alarm Rate) (Green and Swets, 1966; Smith, 1970). However, this correction can give misleading results when high hit rates are accompanied by high false alarm rates. This is illustrated by the extreme case in which a perfect hit rate, 1.0, is accompanied by a false alarm rate of .99, a situation which results in the same perfect score of 1.0 as a hit rate of 1.0 and a false alarm rate of 0.

To better correct for the effect of false alarms, the following formula may be used: Performance = Hit Rate - (Hit Rate * False Alarm Rate). This calculation yields scores from 0 (failure to detect or discriminate) to 1.0 (perfect detection or discrimination without any false alarms). Unlike the classic method, a score of 1.0 can result only from a hit rate of 1.0 and a false alarm rate of 0. In practice, the scores resulting from this formula rarely reach 1.0 because it is desirable to keep the false alarm rate greater than zero to ensure that the animal is sufficiently attentive. Similarly, a score of 0 is usually not reached because an animal unable to detect or discriminate the stimulus will, on average, have a hit rate equal to its false alarm rate, which can give a performance score as high as .25 (i.e., hit and false alarm rates both equal to

.50). Because this formula works well for a wide range of hit and false alarm rates, it is the preferred formula. A detailed comparison of this formula with other measures can be found in Heffner and Heffner, 1988.

Threshold is defined as the stimulus value yielding a performance of .50, which is derived by interpolating if necessary. However, it is important to reduce the stimulus value to a level at which performance falls to statistical chance ($p > 0.01$) in order to rule out the possibility that an animal is using some other cue to perform the discrimination. For example, a sound localization task in which an animal is required to discriminate the locus of two loudspeakers can be confounded if an animal learns to distinguish the speakers by the quality of their sound. Thus, the angle of separation between the speakers must be reduced until performance falls to chance in order to demonstrate that the animal is indeed discriminating locus.

The probability of a particular score can be determined using the binomial distribution (Hays, 1963). This is done using the formula:

$$p(X \geq r) = \sum_{x=r}^{N} \binom{N}{r} p^r q^{N-r} \qquad (1)$$

This formula gives the probability of observing a hit rate, X, equal to or greater than the observed hit rate, r, where N is the number of warning trials, p is the false alarm rate, and q is the *correct rejection* rate, i.e., 1-False Alarms. The result is the probability of obtaining a hit rate equal or greater than that observed, given the observed false alarm rate for that stimulus level.

3 Discussion

The following points can be made regarding the conditioned avoidance procedure. First, the basic training and conditioning can be accomplished in a relatively short time. Because licking is a natural response, mammals typically require no special training to maintain steady contact with the reward spout. Furthermore, once an animal is acclimated to the testing situation, it can be trained within the first session to break contact reliably when an easily detectable or discriminable warning stimulus is presented. As with all procedures, training an animal to attend carefully to stimuli near threshold requires additional practice.

Second, the results obtained with conditioned avoidance have been shown to be highly replicable. Not only is there less variation between subjects than often found when using a purely positive reward procedure (e.g., Heffner and Heffner, 1984a), but comparisons between data obtained by different laboratories show good agreement (cf., Heffner et al., 1994; Kelly and Masterton, 1977).

Third, this procedure can be applied to a wide variety of animals and tests. It has been used with over 30 species of mammals, as well as birds, to assess sensory, perceptual, and cognitive abilities in any test involving two choices (e.g., Heffner & Heffner, 1990b, Smith, 1970). Not only is it an ideal procedure for difficult to test animals (Heffner & Heffner, 1984a), but unlike simple fear conditioning (LeDoux et al, 1984), it works well with animals brain damaged in a wide variety of ways (Heffner & Heffner, 1984b; Kelly & Judge, 1985).

Fourth, conditioned avoidance does not appear to result in "experimental neurosis," a condition which refers to the development of long-standing behavioral disturbances in animals in certain test situations (e.g., Deese, 1958). These disturbances, which include struggling on the part of the animal during testing, have been observed in conditioning experiments involving positive reward, as well as shock, and may appear when an animal is subjected to lengthy testing on a difficult discrimination. The fact that such behavior has not been observed in conditioned avoidance may be due to the fact that the animal can terminate the experiment at any time by failing to return to the reward spout. As a result, an animal cannot be subjected to prolonged unwarned shocks (as when the stimulus is below threshold) beyond its capacity to tolerate them.

Finally, as noted by Smith (1970), a procedure that combines aversive control with positive reinforcement gives good control over an animal's performance (i.e., its hit and false alarm rates). Too low a hit rate can be corrected by increasing the shock level while too high a false alarm rate can be corrected by reducing the shock level and/or increasing the rate at which the reward is delivered. In this way, an animal's behavior can be adjusted to yield its best performance. A procedure which relies solely on positive reward, on the other hand, may lack sufficient punishment for errors. Such procedures usually rely on an "error-time-out" in which testing is momentarily halted following a miss or false alarm. However, the temporary lack of opportunity to obtain access to a reward is not always sufficient punishment for errors and some animals will not perform at optimal levels in a positive reward procedure, especially when the discrimination becomes difficult. While this is not always the case, it should be kept in mind when the results of such tests yield variable or unusually poor performance.

4 Acknowledgements

We are indebted to Bruce Masterton and Jack Kelly for their comments and Gim Koay, Laura Boester, and Robyn Gandy for their contributions. Supported by NIH Grant NS30539.

5 Suggested Readings

Heffner, R.S., & Heffner, H.E. (1988). Sound localization in a predatory rodent, the northern grasshopper mouse (*Onychomys leucogaster*). *Journal of Comparative Psychology, 102*, 66-71.

Masterton, R.B., & Granger, E.M. (1988). Role of the acoustic striae in hearing: Contribution of dorsal and intermediate striae to detection of noises and tones. *Journal of Neurophysiology, 60*, 1841-1860.

Thaw, A.K., & Smith, J.C. (1992). Conditioned suppression as a method of detecting taste thresholds in the rat. *Chemical Senses, 17*, 211-223.

6 References

Berkley, M.A., Crawford, F.T., & Oliff, G. (1971). A universal food-paste dispenser for use with cats and other animals. *Behavior Research Methods & Instrumentation, 3*, 259-260.

Bucci, T.J. (1992). Dietary restriction: Why all the interest? An overview. *Lab Animal, 21*, 29-34.

Deese, J. (1958). *The psychology of learning* (2nd ed.). New York: McGraw-Hill.

Desimone, R., Olson, C., & Erickson, R. (1992). The controlled water access paradigm. *ILAR News, 34*, 27-29.

Estes, W.K., & Skinner, B.F. (1941). Some quantitative properties of anxiety. *Journal of Experimental Psychology, 29*, 390-400.

Green, D.M., & Swets, J.A. (1966). *Signal detection theory and psychophysics.* New York: Wiley.

Hays, W.L. (1963). *Statistics for psychologists.* New York: Holt, Rinehart & Winston.

Heffner, H.E., & Heffner, R.S. (1984a). Sound localization in large mammals: Localization of complex sounds by horses. *Behavioral Neuroscience, 98*, 541-555.

Heffner, H.E., & Heffner, R.S. (1990a). Effect of bilateral auditory cortex lesions on absolute thresholds in Japanese macaques. *Journal of Neurophysiology, 64*, 191-205.

Heffner, H.E., & Heffner, R.S. (1990b). Role of primate auditory cortex in hearing. In W.C. Stebbins & M.A. Berkley (Eds.), *Comparative perception—volume II: Complex signals* (pp. 279-310). New York: Wiley.

Heffner, H.E., Heffner, R.S., Contos, C., & Ott, T. (1994). Audiogram of the hooded norway rat. *Hearing Research, 73*, 244-247.

Heffner, R.S., & Heffner, H.E. (1984b). Hearing loss in dogs after lesions of the brachium of the inferior colliculus and medial geniculate. *Journal of Comparative Neurology, 230*, 207-217.

Heffner, R.S., & Heffner, H.E. (1988). Sound localization in a predatory rodent, the northern grasshopper mouse (*Onychomys leucogaster*). *Journal of Comparative Psychology, 102*, 66-71.

Heffner, R.S., & Heffner, H.E. (1992). Hearing and sound localization in blind mole rats (*Spalax ehrenbergi*). *Hearing Research, 62*, 206-216.

Heffner, R.S., & Heffner, H.E. (1993). Degenerate hearing and sound localization in naked mole rats (*Heterocephalus glaber*), with an overview of central auditory structures. *Journal of Comparative Neurology, 331*, 418-433.

Hutson, K.A., & Masterton, R.B. (1986). The sensory contribution of a single vibrissa's cortical barrel. *Journal of Neurophysiology, 56*, 1196-123.

Kelly, J.B., & Judge, P.W. (1985). Effects of medial geniculate lesions on sound localization by the rat. *Journal of Neurophysiology, 53*, 361-372.

Kelly, J.B., & Masterton, R.B. (1977). Auditory sensitivity of the albino rat. *Journal of Comparative and Physiological Psychology, 91*, 930-936.

LeDoux, J.E., Sakaguchi, A., & Reis, D.J. (1984). Subcortical efferent projections of the medial geniculate nucleus mediate emotional responses conditioned to acoustic stimuli. *Journal of Neuroscience, 4*, 683-698.

Masterton, R.B., Granger, E.M., & Glendenning, K.K. (1994). Role of acoustic striae in hearing: mechanism for enhancement of sound detection in cats. *Hearing Research, 73*, 209-222.

Orlans, F.B. (1991). Prolonged water deprivations: A case study in decision making by an IACUC. *ILAR News, 33*, 48-52.

Petersen, M.R., Prosen, C.A., Moody, D.B., & Stebbins, W.C. (1977). Operant conditioning in the guinea pig. *Journal of the Experimental Analysis of Behavior, 27*, 529-532.

Poling, A., Nickel, M., & Alling, K. (1990). Free birds aren't fat: Weight gain in captured wild pigeons maintained under laboratory conditions. *Journal of the Experimental Analysis of Behavior, 53*, 423-424.

Ray, B.A. (1970). Psychophysical testing of neurologic mutant mice. In W.C. Stebbins (Ed.), *Animal psychophysics: The design and conduct of sensory experiments* (pp. 99-124). New York: Appleton-Century-Crofts.

Schmidt-Nielsen, K. (1979) *Animal physiology: Adaptation and environment.* Cambridge: Cambridge University Press.

Smith, J. (1970). Conditioned suppression as an animal psychophysical technique. In W.C. Stebbins (Ed.), *Animal psychophysics: The design and conduct of sensory experiments* (pp. 125-159). New York: Appleton-Century-Crofts.

Thaw, A.K., & Smith, J.C. (1992). Conditioned suppression as a method of detecting taste thresholds in the rat. *Chemical Senses, 17*, 211-223.

Thompson, M., Porter, B., O'Bryan, J., Heffner, H.E., & Heffner, R.S. (1990). A syringe-pump food-paste dispenser. *Behavior Research Methods, Instruments, & Computers, 22*, 449-450.

Methods in Comparative Psychoacoustics
G. M. Klump, R. J. Dooling, R. R. Fay & W. C. Stebbins (eds)

Design and Conduct of Sensory Experiments for Domestic Cats

Bradford J. May, Aileen Y. Huang, Cynthia M. Aleszczyk and Robert D. Hienz

Center for Hearing Sciences, Depts. of Otolaryngology-HNS, Biomedical Engineering, and Psychiatry and Behavioral Sciences, The Johns Hopkins School of Medicine, 720 Rutland Avenue, Baltimore, Maryland 21205, U.S.A.

Summary. This chapter uses the design and conduct of an auditory detection task to illustrate positive reinforcement operant conditioning methods for domestic cats. Issues of early training are outlined where the investigator plays a critical interactive role with the subject. Automated systems for conducting behavioral testing are summarized. Finally, criteria for evaluating optimal performance and stability of auditory behaviors are discussed. Training and testing strategies presented should prove equally useful for non-auditory sensory experiments.

1 Introduction

Cats have natural tendencies to explore and manipulate their environment. They have a keen sense of hearing and vision; most cats also have strong appetites. These traits make the species an ideal subject for sensory experiments, so it is not surprising that cats have been used as experimental animals in a number of psychoacoustic studies. Cats have been conditioned with electrical shocks to respond to bursts of sound by jumping a barrier (Neff and Hind, 1955; Butler et al., 1957; Elliot et al., 1960; Saunders, 1969; Trahoitis and Elliot, 1970; Igarashi et al., 1979) or by breaking contact with a water or food spout (Heffner and Heffner, 1985, 1988; Martin and Webster, 1987). They have been trained with food rewards to signal the perceived location of a sound by approaching a loudspeaker (Casseday and Neff, 1973; Jenkins and Merzenich, 1984). Reflexive behaviors, such as acoustic startles or head orientations, have even been exploited as behavioral measures (Thompson and Masterton, 1978). This chapter describes the design and conduct of positive reinforcement operant procedures for measuring absolute auditory detection in cats (Neff and Hind, 1955; Elliot et al., 1960; Gerkin and Sandlin, 1977; Heffner and Heffner, 1985). The hold-release task is a versatile paradigm that has also been used in our laboratory to measure difference limens for the frequency, intensity, and location of sound stimuli; it is our laboratory's paradigm during electrophysiological experiments in behaving cats as well (suggested reading: May et al., 1991).

2 Free-Field Sound System

The configuration of a typical free-field sound system for the auditory detection task is diagrammed in Fig. 1. Three sources of sound are shown: a digital oscillator for producing pure tones (a_1), a noise generator for producing noise burst stimuli or continuous background noise (a_2), and a 16-bit D/A board for creating complex stimuli such as human vowel sounds (a_3). The appropriate source for the acoustic stimulus is accessed by a stimulus selector (b). The digital oscillator is usually the sole source of stimuli for the auditory detection task, but a versatile sound system should be capable of generating a variety of stimulus conditions. For example, the system in Fig. 1 can measure tone detection in noise by selecting the digital oscillator for audio channel 1 and the noise generator for audio channel 2. After the stimulus selector, the electronic waveform of the stimulus is gated by an electronic switch (c). Stimuli in psychoacoustic tests are shaped with rise/fall times of 10 - 20 msec to minimize switching transients. The level of the stimulus is adjusted by passing the waveform to a programmable attenuator (d). In cats, minimum detectable tone levels range from -20 to 50 dB SPL as a function of stimulus frequency. The programmable attenuator must be capable of producing a 100-dB range of stimulus levels to measure those thresholds. Attenuated waveforms in the two audio channels are combined by a passive mixer (e) and then boosted in amplitude by a power amplifier (f). At this point in the circuitry, the electronic signal enters the anechoic testing environment via an interface panel and drives a free-field loudspeaker (g).

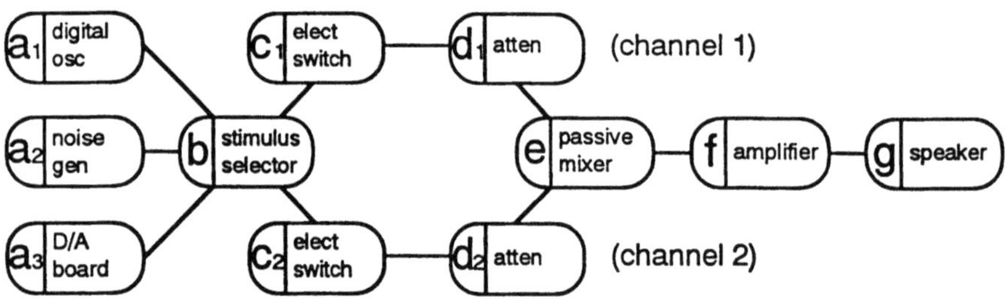

Figure 1. Schematic diagram of the free-field sound system.

Important specifications of components in the free-field sound system are shown in Table I. These components are designed and manufactured within our laboratory or purchased from commercial vendors. The advantage of using "in house" audio equipment for psychoacoustic testing is that the investigator can create a system for generating stimuli that meet the exact specifications of the current experimental protocol. The disadvantage of manufacturing your own equipment is that the process of design, fabrication, and maintenance always involves more time and usually more cost than commercially available components. When a complicated "in house" component fails, repairs are likely to take several days to complete. Alternatively, suppliers of commercial equipment offer maintenance for their own devices and most will provide substitute devices to reduce any "down-time" caused by failure of their equipment. Unless the investigator has the rare combination of expertise, money and patience, the most economical approach to creating a sound system is to purchase one of the high quality and low cost commercial systems that are now available. Vendor and product name for some of the commercial audio components that we have used in our laboratory are shown in Table I.

Table I. Components of sound system for auditory experiments.

	Component	Specifications	Vendor, Product
a_2.	Noise generator	0.02 - 500 kHz Gaussian noise spectral uniformity ±2.5 dB	Elgenco, 602A
a_3.	16 bit D/A convertor	500 kHz mono sampling rate 83 dB signal-to-noise ratio	Tucker-Davis, DD1
c.	Electronic switch	cosine and ramp functions Rise/fall times from 0.1 - 90 msec	Tucker-Davis, SW2
d.	Prog attenuator	±0.05 dB from 0.0 - 99.9 dB 0.1 dB resolution	Tucker-Davis, PA4
f.	Power amplifier	Freq. response ±1 dB from 0.02 - 50 kHz < 0.05% total harmonic distortion	Crown, D-75
g.	Mid range speaker	Freq. response ±3 dB from 0.2 - 15 kHz	Radio Shack, Minimus-7
	High frequency speaker	Freq. response ±3 dB from 10 - 30 kHz	Radio Shack, Tweeter

3 Anechoic Testing Environment

An operant platform for the auditory detection task is shown in Fig. 2. Toward the rear of the platform, a collar and canvas harness maintains the cat in a natural seated posture. The minimal restraint system stabilizes the cat's head orientation in the sound field and keeps the subject engaged in the behavioral task. Above the platform, a video camera provides constant surveillance while the subject is held in restraint. On the floor of the platform near the right front paw, the response lever reacts to a light touch by activating a microswitch which is "sensed" by input/output devices on the computer controlling behavioral testing. When the subject responds to a tone presentation by releasing the lever, a pneumatically activated spout rises from the floor of the platform and delivers liquefied meat paste directly to the cat's mouth.

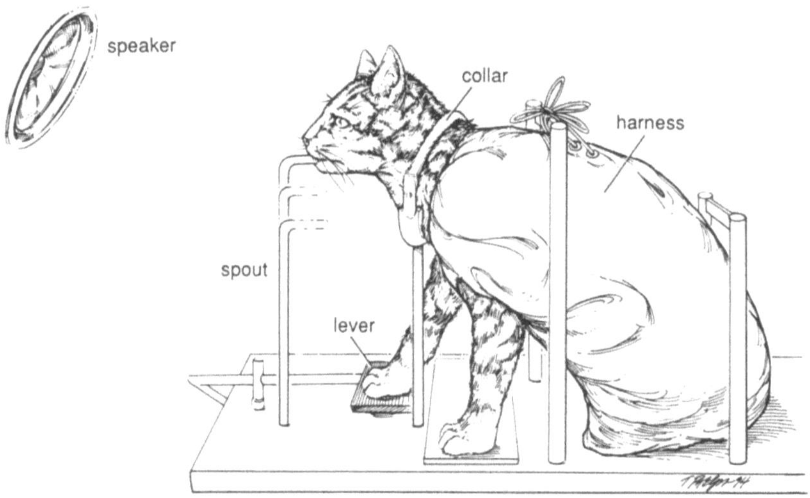

Figure 2. The operant platform.

A free-field loudspeaker delivers the acoustic stimuli. Although shown closer in this figure for artistic reasons, the speaker is actually about 1 m from the cat's normal head position and slightly elevated relative to the interaural axis to reduce acoustic filtering effects of the outer ear (Rice et al., 1992).

Extraneous sounds are eliminated from the testing environment by placing the operant platform inside a double-walled sound-attenuating chamber (Industrial Acoustic Company). The interior surfaces of the chamber (i.e., walls, ceiling and floor) are covered with 3-inch anechoic foam padding (Sonex) to minimize acoustic reflections that distort the frequency response characteristics of the sound system. Anechoic foam of this thickness will absorb sound frequencies above 0.5 kHz and will not significantly reduce the interior dimensions of the testing chamber. As an additional precaution to control acoustic reflections, all surfaces on the operant platform are kept small or covered with anechoic foam, and the platform itself is situated at the center of the chamber, as far as possible from chamber walls.

4 Computerized Testing Routines

Behavioral experiments are conducted by computer in the modern psychoacoustics laboratory. Animal handlers place the cat in the testing apparatus and activate the appropriate computerized testing routine. These programs create the sound stimuli, control the contingencies of reinforcement, and record data until the experimental session is concluded. Automated procedures are more efficient and less prone to experimental bias than interactive procedures.

A particularly effective strategy for the computer programming of psychoacoustic experiments is to design modular code; that is, reducing a large program into smaller subroutines. The major states of the behavioral paradigm shown in Fig. 3 can be thought of as subroutines in a computer program. The lines that connect the states in the diagram represent the flow from one subroutine to another as a consequence of the cat's behavioral responses. Although the logical expressions that interconnect the states of a program change to reflect the reinforcement contingencies of different behavioral tasks, all programs will conserve basic functions such as turning on the tone, waiting for the lever release, or activating the feeder. Libraries of subroutines for these functions eliminate unnecessary duplication of effort and ensure programming consistency.

5 The Hold-Release Paradigm

The hold-release paradigm is an effective operant task for measuring auditory detection in cats. The reinforcement cycle for the task is shown in Fig. 3; values of task parameters are shown in the right column of Table II. Each trial sequence in the auditory detection task begins and ends with an intertrial interval (a). The 1-sec delay imposed by the intertrial interval affords time for the cat to consume food and for the computer to record behavioral data before setting up the next trial sequence. After the intertrial interval, the cat depresses the response lever to start a variable hold (b), which is a period of silence that lasts from 5 - 10 sec. If the cat holds the lever down for the entire duration of the variable hold, a detection trial interval is presented (c). The onset of the detection trial is marked by the presentation of a tone burst. The cat indicates detection of the tone by releasing the lever. Releases during the 3-sec detection trial intervals are reinforced with food rewards (d).

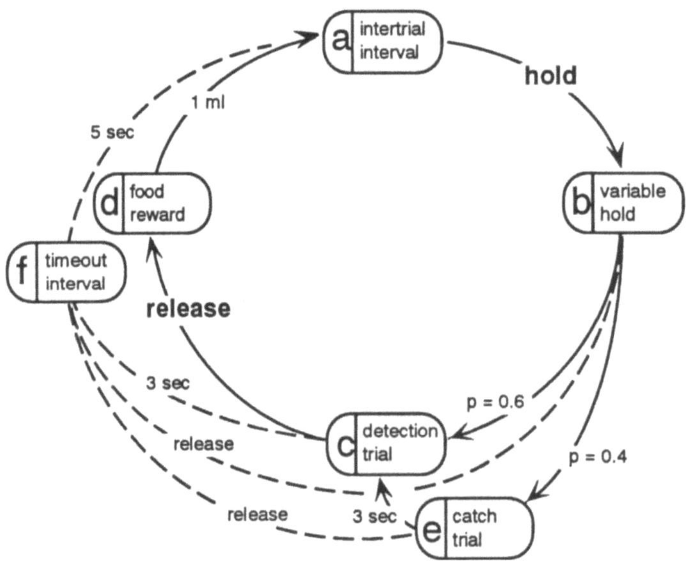

Figure 3. The hold-release paradigm.

A blend of canned cat food and tap water serves as the reward for correct lever releases in the auditory detection task. Cats seem to prefer food rewards made with less expensive and more fatty brands of cat food. The liquefied meat paste is delivered in 1-ml allotments and is only available while cats are performing the behavioral task. Unlimited water and a daily ration of dry chow are provided in the home cage. Although exact rations are determined on an individual basis, the amount of dry food in a cat's diet is decreased gradually from approximately 40 to 10 g as the young animal grows to adulthood. Cats maintained on this moderate schedule of food deprivation show normal healthy growth and will work continuously in daily testing sessions for 100 - 200 ml of the highly preferred liquefied meat.

A well trained and slightly food-deprived cat will attempt to maximize its food reinforcement by rapidly cycling from contingencies (a) through (d) of the hold-release paradigm, as indicated by unbroken lines in Fig. 3. However, cats occasionally will make errors as they perform the operant procedure. Three pathways for potential errors are indicated by dashed lines in the figure. One error results when a cat releases the lever during a variable hold; a second occurs if a cat fails to release the lever during a detection trial interval. Error responses are punished with a 5-sec timeout interval (f). Cats avoid a timeout because this delay postpones the delivery of the next food reward.

Table II. Values for training and testing parameters of the hold-release paradigm.

	Behavioral Parameters	Shaping Values	Training Priorities	Testing Values
a.	Intertrial interval	0 sec	2	1 sec
b.	Variable hold	0 sec	4	5 - 10 sec
c.	Detection trial	until release	1	3 sec
d.	Food reward	0.5 ml	3	1 ml
e.	Catch trial	p = 0.0	5	p = 0.4
f.	Timeout	0 sec	6	5 sec

The basic premise of the auditory detection task is that a cat releases the lever during a detection trial interval because it hears a tone. Unfortunately, the acquisition of psychophysical data is not quite so straightforward because a cat may also obtain food rewards by serendipitous releases of the lever during presentations of inaudible stimuli. Good experimental design can discourage guessing by making it an ineffective response strategy. For example, if the duration of the variable hold is kept long relative to the duration of the detection trial interval, most guesses will occur during the variable hold and will result in a timeout. Even when this precaution is taken, the experimental design should monitor how often guesses contribute to correct releases. That monitor is provided by catch trials (e), which are identical to detection trial intervals with the important exception that no tone is presented during the 3-sec catch trial interval. Therefore, if the cat is attempting to obtain food by timing the release of the response lever, it will respond with equal probability to catch trial intervals and inaudible detection trial intervals. Such false alarms are treated as response errors and punished with a timeout. Alternatively, if the cat rejects the catch trial by continuing to hold the lever, a detection trial is presented and the cat is given the opportunity to obtain food by releasing the lever. Modern measures of psychophysical performance are based not only on the probability of correct responses to detection trials but also on the probability of false alarms to catch trials. Approximately 20 - 40% of all testing cycles involve catch trials to ensure a reasonable sample of false alarms for these analyses.

In addition to the auditory detection task, the hold-release paradigm is equally effective in other sensory tasks where measures of stimulus detection or discrimination are desired. For example, with slight modification, the behavioral contingencies shown in Fig. 3 can measure a cat's ability to discriminate changes in sound intensity (Rosenzweig, 1946; Saunders, 1969; Igarashi et al., 1979). In the resulting intensity discrimination task, tone bursts of fixed frequency and intensity are presented during the 5 - 10 sec variable hold (b). As in the detection task, the cat must maintain the lever press thoughout the variable hold to avoid a timeout and to obtain a trial. Detection trials (c) are replaced by intensity discrimination trials which represent changes in stimulus level relative to tones presented during the variable hold. Cat are trained with food rewards to respond to such changes in level by releasing the lever. No change in level is presented during catch trials (e). By contrasting the relevant stimulus feature in discrimination trial intervals, the hold-release paradigm can also measure the discrimination of sound frequency (Butler et al., 1957; Elliot et al., 1960) or location (Casseday and Neff, 1973; Martin and Webster, 1987; Heffner and Heffner, 1988).

6 Outline of Early Training

Training begins when cats are about 4 - 6 months old. Juvenile cats are fast learners in psychophysical procedures but they can also be sensitive to changes in their environment. If cats are purchased from a commercial vendor, they should be given a few days to adapt to their new housing conditions and the presence of other animals in the laboratory. Once a cat has become accustomed to its new world, training begins with the assignment of a primary trainer.

A cat's experiences during early training exert a strong influence on how quickly the animal learns a task and, once trained, how well it performs as a psychophysical observer. Although there are no specific rules about how the human trainer should interact with an animal during the first days of training, it is clear that some people have a special rapport for training animals. When that rapport appears to be lacking in the primary trainer, the following general guidelines may enhance the training process.

The first goal of the trainer must be to adapt a cat to its testing environment because a frightened cat is neither motivated nor attentive enough to learn the operant procedure. Positive experiences, consistency and repetition in the training routine offer the most direct route to this goal. All training should be performed by the same person, in the same testing environment, and at the same time of the day. It may be occasionally necessary for the trainer to remain in the anechoic chamber reassuring the cat during the first few days of training, but once a cat sits quietly in the restraint harness consuming liquefied food, further training should be performed under computer control with the subject alone in the testing apparatus.

Automated training is performed by the method of successive approximations. That is, a complex operant response like the hold-release task is shaped by training the cat to perform a chain of less complex responses. The first approximation for the hold-release task is to reinforce the animal with food each time it presses and releases the response lever. These contingencies of reinforcement are created by conducting the hold-release task with the shaping parameters shown in Table II. With shaping parameters in effect, the reinforcement cycle shown earlier in Fig. 3 omits the intertrial interval (a), the variable hold (b), the catch trial (e) and the timeout (f); consequently, every lever press produces an instantaneous and continuous detection trial interval (c), which is marked at onset by a clearly audible tone. Cats end the detection trial and gain a food reward (d) by releasing the lever. The allotment of liquefied meat paste is reduced to 0.5 ml during shaping to prevent cats from quickly satiating as they learn the abbreviated reinforcement cycle.

When cats reliably perform the sequence of pressing and releasing the lever for food, shaping parameters are gradually adjusted to final testing values. Suggestions for the sequence of parameter changes are given as training priorities in Table II. The sequence begins by introducing those contingencies that have small effects on behavior and then builds toward more difficult changes. For example, if shaping has produced a cat that briefly holds the lever before releasing it, the detection trial interval can be quickly decreased to 3 sec (priority 1) and the intertrial interval increased to 1 sec (priority 2). It is also unlikely that this pattern of behavior will be negatively affected by increasing the amount of the food reward from 0.5 - 1.0 ml (priority 3). In contrast, the releasing behavior may be lost by rapidly introducing a 5 - 10 sec variable hold (priority 4) because the cat will tend to release the lever before the start of a detection trial. Several weeks of training may be needed to gradually lengthen the variable hold to its final parameters. It is during this stage of training that cats learn to base lever releases on the detection of a tone.

When cats have been trained to maintain the lever response for the duration of the variable hold and to release the lever before the end of a detection trial interval, they are performing the basic reinforcement cycle of the hold-release paradigm. Now, catch trials are added to the testing procedure to determine how well behavioral responses are under stimulus control (priority 5). Catch trials are 3-sec intervals of silence that are interjected between the variable hold and detection trials. In essence, the addition of this behavioral contingency is equivalent to a slight increase in the average length of the variable hold. If cats are basing the lever release on the detection of a tone, their behavior is usually unaffected by introduction of the occasional catch trial. The final parameter change involves increasing the duration of the timeout (priority 6). Gradual changes should be made in the timeout parameter, especially if cats are experiencing difficulty with the task, as indicated by high false alarm rates, early lever releases, or failure to release the lever during detection trials. As the duration of the timeout builds, cats learn to reduce error responses in order to avoid delays in the reinforcement cycle. A 5-sec timeout produces the desired effect in most cats.

7 Evaluating Optimal Performance

A cat's hearing sensitivity is estimated in the auditory detection task by observing the probability of correct lever releases across a range of stimulus levels. In Fig. 4, effects of stimulus level on the operant response are illustrated by a representative cat's releases to five levels of a 16-kHz tone (open symbols). The level of the tone was randomly selected from these five values for presentation during 671 detection trials using the psychophysical procedure known as the method of constant stimuli (suggested reading: Moody et al., 1976). The leftmost data point indicates the cat's rate of false alarms. The solid line fit to these data is known as a psychometric function.

Traditionally, in psychoacoustic studies of human subjects the absolute threshold is defined as the stimulus level that elicits 50% correct releases, assuming that the psychometric function asymptotes near 0% correct releases at low stimulus levels and near 100% correct releases at high stimulus levels. The traditional detection theshold for this cat is 4.6 dB SPL which is indicated on the abscissa of the figure by the reflected line (a). Animal subjects rarely yield "ideal" psychometric data such as the dashed function in Fig. 4. Cats with very low false alarm rates may exhibit a percentage of correct releases that does not reach 100% at even the highest stimulus levels. On the other hand, a high rate of false alarms may inflate the percentage of correct releases at low stimulus levels as it does in this example.

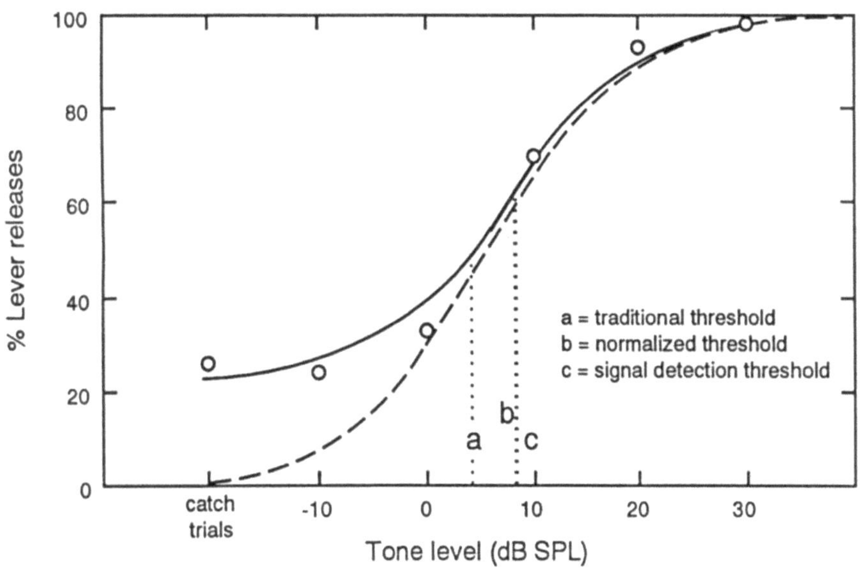

Figure 4. Psychometric function for the auditory detection task.

Different strategies exist for dealing with "nonstandard" psychometric functions. The influence of false alarm rates on threshold estimates may be considered minimal, and therefore ignored, if false alarm rates are less than 15% and constant over time (Stebbins, 1970). Alternatively, psychometric functions with nonzero false alarm rates may be normalized by placing the criterion for the detection threshold halfway between the false alarm rate and 100% correct responding. This "corrected" estimate has been shown to result in slightly higher but more stable thresholds than traditional methods (Hienz et al., 1980). The normalized threshold for data in Fig. 4 is 8.1 dB SPL which is indicated by the reflected line (b). Another alternative to traditional definitions of threshold is to minimize the influence of false alarms by setting the criterion for threshold in terms of signal detection theory (suggested reading: Green and Swets, 1966). Signal detection thresholds like the d' index of sensitivity are less affected by a cat's response bias because they are based on the combined pattern of correct releases and false alarms. Criteria for evaluating optimal performance in behavioral studies range from d' values of 1 to 2. The data in Fig. 4 produce a signal detection threshold (d' = 1) when the sound pressure level of the tone is 8.3 dB SPL, as indicated by the reflected line (c). One point should be clearly evident from these various techniques for analyzing psychophysical performance: "threshold" is not an immutable sensory barrier; it is defined by the context in which it is measured.

In order to measure a sensory threshold, the experimenter must maintain days if not weeks of consistent performance from behavioral subjects. One technique for achieving this objective is to present a number of easily detectable test stimuli during detection trial intervals. Even though such stimuli lend little to the estimate of threshold, they increase the amount of food earned in the experimental session and therefore prevent the subject from ceasing appropriate behavior when too few detection trials result in reinforcement.

A common question asked in any animal psychophysical experiment is "How do I know when a subject has reached its threshold level of performance?" While there is no easy answer to this question, a number of guidelines can help in making such a judgement. One good indicator of threshold is how well does a cat's performance compare with that of other subjects. When comparing performance within the same psychoacoustic study, it is important to use a threshold estimate that takes into account differences in response bias across subjects (e.g., a normalized threshold or one based on d' values). Moreover, comparisons of performance across psychoacoustic studies can be difficult because experiments often use different stimulus conditions or different behavioral paradigms. Even when tested under the same experimental conditions, not all cats perform equally. In this case, graphs of daily threshold estimates and false alarm rates can be examined for upward or downward trends, or other changes in stability. Consistent, abnormally high thresholds are usually signs of poor hearing; whereas, dramatic shifts in threshold most likely indicate poor behavioral control.

8 Concluding Remarks

Studies of auditory physiology are a rich source of questions about auditory perception because the functional significance of a hypothetical sensory process can only be validated with behavioral evidence. Behavioral validation is typically provided in one of three ways. In traditional psychophysical paradigms, hypothetical neural mechanisms are tested by measuring behavioral performance under the same stimulus conditions that produced the relevant neurophysiological results (Pickles, 1979; suggested reading: Hienz et al., 1993). The existence of the sensory process is supported if behavioral responses are predicted by neural responses. In behavior/ablation paradigms, contributions of neural mechanisms are revealed by behavioral deficits that result from selective damage to the nervous system (Trahoitis and Elliot, 1970;

Thompson and Masterton, 1978; Igarashi et al., 1979; Jenkins and Merzenich, 1984). Finally, the significance of neural information may be evaluated by simultaneously sampling neural and behavioral responses to sensory stimuli (May and Sachs, 1992).

Natural auditory behaviors also aid the interpretation of physiology. For example, the cat's sound localization behaviors provide a conceptual framework for understanding the response patterns of auditory neurons in the dorsal cochlear nucleus (Spirou et al., 1993), inferior colliculus (Thompson and Masterton, 1978) and auditory cortex (Jenkins and Merzenich, 1984). Important insights into auditory specialization for human speech perception have been gained from behavioral and physiological studies of cats. Little is known, however, about how cats perceive and process their own vocal signals. Futurc studies of the natural communication behaviors that are associated with these and other biological sounds offer an exciting new direction in the design and conduct of sensory experiments for domestic cats.

9 Acknowledgements

The authors acknowledge the important contributions of William Stebbins and David Moody in the development of our psychophysical methods. The behavioral apparatus shown in Fig. 2 was fabricated at Johns Hopkins in the Biomedical Engineering machine shop of Howard Conner. Major electronic components were designed by Eric Young and Jeremy Rice and were built by Ron Atkinson.

10 Suggested Reading

Green, D.M. & Swets, J.A. (1966). *Signal detection theory and psychophysics*. Huntington: Krieger.

Hienz, R.D., Sachs, M.B. & Aleszczyk, C.M. (1993). Frequency discrimination in noise: comparison of cat performances with auditory-nerve models. *Journal of the Acoustical Society of America, 93*, 462-469.

May, B.J., Aleszczyk, C.M. & Sachs, M.B. (1991). Single-unit recording in the ventral cochlear nucleus of behaving cats. *Journal of Neuroscience Methods, 40*, 155-169.

Moody, D.B., Beecher, M.D. & Stebbins, W.C. (1976). Behavioral methods in auditory research. In Smith, C.A. & Vernon, J.A. (eds.): *Handbook of Auditory and Vestibular Research Methods*. Springfield: Thomas, pp.439-497.

11 References

Butler, R.B., Diamond, I.T. and Neff, W.D. (1957). Role of auditory cortex in discrimination of changes of frequency. *Journal of Neurophysiology, 20*, 108-120.
Casseday, J.H. and Neff, W.D. (1973). Localization of pure tones. *Journal of the Acoustical Society of America, 54*, 365-372.
Elliot, D., Stein, L. and Harrison, M. (1960). Determination of absolute-intensity thresholds and frequency difference thresholds in cats. *Journal of the Acoustical Society of America, 32*, 380-384.
Gerkin, G.M. and Sandlin, D. (1977). Auditory reaction time and absolute threshold in cat. *Journal of the Acoustical Society of America, 61*, 602-607.
Heffner, R.S. and Heffner, H.E. (1985). Hearing range of the domestic cat. *Hearing Research, 19*, 85-88.
Heffner, R.S. and Heffner, H.E. (1988). Sound localization acuity in the cat: effect of azimuth, signal duration, and test procedure. *Hearing Research, 36*, 221-232.
Hienz, R.D., Sinnott, J.M. and Sachs, M.B. (1980). Auditory intensity discrimination in blackbirds and pigeons. *Journal of Comparative and Physiological Psychology, 94*, 993-1002.
Igarashi, M., Cranford, J.L., Allen, E.A. and Alford, B.R. (1979). Behavioral auditory function after transection of crossed olivo-cochlear bundle in the cat: pure-tone intensity discrimination. *Acta Otolaryngology, 87*, 429-433.
Jenkins, W.M and Merzenich, M.M. (1984). Role of cat primary auditory cortex for sound-localization behavior. *Journal of Neurophysiology, 52*, 819-847.
Martin, R.L. and Webster, W.R. (1987). The auditory spatial acuity of the domestic cat in the interaural horizontal and median vertical planes. *Hearing Research, 30*, 239-252.
May, B.J. and Sachs, M.B. (1992). Dynamic range of neural rate responses in the ventral cochlear nucleus of awake cats. *Journal of Neurophysiology, 68*, 1589-1602.
Neff, W.D. and Hind, J. (1955). Auditory thresholds of the cat. *Journal of the Acoustical Society of America, 27*, 480-483.
Pickles, J.O. (1979). Psychophysical frequency resolution in the cat as determined by simultaneous masking and its relation to auditory-nerve resolution. *Journal of the Acoustical Society of America, 66*, 1725-1732.
Rice, J.J., May, B.J., Spirou, G.A. and Young, E.D. (1992). Pinna-based spectral cues for sound localization in cat. *Hearing Research, 58*, 132-152.
Rosenzweig, M.R. (1946). Discrimination of auditory intensities in the cat. *American Journal of Psychology, 59*, 127-136.
Saunders, J.C. (1969). Behavioral discrimination of click intensity in cat. *Journal of the Experimental Analysis of Behavior, 12*, 951-957.
Spirou, G.A., May, B.J., Wright, D.D. and Ryugo, D.K. (1993). Tonotopic organization of the dorsal cochlear nucleus of the cat. *Journal of Comparative Neurology, 329*, 36-52.
Stebbins, W.C. (1970). *Animal psychophysics: the design and conduct of sensory experiments*. New York: Appleton-Century-Crofts.
Thompson, G.C. and Masterton, R.B. (1978). Brain stem auditory pathways involved in reflexive head orientation to sound. *Journal of Neurophysiology, 41*, 1183-1202.
Trahiotis, C. and Elliot, D.N. (1970). Behavioral investigation of some possible effects of sectioning the crossed olivocochlear bundle. *Journal of the Acoustical Society of America, 47*, 592-596.

Methods in Comparative Psychoacoustics
G. M. Klump, R. J. Dooling, R. R. Fay & W. C. Stebbins (eds)

Methods in Directional Hearing

C. H. Brown

University of South Alabama, Department of Psychology, 326 Life Sciences Classroom Building, Mobile, Alabama 36688, USA

Summary. Testing and training behavioral procedures using operant conditioning positive reinforcement methods are described for measuring minimum audible angles in animals. These procedures may be adopted to measure the acuity of localization for both azimuth and elevation.

1 Introduction

The early detection of the movement and activities of other organisms is often revealed acoustically, and it is likely that most vertebrates have evolved mechanisms to swiftly, accurately and effortlessly determine the location of the source of most sounds. The location of a sound source is specified in three-dimensional space relative to the orientation of the listener. Hence, the location of two different sound sources may vary in respect to their azimuth: the horizonal coordinate of their locations, elevation: the vertical coordinate of their locations, or in reference to their respective distances to the listener. This chapter will address the methodological challenges for measuring the perception of sound source azimuth and elevation in terrestrial vertebrates. The measurement of the perception of sound proximity or distance is beyond the scope of this chapter, and readers are referred to Brown (1994) for a recent review.

In most vertebrates the precise localization of the source of a sound is dependent upon the comparative analysis of the signal incident at both ears. If the listener is oriented such that the location of the source is off to one side, the pressure wave will necessarily arrive at the "near" ear before it arrives at the "far" ear. For each additional centimeter the sound wave must travel to reach the far ear, the signal will arrive 29 microseconds later than it will at the near ear. Interaural differences in *time-of-arrival* of the signal at each ear constitute one of the principal cues for directional hearing, and the geometry relevant for this cue is idealized in Figure 1. The wave front of a distant source is approximated by a plane, and an interaural sound propagation distance difference (Δ-d) will occur for all sound locations other than those which lie on the median plane. For a sound source at azimuth x, the additional distance that the sound must travel to reach the far ear (left ear in figure 1) is given by the sum of the linear distance r(sin x) and the curvilinear distance r(x). The propagation distance difference (Δ-d) in the sound pathlength for the two ears is expressed by Equation 1.

$$\Delta\text{-d} = r(x + \sin x) \qquad \text{Eq. 1}$$

where Δ-d is the distance difference in cm, r is the radius of the listener's head in cm, and angle x is measured in radians.

The pathlength difference between the near and far ears produces an interaural difference in the time-of-arrival of corresponding points in the waves incident at both ears. This time-of-arrival difference (Δ-t) is calculated by dividing the distance difference by the velocity of sound (nominally 343 m/second in air). The corresponding relationship between Δ-t and azimuth is expressed by equation 2.

$$\Delta\text{-t} = r(x + \sin x)/3.43 \times 10^4 \qquad \text{Eq. 2}$$

where; Δ-t is the temporal difference in time-of-arrival in microseconds, r is the radius of the listener's head in cm, and the angle x is measured in radians.

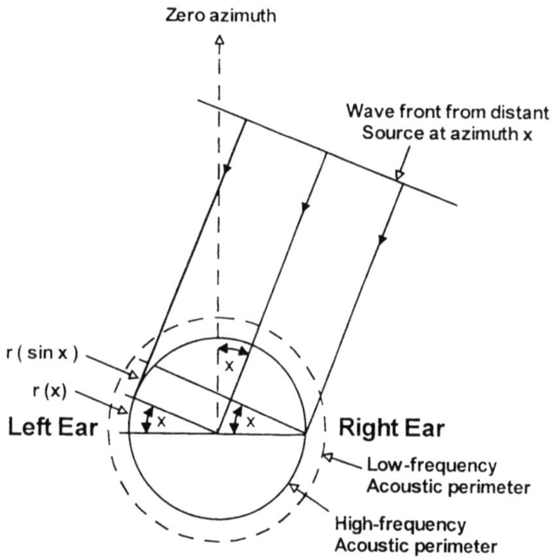

Figure 1. Geometrical basis for the generation of interaural time differences.

From inspection of Figure 1 it is apparent that the magnitude of the interaural time-of-arrival cues for sound localization increase as the position of the sound source defined by angle x approaches π/2 radians or 3π/2 radians (90° or 270°). It is also apparent that for any given azimuth, Δ-t will increase as head size increases. That is, species with relatively large heads will have greater interaural-differences in time-of-arrival than will species with small heads.

The advantage conferred to species with large heads for interaural time-of-arrival differences also applies to interaural-differences in *sound-pressure-level*, the second principal cue for directional hearing. Near and far ear differences in signal level arise because the head casts an "acoustic shadow", and the signal level at the distant ear will be perceptibly less than that at the near ear. It should also be emphasized that the size and orientation of the pinnae can exaggerate the magnitude of both interaural-level and time-of-arrival cues. Many mammals have large mobile pinnae. These pinnae act as a funnel and they can significantly increase the magnitude of interaural differences in the level of the signal at both ears. Interaural-level differences are frequency dependent, and in general the greater the frequency of the signal, the greater the magnitude of interaural level differences (Musicant, Chan & Hind, 1990).

Interaural differences in time-of-arrival are produced for both the leading edge of the wave front, and for on-going time or phase differences in the sound waves. Though the acoustic radius for high-frequency sounds and for the leading edge of the sound wave is determined by the radius of the skull, Kuhn (1977; 1987) showed that the "effective acoustic radius" of the head for on-going time or phase differences is larger than the skull perimeter when low-frequency sounds are presented. If the wavelength of the sound is less than or equal to the diameter of the skull (a high-frequency sound) the acoustic radius is governed by that of the skull; if the frequency of the signal is decreased so that the wavelength grows up four times the diameter of the skull the effective acoustic radius expands. In human subjects, the effective acoustic radius for low-frequency sounds is about 150% of that for high-frequency signals (Kuhn, 1977), and in non-human organisms with pronounced prognathism and large pinnae it is likely that the effective acoustic radius for low-frequency sounds is significantly greater. These observations indicate that species differences in the anatomy of the head and pinnae may influence directional hearing. Furthermore, the acuity of directional hearing and the underlying cues upon which it is based is frequency dependent. Thus, the comparison of directional hearing abilities for different species should be judged in reference to the test stimuli employed, and the nature of the cues available for processing.

Several additional issues relevant to the design of studies of sound localization may also be illustrated by Figure 1. First, many acoustic sources emit sound as a point source, and the shape of the wavefront of an expanding spherical wave will more closely approximate that of a plane surface as the propagation distance increases. Most studies of directional hearing have been conducted in special sound rooms where standard room dimensions commonly limit propagation distances to 2 or 3 meters or less. The wavefronts generated in these investigations are usually regarded as approximately plane surfaces. Second, hearing must be measured in a free-field environment where reflections of the wave from environmental surfaces do not exist, or are so greatly attenuated that they do not significantly contribute to, or interfere with the perception of sound localization. Third, measurements must be conducted in a quiet environment where other sound sources do not obstruct the perception of the target signal. These two requirements have resulted in using anechoic or semi-anechoic rooms for the test environment. In an anechoic room, the walls, floor and ceiling of the room are covered with wedges constructed of a sound absorbent material. The shape of the wedge is designed so that the angle of reflection will direct any unabsorbed sound to the surface of another wedge where it has an additional chance to be absorbed. In these rooms, the subject is suspended over the bottom wedges on a wire or grid floor which is designed to be nearly acoustically transparent. However, because animals may urinate or defecate in the test environment, it may be necessary to place a sound absorbent rug under the animal.

2 Sound Localization Paradigms

Many animals will reflexively orient towards the origin of a sound, and orientation paradigms have been developed to measure the acuity of directional hearing. With these procedures a head turn or body turn is used as an index of perceived sound direction, and the acuity of localization is defined as the average discrepancy between the subject's midline and the azimuth of the sound source (Knudsen & Konishi, 1978; Brown, 1982; Makous & Middlebrooks, 1990; Wagner, this volume). In addition to orientation procedures, some animals readily exhibit a phonotaxic response and will move towards the perceived origin of a sound, and the accuracy of approach has also been used as an index of the acuity of localization (Casseday & Neff, 1973; Feng, Gerhardt & Capranica, 1976; Waser, 1977; Klump, this volume). In both orientation and phonotaxis approaches, some subjects are apt to habituate and cease to respond unless food or some other reinforcement is used to maintain this behavior. In certain instances, phonotaxis is restricted to specific classes of stimuli and may be dependent upon the hormonal state of the listener. In tree frogs, for example, females will approach broadcast exemplars of their species-specific mating call, but not synthetic sounds that depart from species typical parameters and not after the breeding season (Feng et al, 1976; see Gerhardt, this volume).

In both approach and orientation paradigms, localization is made in reference to the subject's physical orientation in space, and not in reference to an external acoustic marker. These procedures have been categorized as egocentric procedures (Brown & May, 1990) because the response is assessed in reference to the orientation of the listener's self (ego). Observed differences in sound localization acuity determined by egocentric methods may be due to differences in the resolution of perceptual systems, or due to limitations in the precision of orientation of motor systems, and egocentric methods do not permit the independent assessment of the performance of the perceptual or motor component. Consequently, otocentric, or ear-centered approaches (Brown & May, 1990), have been developed to directly determine the acuity of the perceptual system. In the otocentric approach, animals are trained to operate response levers to report the detection of a change in sound source location relative to an external acoustic marker (Brown et al. 1978; Heffner & Heffner, 1988), and these procedures complement those developed by Mills (1958) to measure the minimum audible angle in humans.

3 Moving Speakers, Matching Speakers, and Acoustical Artifacts

In both otocentric and egocentric paradigms the experimental design typically requires the technical ability to present signals from different spatial locations with short interstimulus intervals between signal presentations. This can be done in one of two ways: moving a single speaker from one location to the next, or using switching circuits to present the signal from speakers positioned at different locations. Both of these approaches have advantages and disadvantages. Because no two speakers emit acoustically identical sounds some investigators have elected to reposition a single transducer between trials. The advantage of this approach is that the signal is identical irrespective of the position from which it is broadcast. However, it is difficult to move a speaker quickly and silently without introducing other artifacts which could be used to detect the change in the position of the sound source. A masking noise can be presented during the interstimulus interval to help obscure any mechanical sounds which may cue changes in the speaker location.

Multiple speaker arrays and switching circuits permit rapid and silent changes in source location, but the possibility of the artifact of sound "quality" differences between speakers must be addressed. If the test signals are pure tones, then the output of different speakers can easily be matched within 1 dB. Furthermore, programmable attenuators can be used to randomly vary the level of the signal pulse by pulse irrespective of the speaker emitting the sound, and this strategy virtually insures that any differences in signal level, characteristic of different transducers, cannot be used as a cue to detect changes in location (Brown et al, 1978). If the test signal is a click, tone pip, or any other spectrally complex signal, then the demand for matching transducers becomes more challenging. Candidate speakers should be calibrated with a tone swept across the target frequency range in an anechoic room to identify speakers which have nearly identical response characteristics. The third-octave calibration curves provided by some speaker manufactures are inadequate for matching transducers. If the test signals are clicks or bursts, investigators need to be concerned about the possibility of speaker differences in ringing or the rise- and fall-time of the speaker response. Once a set of "matched" transducers has been identified by their physical response characteristics, it is usually possible to confirm the validity of the match through perceptual tests. In nearly every case where the range of human and animal hearing overlap, human listeners are able to hear finer distinctions between sounds than are most animal subjects. Thus, if the test stimuli are in the range of human hearing, perceptual tests may be conducted with human listeners to validate the match. Because the acuity of vertical localization is usually less than that for azimuth, it may be possible to show that listeners can detect changes between speaker locations when the speakers differ in azimuth, but not when the speakers have the same azimuth and are placed directly above one another. If this test can be met, then it is very likely that any differences between speakers is imperceptible to other subjects. Once a set of speakers have been selected, investigators often rotate the speakers across positions over the course of a study. This strategy prevents the formation of an association between location and speaker quality, and this procedure also provides the opportunity for investigators to show that the subject's response is dependent upon location and independent of which speaker occupied any location. Furthermore, it is possible to randomize signal level, as well as randomize subtle sound quality variations for each signal presentation. That is, the increased availability of high-speed computers for experimental control, and improvements in digital audio, make it possible to test listeners, not with just a single exemplar of the stimulus, but with a number of exemplars selected randomly out of a family of signal variations designed to mimic minor speaker differences. Alternatively, it may be possible to digitally equalize the signal for each speaker so that the signal, as broadcast, compensates for any differences between transducers. These strategies would make it very difficult for transducer differences to be a significant artifact for sound localization studies.

Because the auditory system is so sensitive to subtle acoustical changes, attention must be directed to the possibility of an artifact due to position dependent reflections from the apparatus used to orient and position subjects, train subjects, or measure the subject's response. The apparatus should be designed so that the subject's head is located in the center of an arc (or surface of a sphere) along which sounds will be presented, and the apparatus used to position the subject should be acoustically transparent, or present the smallest manageable surface areas. For example, a special primate chair has been designed to minimize the reflective surfaces around the subject's head for sound localization experiments (Brown et al, 1978). In the development of this chair, the reward delivery system was changed so that food pellets were not delivered to the subject's mouth, but to the subject's hand (monkey's would then feed themselves during the intertrial interval), and the size of the apparatus used to position the subject's head was pared to the bare essentials resulting in a 95% reduction in the surface area of reflective materials. By calibrating the test apparatus in the position occupied by the subject for each speaker location, problem reflections can be identified, and sound absorbing foam can be placed to correct some problems or the apparatus can be redesigned. Calibrations made through an omnidirectional microphone placed at the position occupied by the subject's head should reveal no differences between the various broadcast positions. With the appropriate attention to these details, the perception of sound position has been found to be a very robust phenomenon, and measurements of directional hearing show very high agreement both across laboratories and with different procedures (Heffner & Heffner, 1988; Brown et al, 1978; Brown & May, 1990).

4 Training Animals to Detect a Change in Sound Position

Operant conditioning procedures have been successfully employed to train a variety of birds and mammals to emit a behavioral response to indicate the perception of visual or acoustic stimuli (Stebbins, 1970). Operant conditioning procedures functionally transcend the human-animal language barrier, and have been employed to "teach" animals to detect signals, discriminate between signals, and to classify or categorize signals as exemplars of two or more classes. Sound localization studies essentially require animals to discriminate between sounds on the basis of differences in source location. Food reward has been used to train monkeys to make contact with a response-sensitive key as an observing response, and to terminate contact when sound is presented. By varying the amplitude of the signal, the subject's threshold for hearing can be determined. This basic hold-release procedure can be modified to convert it from a detection task into a discrimination task suitable for measuring directional hearing. Monkeys were first trained to contact a response-key illuminated from within by a flashing light. When contact was made the light was illuminated steadily and a very audible sound was presented as a series of 300 msec pulses at a rate of 1.5 pulses per second from a speaker positioned at an azimuth of about 45° to the subject's right. When the subject released the key, food reward was delivered to a feeder cup positioned by the response key. After a brief 2-second intertrial interval (iti) the key light started flashing again signalling the availability of the next trial. If the monkey contacted the key during the intertrial interval the iti clock was reset. Thus the subject learned to refrain from contacting the key, unless the key light was flashing. Over subsequent sessions, monkeys were trained to hold the key for progressively longer intervals until the sound was presented. If the subject released the key prior to sound onset, it entered the monkey into a brief time-out during which the experiment was suspended for several seconds. The hold interval preceding sound onset was gradually lengthened from 50 ms to 8-seconds, and after the monkeys learned to wait for a fixed 8-second interval, the hold interval was changed to a variable interval (1.0 to 10.0) seconds preceding sound presentation. At this stage of training the signal parameters were also modified so that the subject had to release the response key within a 2-second response interval during which three 300-msec pulses were presented. If the subject failed to release the key during this interval, the subject was not punished and the sequence recycled to the beginning of the variable hold state. At this stage in training the procedure is much like the detection paradigm used to measure sound detection thresholds. The procedure was then converted into a sound location discrimination task by presenting very quiet 300 msec pulses of sound during the variable hold interval from a different speaker positioned at 0° azimuth (directly on the monkey's midline). Initially the level of the signal presented at 0° azimuth was barely audible. Hence, the discrimination could be made according to either sound level or sound location. Over subsequent sessions the level of the signal at the reference location, 0° azimuth, was adjusted to equal that at the comparison position (45° azimuth). Other comparison positions were then tested, and the monkey's learned that reporting the detection of any change in sound location from the reference position of 0° azimuth was reinforced. Random variations in signal level (+ or - 3 dB steps, over a 21 dB range) were introduced to insure that level differences between speakers could not serve as a cue for localization. In order to measure the subject's rate of guessing, 20% of the trials were catch-trials. During catch-trials, the signal did not change location, and releases entered the subject into the time-out. Note that in the sequence of training stages involved in the development of this procedure, the discrimination task was acquired through the operant conditioning method of chaining, where the final component of the chain was learned first (releasing the key when the sound was presented from a location off the subject's midline), and each preceding component of the chain was learned in reverse order. The procedure was finally converted from a sound detection task, to a location-dependent discrimination task, by fading-in the level of a repeating background signal at the referent 0° azimuth position. Once the monkeys learned this procedure, a nonverbal communication link had been established between investigator and subject, and the acuity of directional hearing could be tested for any stimulus of interest, impediments to localization could be studied, and the referent position could be shifted to any candidate location.

5 Measuring the Minimum Audible Angle

Figure 2 shows psychophysical functions for three monkeys for the localization of an 8 kHz tone. The referent position was 0° azimuth the percent of trials detected at each comparison position increased with azimuth. The subject's catch-trial rate was very low, and this rate is displayed over the 0° point.

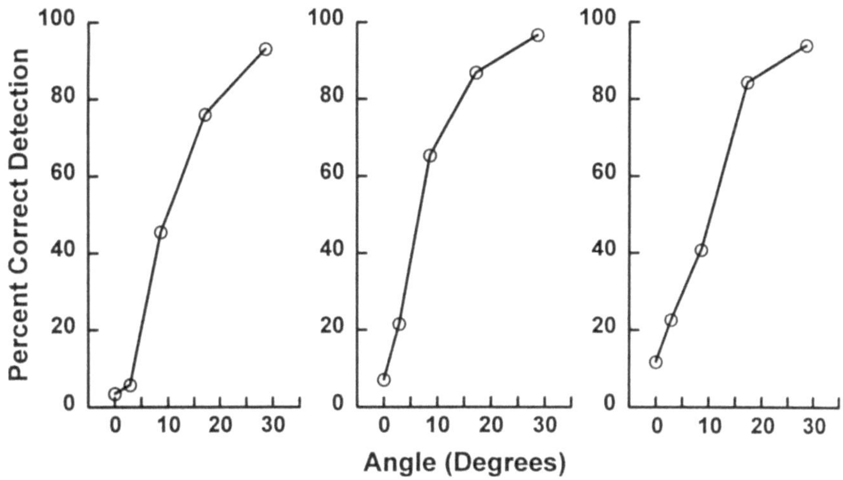

Figure 2. Psychophysical localization functions for three monkeys. The test signal was an 8 kHz tone.

The results show that the monkeys had learned to maintain key contact until the location of the sound pulses changed from the referent location to one of the comparison positions, and that the percent of trials detected increased monotonically approaching 100% correct at 30° azimuth. Psychophysical functions were determined by the method of constant stimuli: the comparison stimuli were presented in random order, and the subject was required to discriminate between sounds presented at the standard locus and all comparison positions. The functions were derived by pooling data over eight or more experimental sessions in which about 3,000 trials were presented. From these functions the minimum audible angle, or subject's threshold (the 50% detectability locus), could be interpolated from the proportions of correct detections for each comparison azimuth.

Figure 3 displays the minimum audible angle as a function of tone frequency. At low-frequencies (signals below 1,000 Hz) thresholds for each monkey are nearly coincident and decrease as stimulus frequency is increased. At stimulus frequencies greater than 1,000 Hz, minimum audible angles fluctuate over frequency and vary widely between monkeys. This pattern of results is consistent with the duplex theory of sound localization which states that the localization of low-frequency tones is governed by a mechanism sensitive to interaural temporal differences in the waveforms, while localization for high-frequency tones is determined by a second mechanism sensitive to interaural level differences. Individual differences in the shape of one's pinnae would result in idiosyncratic frequency dependent transfer functions for the near and far ear at each comparison azimuth, and this would account for the idiosyncratic variations in thresholds for tones above 1,000 Hz.

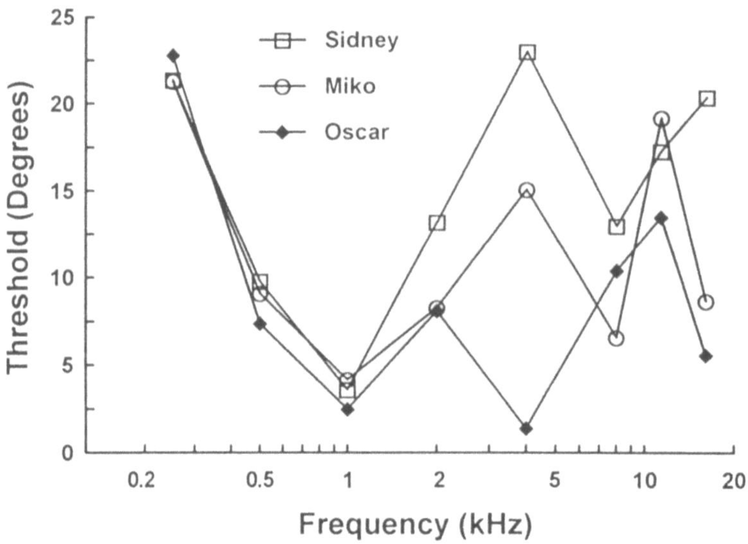

Figure 3. Pure-tone thresholds as a function of frequency. The ordinate is threshold in degrees and the abscissa is signal frequency.

By positioning the location of the reference speaker to a different azimuth it is possible to use the same methods to study how the acuity of localization changes for reference locations off the midline (Brown, Beecher, Moody & Stebbins, 1980). Similarly, the same procedure may be adopted to study vertical localization by rotating the speaker arc 90° so that the speaker locations differ in elevation (Brown, Schessler, Moody & Stebbins, 1982). Figure 4 shows vertical minimum audible angles for bands of noise as a function of the high-frequency cutoff of the noise band. The results show that monkeys exhibit high acuity vertical localization if the signal has sufficient broadband high-frequency information, if the high- frequency content is removed by reducing the cutoff frequency of a low-pass filter, the precision of vertical localization decreases. These findings are consistent with the overall patterns of measurements of pinna transformation functions (Musicant et al, 1990).

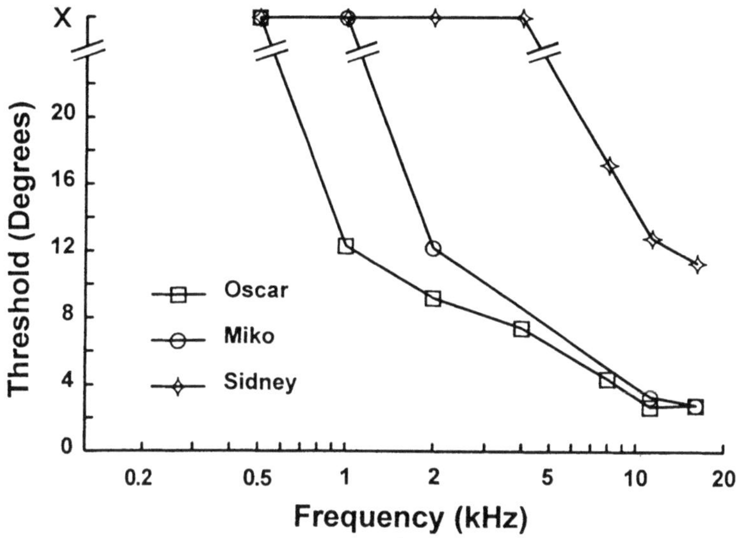

Figure 4. Vertical minimum audible angles for bands of noise, as a function of the high-frequency cutoff of the noise. The low-frequency limit of the bands was 125 Hz. An X indicates that the subject was unable to localize the signal.

Sound localization has been one of the most persistently studied areas in auditory perception and physiology. Directional hearing is important for auditory image analysis, the perception of signals in noise, as well as for sound localization per se (Brown & May, 1990). However, most research has focused on the processing of acoustically simple synthetic signals in quiet echo-free environments, and few researchers have studied how directional hearing may become compromised in complex reverberant, noisy environments, or how the aging process or assaults to the auditory system may influence this fundamental capacity (Brown, 1984). Thus, many questions in directional hearing have yet to be studied, and it is hoped that the methods described here will be adopted by other investigators to gain a fuller understanding of these phenomena.

6 Acknowledgements

I thank J. Sinnott and R. Roush who commented on an earlier version of this manuscript. This work was supported by NIH grant DC00164.

7 Suggested Reading

Yost, W. & Gourevitch, G. (1987). *Directional Hearing.* New York: Academic Press.

8 References

Brown, C.H. (1982). Ventroloquial and locatable vocalizations in birds. *Zeitschrift für Tierpsychologie, 59,* 338-350.

Brown, C.H. (1984). Directional hearing in aging rats. *Experimental Aging Research, 10,* 35-38.

Brown, C.H. (1994). Auditory localization. In: A. Popper & R. Fay (Eds.) *Comparative Mammalian Hearing (Springer Series in Auditory Research, Vol. 4).* New York: Springer-Verlag (in press).

Brown, C.H., Beecher, M.D., Moody, D.B. & Stebbins, W.C. (1978). Localization of pure tones in Old World monkeys. *Journal of the Acoustical Society of America, 63,* 1484-1494.

Brown, C.H., Beecher, M.D., Moody, D.B. & Stebbins, W.C. (1980). Localization of noise bands by Old World monkeys. *Journal of the Acoustical Society of America, 68,* 127-132.

Brown, C.H., Schessler, T., Moody, D.B. & Stebbins, W.C. (1982). Vertical and horizontal localization in primates. *Journal of the Acoustical Society of America, 72,* 1804-1811.

Brown, C.H. & May, B.J. (1990). Sound localization and binaural processes. In: M.A. Berkley & W.C. Stebbins (Eds.) *Comparative Perception, Volume I, Basic Mechanisms.* New York: John Wiley & Sons (pp. 247-284).

Casseday, J.H. & Neff, W.D. (1973). Localization of pure tones. *Journal of the Acoustical Society of America, 54,* 365-372.

Feng, A.S., Gerhardt, H.C. & Capranica, R.R. (1976). Sound localization behaviour of the green tree frog (*Hyla cinerea*) and the barking tree frog (*H. gratiosa*). *Journal of Comparative Physiology, 107,* 241-252.

Heffner, R.S. & Heffner, H.E. (1988). Sound localization acuity in the cat: effect of signal azimuth, signal duration, and test procedure. *Hearing Research, 36,* 221-232.

Knudsen, E.I. & Konishi, M. (1978). A neural map of auditory space in the owl. *Science, 200,* 795-797.

Kuhn, G.F. (1977). Model for the interaural time differences in the azimuthal plane. *Journal of the Acoustical Society of America, 62,* 157-167.

Kuhn, G. F. (1987). Physical acoustics and measurements pertaining to directional hearing. In: W. Yost, & G. Gourevitch (Eds.) *Directional Hearing.* New York: Academic Press (pp. 3-25).

Makous, J.C. & Middlebrooks, J.C. (1990). Two-dimensional sound localization by human listeners. *Journal of the Acoustical Society of America, 87,* 2188-2200.

Mills, A.W. (1958). On the minimum audible angle. *Journal of the Acoustical Society of America, 30,* 237-246.

Musicant, A.D., Chan, J.C.K. & Hind, J.E. (1990). Direction-dependent spectral properties of cat external ear: new data and cross-species comparisions. *Journal of the Acoustical Society of America, 87*, 757-781.

Stebbins, W.C. (1970). *Animal psychophysics: the design and conduct of sensory experiments.* New York: Appleton-Century-Crofts.

Waser, P.M. (1977). Sound localization by monkeys: a field experiment. *Behavioral Ecology and Sociobiology, 2*, 427-431.

G. M. Klump, R. J. Dooling, R. R. Fay & W. C. Stebbins (eds)

Psychoacoustic Studies in Bats

Sabine Schmidt

Zoologisches Institut der Universität München
Luisenstraße 14, 80333 München, Germany

Summary. Two examples of psychoacoustic experiments with bats are discussed: determination of masked auditory thresholds in *Tadarida brasiliensis*, and pitch classification by *Megaderma lyra*. In both cases, a two-alternative, forced choice procedure is combined with a probit analysis of data.

1 Introduction

Echolocating bats are adapted to perceive their environment from acoustical cues in the ultrasonic range, which makes their auditory system an interesting object of study for comparative psychoacoustics. Previous behavioural studies on bat audition centred on a characterization of their active sonar system, e.g. its ability to determine target distance, velocity or object properties (cf. Schnitzler and Henson, 1980). In most of these experiments, some kind of a two-alternative, forced choice (2-AFC) procedure was used (as first applied to bats by Simmons and Vernon, 1971 and Simmons, 1973). Apart from audiograms, only few data are available from passively listening bats in standard psychoacoustic paradigms (cf. Fay, 1988).

In our psychoacoustic experiments with bats, we combined a 2-AFC procedure with a probit analysis of data (Finney, 1971) in order to determine various discrimination thresholds, and to describe the classification of acoustical sensations by bats. Here we give two examples to show how this approach can be realized for different questions and bat species, and to demonstrate the range of knowledge accessible by these methods.

2 Example 1: Masked Auditory Thresholds in *Tadarida brasiliensis*

In the first experiment, *T. brasiliensis* was trained to recognize the presence of short pure tone pulses in longer broadband noise masker pulses. The ability to discriminate masker-plus-pure tone pulses from masker-only pulses was determined as a function of pure tone level for different pure tone durations.

2.1 Tadarida brasiliensis

T.brasiliensis, the Mexican free-tailed bat, is a small Microchiropteran species common in the central and southwestern United States, Mexico and the Caribbean. The fast flying animals use a broadband echolocation system to hunt for insects in open spaces (Simmons et al., 1978). As in many other species, the prey is captured with the help of the wings and a pouch formed by the interfemoral membrane extending between the hind legs. During the day, the bats roost in caves and buildings; they often shelter in narrow crevices, skilfully crawling backward and forward.

The six bats used in our experiments were tested on six days a week. During the period of the experiments, they were fed pieces of mealworms only in the experimental sessions. As food deprivation is critical for these small animals, their weight was checked daily and kept at their individual pre-experimental weight of 9-12 g.

2.2 Training

Our 2-AFC experiments were adapted to the natural behaviour of *T.brasiliensis* in two respects. First, since these bats cannot attain their normal flight speed and manoeuvrability in the experimental chamber, the animals were trained to crawl rather than fly toward the sound source to indicate a choice. Second, because these animals normally gather food from their interfemoral pouch, they have problems picking up a mealworm from a feeding dish. Therefore, we delivered food reward by forceps directly to the bat´s mouth.

The setup used in our experiments is shown in Fig.1. The training procedure can be divided into three stages, i.e. learning to move and feed on the experimental platforms, associating acoustical stimuli with a food reward, and discriminating the rewarded from unrewarded stimuli.

In the first stage, which took only a few days, the bats learned to crawl from the experimenter´s hand onto the starting platform, then to move onto the response platform at which a mealworm was offered, and finally to climb back to the experimenter´s hand after having eaten the reward.

In the second stage, a sequence of pure tone pulses (frequency: 40 kHz; level: 75 dB SPL; duration: 100 ms, repetition period: 1.2 s) was presented from one of the speakers and the reward was *simultaneously* offered at the response platform pointing towards the active speaker. After about two weeks, the bats associated the tone sequence with the reward and moved to the corresponding response platform *before* the reward was delivered.

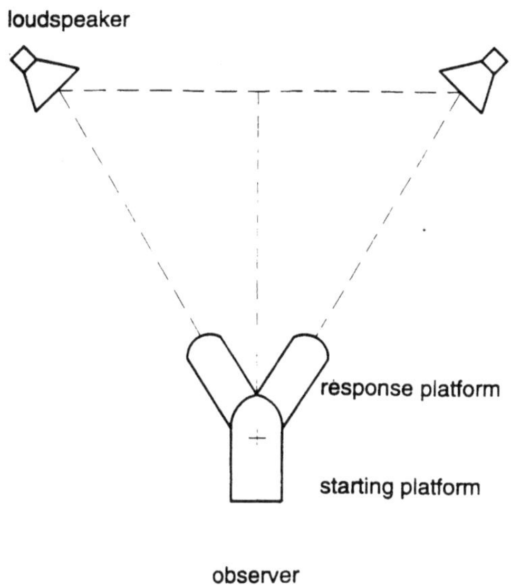

Figure 1. Setup used to determine thresholds in *T.brasiliensis* (drawing not to scale). A central starting platform was connected to two response platforms pointing in the direction of two loudspeakers. The loudspeakers were mounted 22° to the left and right of the starting platform and 7° above it, at 167 cm distance from the head position (cross) of the bat while comparing the stimuli. In order to reduce reflections from the floor and walls, the whole setup was mounted on rods about 1 m above the floor of a sound and echo attenuated chamber. The observer ran the experiments from behind the experimental platforms.

Finally, masker pulses (wide band noise: 10-100 kHz; level: 30 dB SPL; duration: 300 ms) were added so that masker-only pulses (unrewarded stimulus) from one speaker alternated with masker-plus-signal pulses (rewarded stimulus; masker onset preceeding signal onset by 200 ms) from the other speaker. Thc period of the pulse sequences turned out to be critical. Successfully trained bats turn their head towards the active speaker, and thus cannot follow the presentation of the stimuli if the alternation is too rapid. On the other hand, if the alternation cycle is too long, the bats tend to leave the starting platform before they have listened to both stimuli. In our experiments, a period of 1.2 s for one alternation cycle proved most efficient.

After six weeks of training, a stable discrimination performance of about 95 % correct choices was reached.

All trained bats stopped echolocating during stimulus presentation.

2.3 Experimental Procedure and Data Collection

Stimulus presentation started when the bat crawled onto the starting platform and stopped when it had moved completely onto one of the response platforms. The side of the rewarded stimulus was selected according to a pseudorandom schedule in which a bat using a position or alternation strategy would score 50 % correct (chance level), and in which the same side was selected not more than four times in sequence, in order to prevent the formation of side preferences.

The performance of the bats was determined for 6-7 pure tone levels around the expected threshold. A modified method of constant stimuli was used, presenting tones at all tested levels in each experimental session. As the bats refuse to respond to low tone levels directly after the presentation of loud stimuli, the stimuli were successively attenuated in steps of 5 dB after every five trials. About 30-55 trials were typically run in one session. At least 30 decisions of a bat were collected at each pure tone level across several sessions. The value of each datum point was calculated as the percentage of correct choices at a given pure tone level and duration.

Pure tone durations between 2 ms and 400 ms were tested in independent experimental series. For pure tone durations above 100 ms, masker duration was set to 600 ms; at these long stimulus durations, the longer alternation period of 2 s was accepted by the bats.

2.4 Threshold Determination

The *threshold* for the perception of a stimulus (in our case, the masked pure tone) is based on the concept that there is some value of the stimulus parameter (here the pure tone level) above which the animal perceives the stimulus and reacts accordingly (leading to 100 % correct choices), whereas it is quite unable to perceive stimuli below their critical level, and chooses completely at random in this case (50 % correct choices).

In actual experiments, we observe no such transition from discriminable to indiscriminable stimuli, but rather a gradual decrease in the percentage of correct choices. To account for this fact, we assume that the ideal threshold is statistically smeared out by a large number of unknown random factors, e.g., the motivation or the attention of the animal. By the Central Limit Theorem of probability theory, a function of many independent random parameters will be normally distributed irrespective of their individual probability distributions. Thus we are led to expect that the observed performance will be normally distributed around the ideal threshold.

Accordingly, we approximate data points by Gaussian error functions, varying their centre and steepness, and interpret the centre of the optimal fit curve (where the density of the corresponding normal distribution is maximal) as the most probable threshold value (Finney, 1971). Apart from the threshold itself, this method also provides a standard error for the threshold value which is calculated from the quadratic deviation of the data points from the fit curve. The quality of the approximation can be assessed using a chi^2-test.

In order to determine the masked auditory threshold at a given pure tone duration, we used Gaussian error functions between the 50 % chance level and 95 % correct choices (i.e. the upper asymptote was set to the average performance of *T.brasiliensis* measured in simple tasks). The centre of the fit curve, then, correspnds to 72.5 % correct choices. Two examples of threshold determinations are shown in Fig.2.

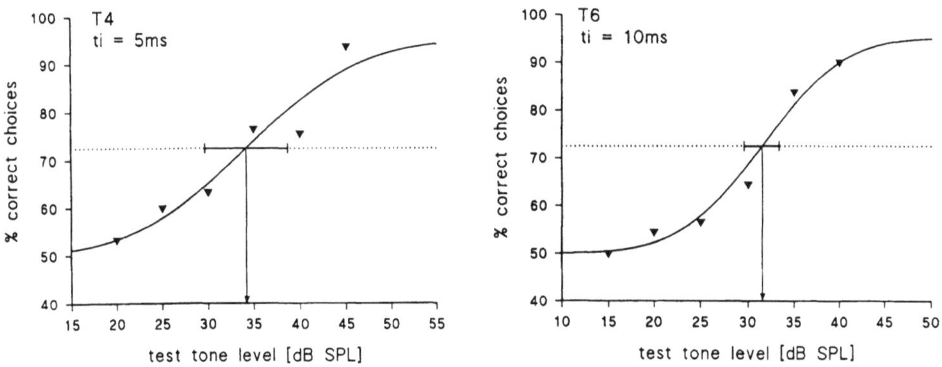

Figure 2. Threshold determination by probit analysis. The triangles represent the performance [% correct choices] of *T.brasiliensis* in discriminating 40 kHz pure tone pulses in masking noise as a function of test tone level [dB SPL] at a fixed test tone duration t_i. These data points are approximated by a Gaussian error function (continuous line). The dotted line marks the threshold criterion. The threshold is indicated by a vertical arrow. The error bar represents the standard error of the threshold value calculated from the residual deviations of the data points from the fit curve. Both fit curves are good representations of their respective data points (chi^2-test, $p < 5$ %).

2.5 Survey of Similar Threshold Experiments

The above threshold determination method has been applied by us to measure a number of basic psychoacoustic properties in *T.brasiliensis*. In addition to the temporal summation function, considering thresholds at different test tone durations (Schmidt & Thaller, 1994), an audiogram and the critical ratio function (Schmidt et al., 1990), as well as frequency modulation thresholds (Bartsch & Schmidt, 1993) were obtained.

Moreover, the audiogram (Schmidt et al., 1984), the critical ratio function (Sedlmeier & Schmidt, 1989) and excitation patterns (Schmidt et al., 1992) in the bat *Megaderma lyra* were determined using a similar method; in these experiments, of course, the setup was modified to fit the behavioral requirements of the latter species (see below).

3 Example 2: Pitch Classification in *Megaderma lyra*

In the following experiments, it was studied how *M.lyra* classifies the pitch of harmonically structured sounds, so-called complex tones, in comparison to the pitch of a pure tone.

3.1 Megaderma lyra

M.lyra, the Indian False Vampire, is a relatively large bat with a highly manoeuvrable flight adapted to hunting for small vertebrates and insects on the ground and in or close to vegetation. The prey is caught during short flights starting and ending at a repeatedly used night perch at which the prey is eaten (Audet et al., 1991). The species uses multiharmonic sonar calls for echolocation.

The five bats used in this study were fed mealworms exclusively during the experimental sessions; in addition, a house mouse was supplied every two weeks. The animals were kept at their normal pre-experimental weights (about 35 g).

3.2 Training

The experimental setup (see Fig. 3) was adapted to the natural hunting behaviour of the species in that the bats had to fly to one of the feeding dishes below the starting position and then return to the perch et the end of each trial to eat the reward.

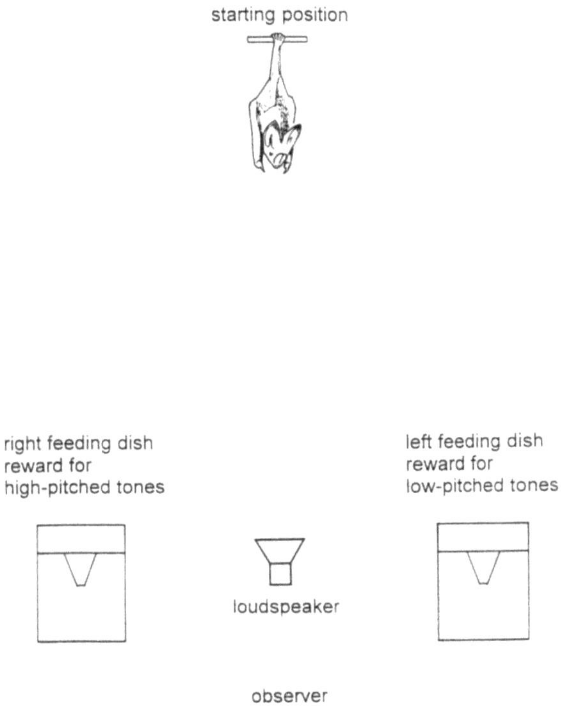

Figure 3. Setup for the pitch classification experiment in *M.lyra*. The loudspeaker for the presentation of the stimuli was positioned at a distance of 190 cm, 40 ° below the starting position of the bat. Two feeding dishes were mounted 70 cm to the left and right of the speaker. The bats classified the pitch of tones as higher or lower than a pure tone reference by flying to the right or left feeding dish, respectively. The observer was either seated on the floor behind the speaker, or watching the experiment through a window from outside the sound-attenuated chamber.

As in example 1, the bats were first trained to fly and feed in the experimental setup, and to associate acoustical stimuli with a food reward. Then, the animals had to learn the classification task. A sequence of at most 10 pairs of pure tones was presented to the bat at the starting position, each consisting of an anchor tone at 23 kHz followed by a test tone of different frequency (duration of all tones: 80 ms; intertone interval: 20 ms; interpair interval: 2 s; level: 60 dB SPL $\pm$ 3 dB random variation). When the bat left the starting position, stimulus presentation was stopped. The bats were rewarded for selecting the left, respectively the right, feeding dish for test tones below, or above, the anchor tone. Finally, the bats were trained to classify complex tones with four harmonics according to their fundamental pitch by adding higher harmonics to the pure tones and gradually increasing their intensities. After about three months of training, data collection started.

3.3 Experimental Procedure and Data Collection

Pitch classification data were collected using two types of stimulus sequences. In one experimental series, the sequence of anchor and test tone pairs was presented as in the training sessions. In the other series, the anchor tones were omitted and the series consisted of the test tones alone (intertone interval: 2 s). In this case, the bats had to refer to a memorized version of the anchor tone, i.e. absolute pitch, in order to solve the classification task.

The experimental procedure was very similar to that described for the threshold determination experiments. Here, the pseudorandom schedule determined whether the fundamental frequency of the test tone was above or below the anchor tone. All test tone frequencies were presented in each session, comprising about 50 trials. In contrast to example 1 (see 2.3), test tone frequency was varied at random, as difficult classification tasks could follow simple ones without affecting the bat's readiness to perform. The decisions at a given test tone frequency were pooled over several sessions. Data points were calculated as the percentage of trials classified as above the anchor tone (high categorization) for a given test tone frequency, or fundamental frequency, respectively.

3.4 Probit Analysis of Classification Data

In a *classification* task, the value of a test stimulus (in this case, test tone frequency or fundamental frequency) has to be categorized as smaller or greater than a reference stimulus. Ideally, all stimuli above the reference, and only these, should be classified as high.

The performance of the animals in actual experiments, however, shows a gradual transition as the stimulus parameter passes the reference value. The steepness of this transition is taken as a measure for the quality of the bat's classifying mechanism. Assuming that this quality is limited by a number of inherent, independent randomizing factors, we are led to approximate the data by Gaussian errror functions (see 2.4, and Finney, 1971). The optimal fit curve, here interpreted as the psychometric function, yields a measure for the steepness of transition, and its centre indicates the actually used reference criterion. Similar to the procedure in example 1 (see 2.4), we set the lower and upper asymptote of the approximation to the average performance for simple classification tasks (cf. Fig.5).

3.5 Survey of Classification Experiments

As a first result, the centre of the optimal fit curve helps to clarify whether the bats use an absolute reference, or a relative criterion, for pitch classification (Fig. 4). If the bats used a relative criterion, i.e. classified the anchor tone-test tone pairs as rising or falling, respectively, one would expect that the data points were distributed around two Gaussian error functions centred at the corresponding anchor tones of 22 kHz and 23 kHz. In contrast, the bats categorized only test tones with a fundamental frequency above 23 kHz as high, regardless of the frequency of the anchor tone, and all data points are well described by the fit curve based

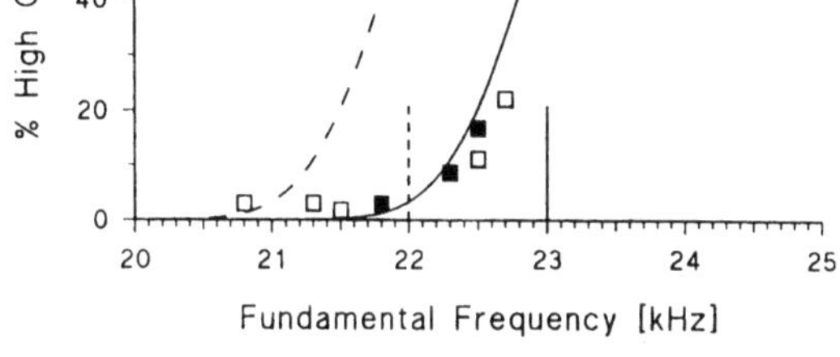

Figure 4. Discrimination reference used by *M.lyra* in pitch classification experiment (data taken from Sedlmeier, 1992). In this experiment, tone pairs consisting of a randomly chosen anchor tone of 23 kHz or 22 kHz followed by a complex test tone were presented to 5 animals previously trained to classify tones relative to a 23 kHz reference. The mean classification performance in the presence of the 23 kHz (filled symbols) and 22 kHz (open symbols) anchor tones is plotted as a function of the fundamental frequency [kHz] of the complex tones. The continuous curve represents the optimal Gaussian error function approximating the performance in the presence of the 23 kHz anchor tone (continuous vertical line). For comparison, a shifted curve, centered at 22 kHz, is also given (dashed lines).

on the data points with 23 kHz anchor tone alone. This reveals that the bats referred to the 23 kHz *memory* reference acquired during training!

In Fig.5., we compare the classification performance relative to the memorized 23 kHz reference (no anchor tones were given), for pure tones and complex tones consisting of four harmonic components. The centres of the two optimal fit curves coincide within their standard errors, which shows that the *fundamental* pitch of the complex tones is used for classification. Moreover, the curves have the same steepness which suggests that the ability to classify pitch is not affected by the presence of the higher harmonics.

An analogous experimental procedure has been used to establish virtual pitch perception in *M.lyra* (Schmidt et al., in press). The method of data analysis has also been applied in an active texture discrimination experiment (Schmidt, 1992).

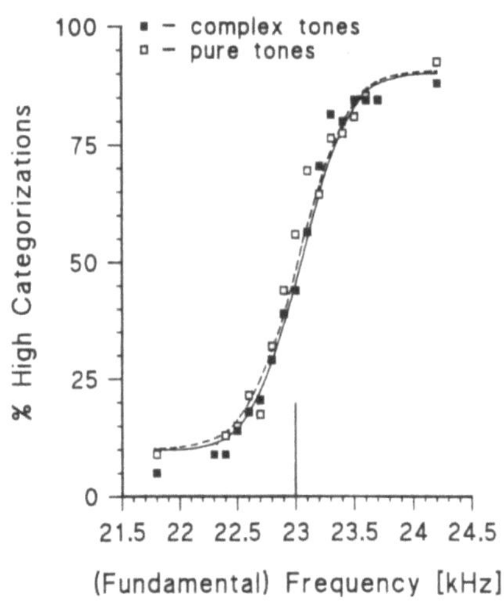

Figure 5. Description of pitch classification by probit analysis. The mean performance of 5 *M.lyra* in categorizing pure tones (open squares) and complex tones (filled squares) as high or low compared to a 23 kHz pure tone reference (marked by the vertical line) is plotted as a function of the frequency of the pure tones or the fundamental frequency [kHz] of the complex tones. Both sets of data points are approximated by Gaussian error functions (dashed line for pure tones, continuous line for complex tones; data taken from Sedlmeier, 1992). The two fit curves are indistinguishable within their standard deviations ($\pm$ 90 Hz for pure tones, $\pm$ 70 Hz for complex tones).

4 Concluding Remarks

Finally we discuss alternatives to the 2-AFC method that have been used in studies of bat psychoacoustics.

Long (Long, 1977; Long & Schnitzler, 1975) applied classical conditioning to determine audiograms and masked thresholds in *Rhinolophus ferrumequinum*, measuring a conditioned heart rate increase in response to the stimulus. While this method provides reliable threshold measurements and shortens the time consuming training phase, it has the disadvantage of inducing considerable stress to the delicate animals, thus limiting the number of possible trials.

In a study on the audiogram, masked thresholds and temporal summation of the echolocating flying fox, *Rousettus aegyptiacus*, Suthers & Summers (1980) trained the bat in a Go/No-Go procedure with a food reward. The thresholds obtained in this way appear rather too insensitive. This may be due to the Go/No-Go procedure. Since many bat species, e.g. *T. brasiliensis*, tend to hesitate in their decisions, they may simply not report a perceived stimulus with this paradigm. In addition, bats tend to be easily frustrated when a correct response is not rewarded. Thus, a procedure in which less than 100% of the correct responses are rewarded in order to increase the possible number of trials per experimental session until the bat is satiated is probably not feasible.

Finally, more elaborate paradigms for measuring thresholds directly as in the staircase method combined with a Go/No-Go procedure (cf. Okanoya, this volume; Niemiec and Moody, this volume) are likely to fail with many bat species (e.g., in *T. brasiliensis*) since individuals of these species typically attain significantly less than 100 % correct choices in simple tasks. Therefore we prefer our statistically safe method of threshold determination, even though it requires a substantially larger number of trials, in bat psychoacoustics.

5 Acknowledgements

This study was funded by the Deutsche Forschungsgemeinschaft, SFB 204/TP8, "Gehör", München. I thank Dr. K.M.Schmidt for his critical comments on this paper.

6 Suggested Reading

Schmidt S. & Thaller J. (1994) Temporal auditory summation in the echolocating bat, *Tadarida brasiliensis*. *Hearing Research, 77*, 125-134.

Simmons, J.A. & Vernon, J.A. (1971) Echolocation: discrimination of targets by the bat, *Eptesicus fuscus*. *Journal of Experimental Zoology, 176*, 315-328.

7 References

Audet, D., Krull, D., Marimuthu, G., Sumithran, S. & Balasingh J. (1991) Foraging behavior of the Indian False Vampire bat, *Megaderma lyra*. (Chiroptera, Megadermatidae). *Biotropica, 23*, 63-67.

Bartsch E. & Schmidt S. (1993) Psychophysical frequency modulation thresholds in a FM-bat, *Tadarida brasiliensis*. *Hearing Research, 67*, 128-138.

Fay, R.R. (1988) *Hearing in Vertebrates. A psychophysics data book*. Winnetka, IL: Hill and Fay Associates.

Finney, D.J. (1971) *Probit Analysis. 3rd edition*. Cambridge, UK: Cambridge University Press.

Long. G.R. (1977) Masked auditory thresholds from the bat, *Rhinolophus ferrumequinum*. *Journal of Comparative Physiology, 116*, 247-255.

Long G.R. & Schnitzler, H.U. (1975) Behavioural audiograms from the bat, *Rhinolophus ferrumequinum*. *Journal of Comparative Physiology, 100*, 211-219.

Schmidt S. (1992) Perception of structured phantom targets in the echolocating bat, *Megaderma lyra*. *Journal of the Acoustical Society of America, 91*, 2203-2223.

Schmidt S., Gerber P. and Pötke M. (1992) Psychoacoustical characterization of auditory filter shape in the bat, *Medaderma lyra*. In: Abstracts of the 15th annual meeting of the European Neuroscience Association, Supplement No.5 to the European Journal of Neuroscience, Oxford University Press, p.141.

Schmidt S., Preisler A. & Sedlmeier H. (in press) Aspects of pitch perception in the ultrasonic range. In: G.A. Manley, G.M. Klump , C. Köppl, H. Fastl & H. Oeckinghaus (Eds.) *Advances in Hearing Research, Proceedings of the 10th International Symposium on Hearing*, Singapore: World Scientific Publishers.

Schmidt S., Thaller J. & Pötke M. (1990) Behavioural audiogram and masked thresholds in the free-tailed bat, *Tadarida brasiliensis*. In: N. Elsner & Roth (Eds.). *Brain - Perception - Cognition. Proceedings of the 18th Göttingen Neurobiology Conference*, Stuttgart: Georg Thieme Verlag, p.146.

Schmidt S., Türke B., & Vogler B. (1984) Behavioural Audiogram from the Bat, *Megaderma lyra* (Geoffroy, 1810; Microchiroptera). *Myotis, 21/22*, 62-66.

Schnitzler, H.U. & Henson W.JR. (1980) Performance of airborne animal sonar systems. In: R.G. Busnel and J.F.Fish (Eds.), *Animal Sonar Systems*, New York and London: Plenum Press, pp.109-181.

Sedlmeier H. (1992) Tonhöhenwahrnehmung beim Falschen Vampir, *Megaderma lyra*. Dissertation, Fakultät für Biologie, LMU, München.

Sedlmeier H. & Schmidt S. (1989) Masked auditory thresholds from the Indian false vampire bat (*Megaderma lyra*). In: N. Elsner & W. Singer (Eds.). *Dynamics and Plasticity in Neuronal Systems. Proceedings of the 17th Göttingen Neurobiology Conference*, Stuttgart: Georg Thieme Verlag, p.292.

Simmons, J.A. (1973) The resolution of target range by echolocating bats. *Journal of the Acoustical Society of America, 54*, 157-173.

Simmons, J.A., Lavender, W.A., Lavender, B.A., Childs, J.E., Hulebak, K., Ridgen, M.R., Sherman, J. & Woolman, B. (1978) Echolocation by free-tailed bats (*Tadarida*). *Journal of Comparative Physiology, 125*, 291-299.

Suthers, R.A. and Summers, C.A. (1980) Behavioral audiogram and masked thresholds of the megachiropteran echolocating bat, *Rousettus*. *Journal of Comparative Physiology, 136*, 227-233.

Methods in Comparative Psychoacoustics
G. M. Klump, R. J. Dooling, R. R. Fay & W. C. Stebbins (eds)

Observer-based Approaches to Human Infant Psychoacoustics

Lynne A. Werner

Department of Speech and Hearing Science, University of Washington
CDMRC, WJ-10, Box 47, Seattle, WA 98195, U.S.A.

Summary. As subjects of psychoacoustical research, human infants, especially very young infants, pose special problems. Their response repertoire is limited; they make few unique responses to sound; and they are available for testing for only short periods of time. Our approach to the testing of infants has been to use existing spontaneous responses, but to use the infants' behavior to judge whether or not a sound has occurred rather than whether or not a response has occurred. This *observer-based approach* has allowed us to assess sensitivity, frequency resolution, and temporal resolution in infants as young as 1 month of age.

1 Introduction

The first rigorous method to assess auditory thresholds in human infants was the *conditioned head-turn procedure*, a derivative of a method developed for pediatric audiology (Moore, Thompson, & Thompson, 1975). This method shares many characteristics of traditional operant discrimination learning paradigms: Infants learn to make a head-turn to the left or to the right when they hear a sound or a change in a sound. Correct head-turns are reinforced with an interesting audiovisual display, typically a mechanical toy bear. This method has been used in several laboratories to examine infants' absolute sensitivity, frequency resolution, speech discrimination, and other auditory capacities (e.g., Kuhl, 1979; Olsho, 1984, 1985; Trehub, Schneider, & Endman, 1980). Unfortunately, conditioned head-turn procedures are not effective for infants younger than about 5.5 months, who do not make crisp, directed head-turns to sound sources.

Younger infants do respond to sound, however. Watrous, McConnell, Sitton, and Fleet (1975), for example, listed many behaviors observed in 3- to 12-month-old infants following presentation of calibrated noisemakers (Table I). An inventory of behaviors like these reveals the difficulty in finding a useable auditory response, particularly for young infants. None of the behaviors occurs on a very high proportion of trials, different infants tend to make different responses, and a single infant's response may change from trial to trial. Moreover, except for startle, none of the observed responses is a unique response to sound and so may occur for a variety of reasons not related to hearing. In fact, procedures based on documenting such responses tended to produce absolute thresholds for young infants that seem unreasonably high (e.g., Hoversten & Moncur, 1969; Weir, 1979).

Table I. Percentage of behaviors of various types elicited from infants of three ages in response to calibrated noisemakers, as reported by Watrous et al. (1975).

Behavior	Age		
	3-5 months	6-8 months	9-12 months
Body startle	0	0.4	2.6
Eye blink	0	1.6	1.9
Eye widening	23.8	12.3	10.5
Eye movement	9.5	10.3	3.8
Brow movement	31.0	11.1	4.5
Head movement	7.1	5.2	0.8
Brief cessation of activity	26.2	14.7	8.6
Initiate or increase activity	31.0	8.3	7.1
Prolonged grimace or cessation of activity	19.0	13.1	15.0
"Listening behavior"	7.1	21.0	13.9
Search behavior	21.4	32.5	20.7
Localization	14.3	56.7	60.9

Progress in the field of infant visual psychophysics suggested a solution to this predicament. Teller and her colleagues (e.g., Teller, 1979) had shown that an adult observer could reliably tell whether a visual pattern was located to an infant's right or left simply by observing the infant's looking behavior. Although eye movements and visual fixation were primary cues, the observer was permitted to use any infant behavior as the basis of his or her judgments. The observer received feedback after each trial. We reasoned that an observer might be able to tell whether or not a sound had been presented to an infant, simply by watching the infant's behavior during a possible sound presentation.

As it turns out, observers are able to reliably detect sounds presented to an infant, even when the sounds are at rather low intensities. This fact has led to the development of a family of techniques known as *observer-based procedures*, which have now been employed in several laboratories to assess auditory function in infants ranging from 2 weeks to 12 months of age (e.g., (Morrongiello, 1988; Olsho, Koch, Halpin, & Carter, 1987; Trehub, Schneider, Thorpe, & Judge, 1991; Werner & Gillenwater, 1990). The important characteristics of these methods as they have been employed in my laboratory are described in the following sections. The method is described for the case of a detection experiment, but the application of the method to discrimination experiments is discussed in a subsequent section.

2 Observer-based Methods

2.1 The Test Layout

The typical layout of the test suite for observer-based procedures is very similar to that used in conditioned head-turn testing of infants (Figure 1). The infant is seated on a parent's lap facing a one-way window and a video camera. The only real restriction on the infant's position is that the observer can see the behavior being used to identify sound presentations. If the observer only needs to see the infant's face, the infant can be tested on the parent's shoulder, facing toward the parent's back, but with the parent sitting or standing facing away from the window and video camera. Lighting from the front is preferable to overhead lighting to allow a clear view of the infant's face.

The parent listens to music over circumaural headphones. Although we usually employ earphone presentation, giving the parent masking sounds makes it very unlikely that the parent could hear the sounds presented to the infant. Of course, it is important to make sure that the infant can't hear the parent's music! The parent is instructed to be neutral and not influence the infant's behavior.

A "visual reinforcer", a mechanical toy as described above, is located about a meter from the infant along a line 45° from midline and to the infant's right. The mechanical toy is enclosed in a dark plexiglass box with internal lighting, so that it can only be seen when it is activated as reinforcement. Several mechanical toys can be contained in a tower or behind a panel, so that different toys can be activated to maintain infant interest. We have also had some success using a "video reinforcer"; video recordings can be tailored to an infant's interests and a video monitor substituted for the plexiglass box.

An assistant sits about the same distance from the infant, but 45° to the left. A low table is placed just in front of the parent and infant. The assistant manipulates simple toys to direct the infant's gaze toward midline, using the table as a support. It takes some practice to become good as an assistant. It is important to perform activities that are interesting enough to keep the infant looking, but not so interesting as to coopt the infant's attention entirely. The assistant also wears circumaural headphones, but listens to the output of a microphone in the control room, allowing him or her to monitor what the observer is doing and the progress of the experiment. Thus, a sort of rhythm is typically established in which the assistant uses the simple toys to bring the infant to midline, the observer starts a trial with a signal to the assistant, a trial takes place, the assistant lets the infant go for a few seconds, then brings him or her back to midline. The assistant avoids making any major changes in activity during a trial, so that a response to such a change will not be mistaken for á response to the sound. The assistant is also responsible for communications between the observer and the parent; for example, the assistant might suggest that it is time for a short break or ask the mother to stop jiggling the baby.

2.2 The Observer

The observer sits at the computer console in a room adjacent to the test booth. Activity in the test booth can be monitored through a one-way window, and a close up view of the infant's face is available on a video monitor. Most observers use both sources of information: It is important to factor out any activity of the parent or of the assistant that may have influenced the infant's behavior, and only the window provides a wide enough view. At the same time, because at least some infants make relatively subtle responses, the video information is also important.

The observer waits for the assistant to bring the baby to a "ready" state. The baby is usually facing at midline, but it is not necessary to keep the baby completely still. Often a low-to-moderate degree of motor activity (e.g., limb movement) provides the best background against which to detect signals. Essentially the observer quickly learns what this infant's ready state looks like and tells the assistant what to shoot for.

When the baby is in the appropriate state, the observer initiates a trial. Either a *signal* or *no-signal* trial occurs according to some random schedule. In the detection experiment, a sound is repeated several times on a signal trial, perhaps 4-8 repetitions of a 500-ms tone with a 50% duty cycle; no sound is presented on no-signal trials. The observer has the trial duration (4-8 s, depending on the experiment) to decide whether or not a signal has occurred. In this respect, the procedure is *go/no-go*: The observer pushes a button if he or she thinks a signal is occurring, but does nothing otherwise. The observer receives feedback on all trials. The current hit and false alarm rates are also displayed throughout the session. The visual reinforcer in the test booth is activated when the observer responds on a signal trial; thus, the baby only receives "feedback" for a hit or a false alarm, not for misses or correct rejections.

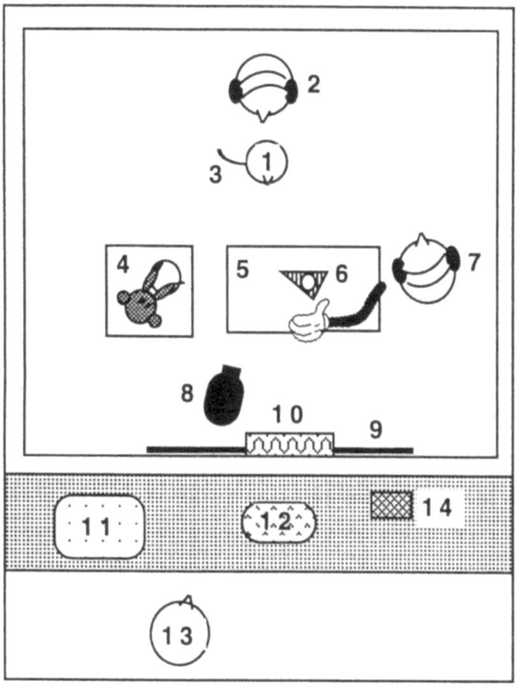

Figure 1. Schematic of laboratory set-up for observer-based test session. 1=infant subject; 2=parent; 3=insert earphone tube; 4=mechanical tory "visual reinforcer"; 5=table; 6=assistant's boring toy; 7=assistant; 8=video camera; 9=one-way window; 10=light; 11=computer monitor; 12=video monitor; 13=observer; 14=speaker monitoring test booth.

As a general rule, trials should be presented in rapid sequence, leaving just enough time for the assistant to bring the baby back to a ready state between trials. This ensures that a reasonable number of trials can be obtained in a session, and that the infant will not become excessively tired of the situation because of long delays between trials with nothing to do but look at the boring assistant's toys. Naturally longer delays must be tolerated at certain times, as when the baby starts to "lose it". If this happens relatively early in the session, a short break outside the booth may help the baby to get back on task. Sessions usually last from 20 to 30 minutes; about 60 total trials may be obtained in a good session.

2.3 Training the Infant-Observer Team

The behaviors that observers use to decide whether or not a signal has occurred are just those listed in Table I. Eye or head movements, changes in facial expression, and changes in motor activity are probably the most commonly noted. Each session begins with a training phase, meant to acquaint the infant with the contingencies in the situation and to allow the observer to learn the behaviors that distinguish signal and no-signal trials for a specific infant. On signal trials during the training phase, the stimulus is presented at a level that is clearly audible to the infant. In our current studies we use a two-part training sequence.

The first part of the training sequence is meant to demonstrate the association between sound and the visual reinforcer to the infant, as well as to give the observer an idea of what this baby will do. Both signal and no-signal trials are presented, and the observer responds if he or she detects a signal, but the visual reinforcer is activated at the end of the trial whether or not the observer responds. In a detection experiment, more signal than no-signal trials are usually presented, and this part of the sequence continues until the observer has been correct on 3 of the last 4 trials, with at least one correct rejection. This is usually a short part of the session, often completed in 5 trials, typically in fewer than 10 trials.

The second part of the sequence begins when the initial criterion is met and the observer is informed when that occurs. Now the probability of signal and no-signal trials is equal. The sequence is completely random with no constraints on the number of trials of one type that can occur in succession (see Spetner & Olsho, 1990). The visual reinforcer is activated only when the observer scores a hit. As far as the infant is concerned, then, reinforcement is contingent on producing a response that is big enough and consistent enough for the observer to see. This part of the sequence continues until the observer has been correct on 4 of the last 5 signal trials and 4 of the last 5 no-signal trials. The average infant-observer team takes about 20 trials to reach this criterion; a general rule is to abandon the session if criterion is not met within 40 trials.

An infant may start out making a response with several components, but only one of these may persist into the test phase of the session. Good observers become sensitive to such changes in the infant's behavior within only 2 or 3 trials. One option when the infant's behavior seems to have changed is to go back to the second training phase to see if the response can be reestablished. If the infant does not complete the test procedure for whatever reason, the entire training procedure is repeated when the infant returns for the next session.

2.4 The Test Phase: Psychophysical Considerations

The actual data collection phase of the experiment occurs once the infant-observer team has reached the final training criterion. The structure of this phase will depend on the performance measure and the psychophysical method. In many experiments, the goal is to obtain a threshold. Adaptive methods and the method of constant stimuli have both been used in conjunction with observer-based procedures effectively. When the two methods have been compared, no difference in mean threshold for either 3- or 6-month-olds has been found, although adaptive thresholds may be more variable (Werner & Marean, 1991). If the goal is to examine performance at one level or in one condition, a larger block of trials at a fixed level is used. For example, *d'* could be estimated for 30 trials, where the level is fixed at 35 dB SPL on signal trials. We have also constructed entire psychometric functions for infants by collecting 4 or 5 blocks of trials at fixed levels in several sessions. We generally expect to be able to complete 30 test-phase trials in a session, although in some experiments we have completed as many as 60 test-phase trials in a session.

It is usually the case during the test phase that the stimulus condition is more difficult than that used in the training phase. During an adaptive run, for example, most of the trials are intentionally close to threshold. The problem is to ensure that when performance gets poorer, it is because the stimulus is hard to hear and not because the infant and observer are tired or bored. One approach to this problem is to insert *probe trials* into the test sequence. Probe trials are signal trials on which the stimulus is presented at a more audible level, perhaps the one used during training. They can be used in different ways. For example, if the observer misses 2 or 3 consecutive signals, probe trials are presented until some criterion, say 2 in a row correct, is reached; then the test sequence resumes where it left off. In a block of trials at a "fixed" level, 5 probe trials can be randomly presented, and data only included if the observer gets at least 4 of these correct.

2.5 Variations on the Observer-based Theme

We refer to the method described in the previous paragraphs as the *observer-based psychoacoustic procedure* (*OPP*). This is the technique that we originally developed and it is distinguished by the use of visual reinforcement in conjunction with the idea of measuring how well observers can detect sounds by using an infant's listening behavior. There are some variations of the method that have been used in other laboratories and under certain circumstances in ours.

In the case of very young infants, for example, it may be better to place the infant in a seat or on a mattress during testing and to observe from the test booth. It is rather easy for an observer to get a 1-month-old to look at a simple toy and then complete a trial without assistance. In addition, the usual mechanical toy visual reinforcer is likely to be ineffective. In our original studies of 2- to 5-week-olds, in fact, we did not use reinforcement (Werner & Gillenwater, 1990). Trehub and her colleagues (1991) have also been able to obtain sensitivity data from 1- to 3-month-olds using the observer-based approach without reinforcement. In recent work, however, we have found that *auditory reinforcement* of the young infant's response may improve threshold estimates. In these studies, we present an audio recording of highly intoned readings from children's books to the infant whenever the observer scores a hit.

It is possible to manipulate the observer's response in ways that we could never manipulate an infant's response. This means that some useful psychophysical procedures become possible. For example, one can ask the observer to give a *confidence rating* on each trial, expressing his or her certainty that a signal was presented (Werner & Gillenwater, 1990). These confidence ratings can then be used to construct receiver operating characteristic (ROC) curves, and to derive nonparametric measures of sensitivity such as the area under the ROC curve, *p(A)*. Given the complex nature of the observer-based methods, it may well be safer to use such estimates of sensitivity than to make assumptions about the distribution of evidence upon which the infant-observer team makes decisions. *Response latencies* can also be converted to a confidence scale and used to construct ROCs and to calculate *p(A)*. We have found that in some cases *p(A)* and *d'* can give fairly different pictures of infants' sensitivity (Werner, Kopyar, & Bargones, 1993).

3 Applications of Observer-based Methods

Olsho, Koch, Carter, Halpin, and Spetner (1988) reported the first large-scale application of OPP to the measurement of infant absolute sensitivity. The 3-month-olds' thresholds from that study are used here to illustrate the results that the method produces. Olsho et al. trained infant-observer teams to detect tones at frequencies ranging from 250 to 8000 Hz when they were presented at 60-70 dB SPL. The tones were 500 ms in duration and were repeated with a 50% duty cycle 10 times on each signal trial. An adaptive technique was used to estimate thresholds. One way to look at the results is in terms of the average psychometric function. The average proportion of detections within a 10-dB "bin" was calculated and averaged across subjects. These functions (Figure 2) illustrate that performance in OPP varies in a sensible way with stimulus intensity: On no-signal trials, the observer said "yes" about 10-20% of the time.

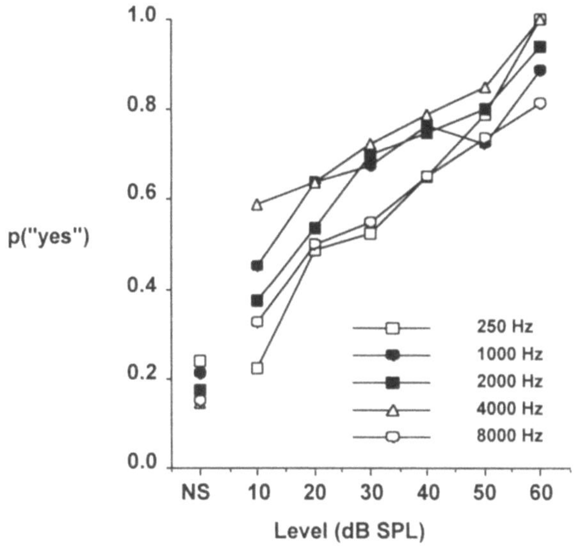

Figure 2. Group psychometric functions for the detection of pure tones at 5 frequencies by 3-month-olds, obtained using OPP. "NS" is the response rate on no-signal trials (i.e., the false alarm rate).

As the signal level increased from 10 to 60 dB SPL, the average psychometric function was more or less monotonic, reasonably steep, and at least at some frequencies even reached 100% detection. These functions are, of course, limited as descriptions of the detection process for 3-month-olds because they are averaged and because different infants may have contributed data at different levels. They do, however, illustrate that the method produces qualitatively reasonable results. Furthermore, the average thresholds obtained for 3-month-olds are much lower than thresholds previously estimated for infants at this age using other techniques (Figure 3).

To this point, I have only discussed using observer-based techniques to measure absolute sensitivity. In fact, any question that can be cast in either a detection or discrimination mold can be addressed using the observer-based approach. For example, to examine gap detection, a continuous noise was presented as a background stimulus, and gaps were introduced in the noise during signal trials (Werner, Marean, Halpin, Spetner, & Gillenwater, 1992). To examine frequency discrimination, 6 tone bursts at one frequency were presented on no-signal trials, while 3 tone bursts at that frequency and 3 tone bursts at a different frequency alternated on signal trials (Olsho, Koch, & Halpin, 1987). The latter situation presents infants with a somewhat different problem: Instead of responding when something happens or changes, they now have to learn that the mechanical toy will come on if they respond to one stimulus, but not when they respond to the other (see also Spetner & Olsho, 1990). However, infant-observer teams are readily able to learn this sort of task.

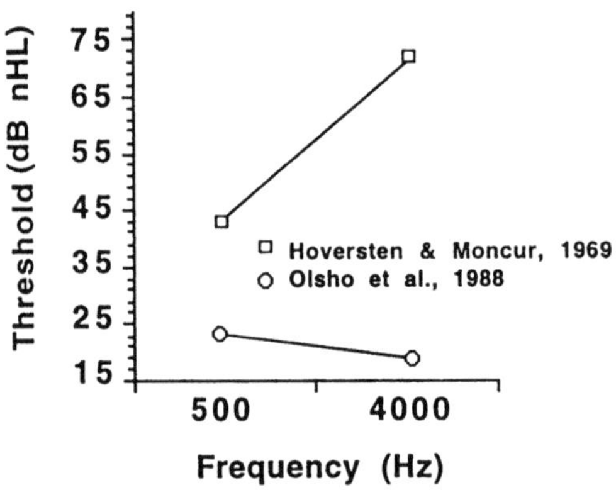

Figure 3. Average threshold for pure tones at two frequencies obtained for 3-month-olds using behavioral observation audiometry (Hoversten & Moncur, 1969) and OPP (Olsho et al., 1988).

4 Advantages and Difficulties of Observer-based Methods

The advantages of the observer-based approach should be evident from the discussion so far. First, thresholds or other measures can be obtained even from infants who make limited and variable behavioral responses, with much more sensitivity than methods which ask whether the infant responded. Second, the method is applicable over a fairly broad age range, so that one can avoid having to assess developmental change by comparing performance of infants who differ in both age and test method. Third, a variety of questions about auditory capacity can be asked using the same procedure.

The most common difficulties associated with the procedure have to do with observers. The first is that being able to "score babies" reliably takes some training. While some observers seem to get the hang of things within a couple of months, others may take as long as a year. Of course, one must address the issue of when to consider an observer "trained", especially when conducting an experiment for which the expected result is not known. In our initial work with these methods, we assumed that if observers got about the same results for 6-month-olds tested in OPP as they got for 6-month-olds tested in the conditioned head-turn procedure, then they were observing reliably. Currently, we consider several factors. Is the observer able to obtain data from a reasonable number of sessions, say 60-70%? Does the observer obtain data similar to that obtained by other observers? Is the variability in the observer's data acceptable? Finally, even if they meet these criteria, some observers are better than others, producing a somewhat lower number of more variable data points. However, we do not believe that these differences among observers are substantial enough to seriously affect the conclusions that we draw from our data. In sum, although there are some difficulties with the method, with care the difficulties can be addressed.

5 Acknowledgements

The work described in this chapter and the preparation of this chapter were supported by NIH grants DC00396 and DC00520 and by the March of Dimes Birth Defects Foundation.

6 Suggested Reading

Olsho, L. W., Koch, E. G., Halpin, C. F., & Carter, E. A. (1987). An observer-based psychoacoustic procedure for use with young infants. *Developmental Psychology*, *23*, 627-640.

Teller, D. Y. (1979). The forced-choice preferential looking procedure: A psychophysical technique for use with human infants. *Infant Behavior and Development*, *2*, 135-153.

Trehub, S. E., Schneider, B. A., Thorpe, L. A., & Judge, P. (1991). Observational measures of auditory sensitivity in early infancy. *Developmental Psychology*, *27*, 40-49.

7 References

Hoversten, G. H., & Moncur, J. P. (1969). Stimuli and intensity factors in testing infants. *Journal of Speech and Hearing Research*, *12*, 677-686.

Kuhl, P. K. (1979). Speech perception in early infancy: Perceptual constancy for spectrally dissimilar vowel categories. *Journal of the Acoustical Society of America*, *66*, 1668-1679.

Moore, J. M., Thompson, G., & Thompson, M. (1975). Auditory localization of infants as a function of reinforcement conditions. *Journal of Speech and Hearing Disorders, 40*, 29-34.
Morrongiello, B. A. (1988). Infants' localization of sounds in the horizontal plane: Estimates of minimum audible angle. *Developmental Psychology, 24*, 8-13.
Olsho, L. W. (1985). Infant auditory perception: Tonal masking. *Infant Behavior and Development, 7*, 27-35.
Olsho, L. W. (1984). Infant frequency discrimination. *Infant Behavior and Development, 7*, 27-35.
Olsho, L. W., Koch, E. G., Carter, E. A., Halpin, C. F., & Spetner, N. B. (1988). Pure-tone sensitivity of human infants. *Journal of the Acoustical Society of America, 84*, 1316-1324.
Olsho, L. W., Koch, E. G., & Halpin, C. F. (1987). Level and age effects in infant frequency discrimination. *Journal of the Acoustical Society of America, 82*, 454-464.
Olsho, L. W., Koch, E. G., Halpin, C. F., & Carter, E. A. (1987). An observer-based psychoacoustic procedure for use with young infants. *Developmental Psychology, 23*, 627-640.
Spetner, N. B., & Olsho, L. W. (1990). Auditory frequency resolution in human infancy. *Child Development, 61*, 632-652.
Teller, D. Y. (1979). The forced-choice preferential looking procedure: A psychophysical technique for use with human infants. *Infant Behavior and Development, 2*, 135-153.
Trehub, S. E., Schneider, B. A., & Endman, M. (1980). Developmental changes in infants' sensitivity to octave-band noises. *Journal of Experimental Child Psychology, 29*, 282-293.
Trehub, S. E., Schneider, B. A., Thorpe, L. A., & Judge, P. (1991). Observational measures of auditory sensitivity in early infancy. *Developmental Psychology, 27*, 40-49.
Watrous, B. S., McConnell, F., Sitton, A. B., & Fleet, W. F. (1975). Auditory responses of infants. *Journal of Speech and Hearing Disorders,40*, 357-366.
Weir, C. (1979). Auditory frequency sensitivity of human newborns: Some data with improved acoustic and behavioral controls. *Perception and Psychophysics, 26*, 287-294.
Werner, L. A., & Gillenwater, J. M. (1990). Pure-tone sensitivity of 2- to 5-week-old infants. *Infant Behavior and Development, 13*, 355-375.
Werner, L. A., Kopyar, B. A., & Bargones, J. Y. (1993.). *Psychophysical measures of auditory sensitivity in human infants*. Paper presented to the Midwinter Research Meeting of the Association for Research in Otolaryngology, St. Petersburg Beach, Florida.
Werner, L. A., & Marean, G. C. (1991). Methods for estimating infant thresholds. *Journal of the Acoustical Society of America, 90*, 1867-1875.
Werner, L. A., Marean, G. C., Halpin, C. F., Spetner, N. B., & Gillenwater, J. M. (1992). Infant auditory temporal acuity: Gap detection. *Child Development, 63*, 260-272.

Birds

Among the vertebrates, birds have evolved highly complex vocal repertoires as part of their most important mode of communication and interaction. The complex vocalizations of birds are appealing to human listeners resulting in a large interest in the study of birdsong. Experiments range from the analysis of behavior to the investigation of neural processes involved in the perception and generation of acoustic signals. The song learning of passerine birds, which shows many parallels to the acquision of language by humans, has served as a model for the study of the ontogeny of sensory and motor systems. These two areas of research render it especially interesting to study the basic perceptual processes underlying bird audition. A second reason for the interest in auditory perception in birds derives from the similarities and differences between birds and mammals in the morphology of their hearing organs. This offers the unique opportunity to investigate the relationship between structure and function in the vertebrate auditory system by comparing auditory capabilities with morphological features of the ear. Not only have birds proven to be good experimental subjects in psychoacoustic studies, but they have also served well in many studies of the neural mechanisms that underlie perception.

The experimental paradigms used in the study of hearing in mammals have been successfully applied to birds. The chapter by Okanoya describes the use of tracking procedures in the psychoacoustic research, and the chapter by Dooling and Okanoya gives an example of the method of constant stimuli. The question of how sounds are localized has been of special interest in birds, since many of their own vocalizations may be difficult to localize given the small distance between the ears. Sound localization performance, measured behaviorally in non-specialized birds, is the topic of the chapter by Klump. Finally, the barn owl has been the premier model for studying the principles by which the brain extracts the spatial information from the sound field. The chapter by Wagner describes various behavioral approaches to the study of sound localization in owls and illustrates how an animal's localization abilities can be investigated at several different levels of analysis.

Methods in Comparative Psychoacoustics
G. M. Klump, R. J. Dooling, R. R. Fay & W. C. Stebbins (eds)

Adaptive Tracking Procedures to Measure Auditory Sensitivity in Songbirds

Kazuo Okanoya

Cognitive and Information Sciences, Chiba University,
1-33 Yayoi-cho, Inage-ku, Chiba-city, 263 Japan

Summary. A combination of an appetitive operant conditioning and a psychophysical tracking procedure was used to measure auditory sensitivity in songbirds. The bird was trained to peck an observation key repeatedly when it did not hear a tone and to peck a report key when it did. By utilizing a two-up, one-down algorithm, a 70.7% hit threshold was tracked. An example experiment comparing auditory sensitivity in canaries of different strains is described. Strength and weakness of the procedure are then discussed. The procedure has been used in several experiments measuring auditory sensitivity, critical ratios, and intensity difference thresholds in songbirds.

1 Introduction

1.1 Hearing in Songbirds

The neuroethological and behavioral studies of vocal learning in songbirds have provided exciting scientific discoveries concerning the interaction between hearing and vocalizations (Nottebohm, et al., 1990). Auditory feedback is necessary to acquire species specific song in all species of songbirds studied so far. In some species, auditory feedback is indispensable to maintain the learned song (Nordeen & Nordeen, 1991; Okanoya et al., 1991). Each species of birds is more capable of discriminating among songs and acoustical characteristics of its own. Basic psychoacoustical data on hearing in songbirds are necessary to supplement these findings.

What we need is a quick and reliable procedure to measure auditory capabilities of songbirds. Since songbirds are usually very small and delicate, typically under 20 g, procedures utilizing negative reinforcement techniques (Dooling et al., 1971) are not desirable because such procedures do not allow prolonged series of experiments. Operant techniques utilizing positive reinforcement had been used to measure auditory sensitivities in larger birds (Hienz, et al., 1977). We have devised a food deprivation procedure and delicate apparatus to utilize appetitive conditioning techniques in smaller songbirds (Park et al., 1985).

1.2 Tracking Procedures

Tracking procedures enable rapid estimation of auditory thresholds. The first application of a tracking procedure in animal psychophysics was by Blough (1956). Blough measured dark adaptation in pigeons using a combination of yes/no task and a simple up-down procedure. Gourevitch et al. (1960) and Stebbins (1970) also applied a similar tracking procedure to measure auditory sensitivity in the rat and the monkey. These earlier studies implemented a Bekesy tracking procedure. However, the Bekesy procedure does not guarantee the subject to be under the stimulus control (Rosenberger, 1970). Also, since the Bekesy tracking procedure does not control false alarm rate and the measure of false alarm could not be obtained, one cannot be sure about the reliability of the threshold.

In this chapter, we avoided these problems by tracking a higher stimulus level and by conducting a post-hoc analysis of false alarm rate and response latencies. With these modifications in the tracking procedure, we measured auditory sensitivity in songbirds.

2 Methods

2.1 Subjects

The procedure described in this chapter has been successfully applied in five species of songbirds including the canary, European starling, song sparrow, swamp sparrow, and the zebra finch (Okanoya & Dooling, 1987a). The procedure was also applicable in two species of parrots including the budgerigar and the cockatiel.

2.2 Food Deprivation

The procedure utilized a positive reinforcement operant technique. Some degree of food deprivation is therefore necessary to control the motivation of the birds. Since monitoring the degree of food deprivation by weight is exceedingly difficult with these small songbirds, we adopted the procedure of restricting food intake for measured periods of time prior to experimental sessions (Park, et al., 1985). In the beginning of the experiment, the bird was deprived of food for 2 hours. The deprivation time was gradually increased from 2 to 8 hours over several days. Eventually, the birds were food deprived overnight and, depending on subjects, were fed for about one to four hours after the experimental sessions were completed for the day. Experimental sessions took place 6 days a week. During food deprivation, the birds were monitored closely for signs of over deprivation.

2.3 Apparatus

The birds were tested in a small wire cage (25 x 25 x 25 cm) which was mounted inside an anechoic chamber (2.5 x 2.0 x 2.5 m). Pure-tone stimuli were presented from a loudspeaker mounted 42 cm directly above the bird's head. A pigeon grain hopper (Coulbourn Instrument, Inc.) was used for food delivery. Two response keys were constructed from microswitches (Microswitch, Inc., Serial #311SM703-T) and light emitting diodes (LEDs). The LED was mounted on the 3-inch arm of each microswitch. The birds reported detection of a pure tone by pecking the LEDs. All aspects of stimulus presentation during both training and testing were controlled by a microcomputer. The audiometric circuits consisted of logic and analog modules from Coulbourn Instruments. Sound field calibration was accomplished by placing the microphone of a General Radio 1982 sound level meter in the position normally occupied by the bird's head during testing.

2.4 Operant Conditioning

The food-deprived bird was first placed in the operant cage with the food hopper fixed at the position where the bird could eat from the hopper anytime. After the bird learned to eat from the hopper, a standard autoshaping procedure was applied to train the bird to peck a response key. Following a 40 s intertrial-interval, one of the LED was illuminated for 10 s. After that, the food hopper and hopper light were activated for 4 s. If the bird pecked the response key while the LED was illuminated, reinforcement followed immediately. The bird was first trained to peck a green LED and then trained to peck a red LED.

2.5 Training the Response Sequence

After the key peck response was established for both of the keys, the training of the response sequence necessary for the detection task was initiated. The green LED was first illuminated.

When the bird pecked the green LED, this LED was immediately extinguished and the red LED was illuminated. When the bird pecked the red LED, it was reinforced by the presentation of food. In this fashion, the bird learned to peck the green LED first and then to peck the red LED to obtain reinforcement.

2.6 Training of the Detection Task

Next, the two LEDs were simultaneously illuminated and the bird was trained to peck the green LED (observation key) repeatedly when it did not hear a tone and to peck the red LED (report key) when it did. A correct response or hit (a peck on the report key during tone presentation) was reinforced with food (Figure 1). A false alarm (a peck on the report key when no tone was presented) was punished with a 4-s timeout period during which the lights were extinguished in the test chamber. Food-deprived birds were tested in daily sessions consisting of 40 trials. The birds were trained with a 1.0 kHz or a 2.0 kHz pure tone presented at 83 dB SPL

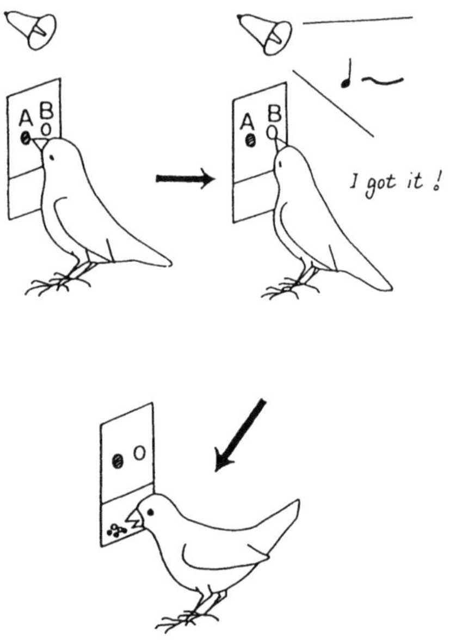

Figure 1. The bird was trained to peck one key (A; observation key) repeatedly when it did not hear a tone and to peck the other key (B; report key) when it did. The correct detection of the tone was reinforced by food.

until they attained a level of 90% correct combined with a false alarm rate of less than 10% over a session of 40 trials. Threshold testing commenced at this point. A false alarm rate was estimated by dividing the number of responses to the report key when no tone was presented by the total number of observing responses. The probability of tone onset after each observing response was initially set at 0.5 and gradually decreased to 0.1 as the bird learned the response sequence.

2.7 Testing Procedures

Thresholds were determined with a tracking procedure. In this phase of the experiment, the probability of tone onset following an observing response ranged from .10 to .25 depending on the subject's response tendencies. If the bird tended to peck the report key often and consistently had a high false alarm rate, the probability of tone presentation for the subsequent session was decreased. On the other hand, if the bird was too conservative and consistently had a very low false alarm rate, the probability of tone presentation was increased for the next session.

For the first 10 trials of each session, the tone intensity was fixed at a level of 10 to 50 dB SPL above the bird's suspected threshold. In essence, these first 10 trials of a session served as a warm-up period. For the next 30 trials, the tracking procedure was implemented with the stimulus level now contingent upon the bird's performance. In this phase, two successive correct response lowered the stimulus intensity, whereas one miss of a signal raised the stimulus level for the next trial. This is similar to the psychophysical procedure described by Levitt (1970) as an Up-Down Transfer (UDTR).707 rule. Because the nature of the UDTR.707 [p2=0.5], the percent correct detection is necessarily 70.7% when the animal is actually tracking its threshold. UDTR.707 was preferred over a simple up-down procedure which tracked 50% threshold. This is because the subject tends to lose stimulus control with the simple up-down procedure.

At the start of tracking, the stimulus intensity changed in 8-dB steps. Within 10 dB of the bird's suspected threshold, a 2-dB step size was utilized. The very first test session of a given frequency was run with the step size fixed at 8-dB to obtain a rough estimate of the threshold. A typical tracking session of 40 trials is shown in Figure 2.

Spontaneous response rates and response latencies were monitored for the last 10 trials of each session. The spontaneous response rate was maintained at a level between about 5 and 15% for each bird by manipulating, on a session-by-session basis, the probability of tone presentations, the length of the timeout period, and the degree of food deprivation. Response latencies were also monitored over the last 10 trials of each session. These latencies were then compared across birds as an additional assurance that no motivational differences existed between the species tested.

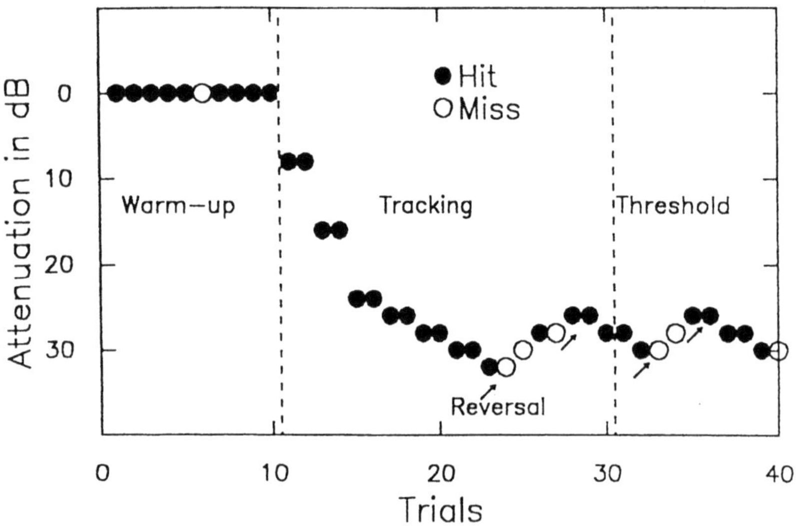

Figure 2. Results of typical session consisting of 40 trials (After the first 10 trials, stimulus intensity is reduced following two correct responses and increased following a miss. Threshold for a session is defined as the mean sound pressure level of the reveal points occurring over the last 10 trials.)

Testing was continued at a given test frequency until the difference between the tracking reversal points over the last 10 trials of a session was less than 4 dB and the false alarm rate was between 5 and 15%. By this criteria, about 70 to 50% of the sessions had to be discarded. The mean of the tracking reversals which satisfied these criteria was taken as a threshold estimate. Threshold estimates from at least three sessions at each test frequency were pooled, and the mean of these served as final threshold value. The order of the test frequencies was randomized, and a different random order was used for each bird. About 2-3 sessions could be run in a day.

3 Example: Strain Differences in Auditory Thresholds in the Canary

The experiment described here was originally reported by Okanoya & Dooling, 1985.

3.1 Subjects

Five canaries of the Belgian Waterslager strain (WS) and 4 canaries of the American Singer and the German Roller (non-WS) strains were used. During this study, all birds were housed in an aviary in the University of Maryland and kept on normal light cycle.

3.2 Methods

The adaptive tracking procedure explained in the previous section was used to measure auditory sensitivity in the canaries.

3.3 Results and Discussion

The average results for the 5 WS canaries and the 4 non-WS canaries are shown in Figure 3. Standard errors of the average were within 3 dB for each of the test frequency and for each of the bird groups. Results for the non-WS birds agree well with previous data on the canary obtained by avoidance conditioning (Dooling et al., 1971), and they were within a range of data obtained for other songbirds (Heinz et al., 1977; Okanoya & Dooling, 1987a). However, results for the WS canaries were between 20 and 40 dB higher (4-5 standard deviations) than the non-WS canaries for the frequencies above 2 kHz.

There were no differences at threshold between the two groups of canaries either in false alarm rate or in response latency. Average false alarm rate and response latency at threshold were 7.8% and 1.07 s for WS canaries and 9.9% and 0.96 s for non-WS canaries. Differences were not significant by a two-way (group by test frequency) ANOVA either in false alarm rate ($df=1,55$, $F=2.35$, $p>0.10$) or response latency ($df=1,55$, $F=2.57$, $p>0.10$). Thus, the motivational and strategic differences did not explain the differences in auditory sensitivity between the two groups of the canaries.

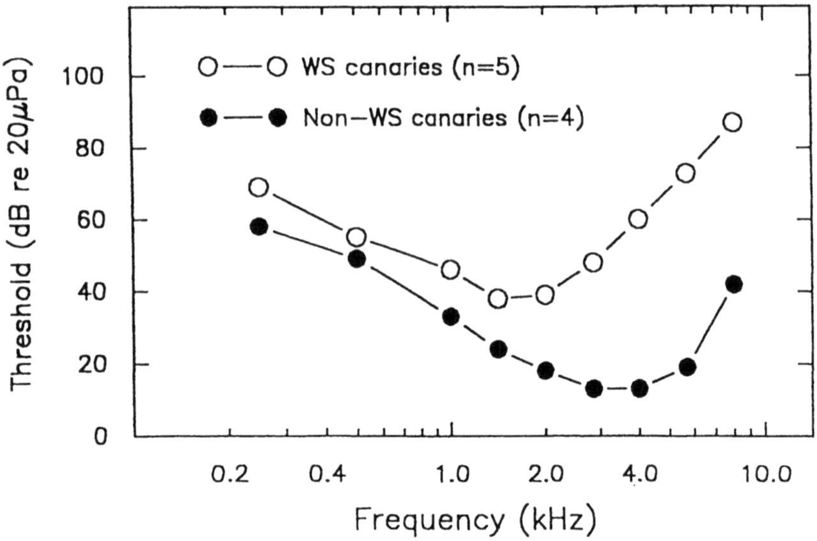

Figure 3. The mean audibility curve (open circles, solid line) of 5 WS canaries and that of 4 non-WS canaries are compared.

The results suggested that the WS strain of the canary has an elevated auditory threshold compared to other strains of canary (Okanoya & Dooling, 1987b). In a subsequent study (Okanoya et al., 1990), by measuring auditory sensitivity of hybrid Waterslager-Roller canaries we showed that this elevated threshold has a genetic basis. Out of 6 hybrid canaries, 4 had the elevated threshold similar to the WS strain and the remaining two had normal thresholds comparable to the non-WS canaries. As an additional evidence, analyses of vocalizations revealed that the peak in the power spectrum of vocalizations of the WS canaries was at 2 kHz whereas that of non-WS canaries was at 5 kHz. The peak in the power spectrum of vocalizations of the hybrid canaries was at 3 kHz, which was in between that of WS and non-WS canaries.

Recently, the result on strain differences in the canary auditory sensitivity was reinforced by reports of pathological changes in the structure of the basilar papilla of WS canaries (Dooling & Gleich, 1993). The number of hair cells in a sample of WS canaries was reduced by about 30% compared to other strains of the canary. Furthermore, the orderly arrangement of hair cell orientation appeared to be severely disturbed in the WS canaries.

Taken together, the present tracking procedure accurately described the differences in auditory sensitivity in these strains of the canary.

4 General Discussion

4.1 Strength and Weakness of the Procedure

The procedure was efficient in obtaining thresholds once the experimenter acquired the skill and intuitions. Birds were generally trained within 2 weeks for the detection task. Testing for 8 frequencies usually completed within a month. Once the bird was trained for the auditory sensitivity task, the bird could be used to measure critical ratios and intensity difference limens (Okanoya & Dooling, 1985; 1987a) without an additional training. The bird seems to acquire the concept of "respond when a change occurs". We were able to use the same birds in discrimination experiments using complex natural sounds (Okanoya & Dooling, 1988).

On the other hand, the procedure certainly required some expertise. For an efficient tracking, the experimenter must be able to manipulate deprivation time, timeout duration, and the probability of tone presentation after the observation response. Furthermore, since the task did not control false alarm rate and response latencies, a post-hoc analysis of these variables is necessary to compare the data. In some cases as much as 70% of the test sessions had to be discarded because they did not satisfy the false alarm and response variability criteria. Thus during the sessions the experimenter was required to closely observe all aspects of the experiment.

4.2 Improvement of the Procedure

In human psychoacoustics, tracking procedures are used in combination with the two-interval forced choice (2IFC) task (Levitt, 1970). With the 2IFC, false alarm rate is not a free parameter as in the yes/no task. Since the 2IFC requires the animal to remember which of the two intervals the sound is presented, the task is difficult to implement in animal psychophysics. Therefore, as in the present procedure, a yes/no task is generally used in tracking procedures in animal psychophysics (Blough, 1956; Stebbins, 1970). Although I introduced the post-hoc analysis of false alarm rate, the problem of manipulating false alarm rate remains. If the duration of timeout and the probability of the tone presentation after observation response could be adaptively changed during a session so that the false alarm rate would remain relatively constant, such modification will greatly reduce the percentage of the sessions that have to be discarded.

5 Acknowledgements

The procedure described in this chapter have been developed at the University of Maryland while the author was supervised by Professor R. J. Dooling as a graduate student. I thank Professor Dooling for his advise during this period. Figure 1 was prepared by T. Yoneda.

6 Suggested Readings

Levitt, H. (1970). Transformed up-down methods in psychoacoustics. *Journal of the Acoustical Society of America, 49*, 467-477.

Okanoya, K. & Dooling, R. J. (1985). Colony differences in auditory thresholds in the canary (*Serinus canarius*). *Journal of the Acoustical Society of America, 78*, 1170-1176.

Okanoya, K. & Dooling, R. J. (1987a). Hearing in passerine and psittacine birds: a comparative study of absolute and masked thresholds. *Journal of Comparative Psychology, 101*, 7-15.

7 References

Blough, D. S. (1956). Dark adaptation in the pigeons. *Journal of Comparative and Physiological Psychology, 49*, 425-430.

Dooling, R. J. & Gleich, O. (1993). Pathological findings in the basilar papilla of Belgian Waterslager canaries. *Abstracts of the 16th midwinter research meeting, Association for Research in Otolaryngology*, **88**.

Dooling, R. J., Mulligan, J. A., & Miller, J. D. (1971). Auditory sensitivity and song spectrum of the canary (*Serinus canarius*). *Journal of the Acoustical Society of America, 50*, 700-709.

Gourevitch, G., Hack, M. H., & Hawkins, J. E. (1960). Auditory thresholds in the rat measured by an operant technique. *Science, 131*, 1046-1047.

Hienz, R. D., Sinnott, J. M., & Sachs, M. B. (1977). Auditory sensitivity of the redwing blackbird (*Agelaius phoeniceus*) and brown-headed cowbird (*Molothrus ater*). *Journal of Comparative and Physiological Psychology, 91*, 1365-1376.

Nordeen, K. W. & Nordeen, E. J. Auditory feedback is necessary for the maintenance of stereotyped songs in adult zebra finches. *Behavioral and Neural Biology, 57*, 58-66.

Nottebohm, F., Alvarez-Buylla, A., Cynx, F., Kirn, J., Ling, C-Y., Nottebohm, M., Suter, R., Tolles, A., & Williams, H. (1990). Song learning in birds: the relation between perception and production. *Philosophical Transactions of the Royal Society London Series B 329*, 115-124.

Okanoya, K. & Dooling, R. J. (1987b). Strain differences in auditory thresholds in the canary (*Serinus canarius*). *Journal of Comparative Psychology, 101*, 213-215.

Okanoya, K. & Dooling, R. J. (1988). Obtaining acoustic similarity measures from animals: a method for species comparison. *Journal of Acoustical Society of America, 83*, 1690-1693.

Okanoya, K., Dooling, R. J., & Downing, J. D. (1990). Hearing and vocalizations in hybrid Waterslager-Roller canaries. *Hearing Research, 46*, 271-276.

Okanoya, K., Yamaguchi, A., & Yoneda, T. (1991). Auditory feedback is necessary to maintain temporal structures of crystallized song in Bengalese finches. *Abstracts of the International Congress of Ethology* Kyoto, Japan.

Park, T., Okanoya, K. & Dooling, R. J. (1985). Operant conditioning of small birds for acoustic discrimination. *Journal of Ethology, 3*, 5-9.

Rosenberger, P. B. (1970). Response-adjusting stimulus intensity. In W. C. Stebbins (Ed.), *Animal Psychophysics: the design and conduct of sensory experiments*. New York: Plenum Press, pp. 41-66.

Stebbins, W. C. (1970). Studies of hearing and hearing loss in the monkey. In W. C. Stebbins (Ed.), *Animal Psychophysics: the design and conduct of sensory experiments*. New York: Plenum Press, pp. 41-66.

Methods in Comparative Psychoacoustics
G. M. Klump, R. J. Dooling, R. R. Fay & W. C. Stebbins (eds)

The Method of Constant Stimuli in Testing Auditory Sensitivity in Small Birds

Robert J. Dooling[1] and Kazuo Okanoya[2]

[1] *Department of Psychology, University of Maryland, College Park, MD 20742, USA*
[2] *Cognitive and Information Sciences, Chiba University, 1-33 Yayoi-cho, Inage-ku, Chiba-city, 263 Japan*

Summary. For well over four decades, birds have proven to be remarkably tractable subjects for operant conditioning work in the laboratory. The method of constant stimuli is a classic psychophysical method which has numerous advantages when used in animal psychophysics and has been used for determining psychophysical threshold across a wide range of animal species. Here we describe an apparatus and a set of software for training and testing small birds for auditory detection and discrimination tests using the method of constant stimuli.

1 Introduction

The method of constant stimuli is one of three classical psychophysical methods for determining threshold developed by Fechner. This procedure is characterized by repeatedly using the same stimuli throughout the experiment. Stimulus values, normally five to seven, are selected to bracket the presumed threshold. In most animal work, these values are selected such that the presumed 50% threshold is located near the lower end of the stimulus range. As described in other chapters in this volume (e.g., May, et.al., this volume), there are compelling advantages to the method of constant stimuli. For one, a greater proportion of test stimuli fall above the animal's threshold compared to the method of limits or various tracking procedures. A second advantage is that, at the end of testing, one obtains a complete psychometric function.

2 Methods and Procedures

2.1 Subjects

We have used these procedures with several species of small songbirds and three species of psittacines. In the experiment described below, four japanese quail (*Coturnix coturnix japonica*) served as subjects. It is probably safe to say that these procedures will probably work with most birds with little or no modification.

2.2 Apparatus

The apparatus used for training and testing the birds has been described previously (Park, Okanoya, and Dooling, 1985). It is basically a modification of standard operant conditioning with a pigeon grain hopper (Coulbourn Instruments) mounted on one wall of the test chamber. The test chamber consists of a small wire cage (15 x 15 x 15 cm) placed in a sound attenuation chamber (IAC-1; 56 x 62 x 78 cm). Small songbirds such as zebra finches and canaries weighing between 10-20 g are unable to trip the standard response keys used for conditioning pigeons so special keys were designed. Instead of standard pigeon response keys, a custom made response panel was designed consisting of sensitive microswitches (Microswitch, Inc. #311SM703-T) with light emitting diodes (LEDs) attached. These keys were mounted on the food hopper just above the hopper opening. A schematic of the response panel is shown in Figure 1. Small birds such as zebra finches weighing as little as 10 g can easily trip these microswitches by pecking at the illuminated LEDs. Other response manipulanda requiring a little or no force can also be used with small birds including reflecting photocells or contact switches.

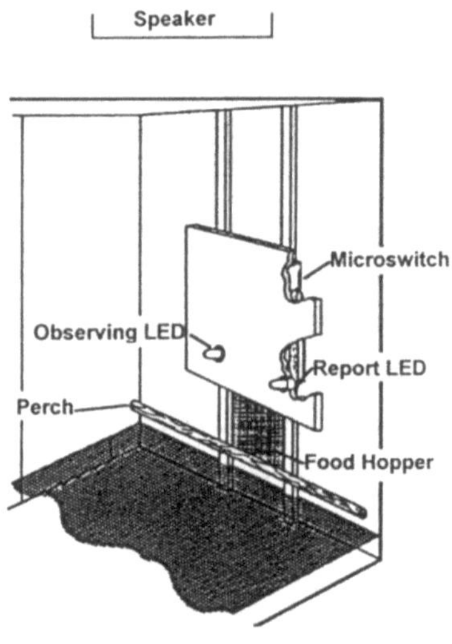

Figure 1. Schematic of the test apparatus showing the arrangement of response switches and LEDs and the food hopper.

In typical experiments, all experimenal events were controlled by an IBM-AT type personal computer. Illumination of the LEDs, activation of the hopper and house lights, and detection of switch closure was done through a standard parallel interface card operating optical relays.

Pure tones were generated in software prior to the experiment and stored on the hard disk. During the experiment, these sounds were output at a sampling rate of 20 kHz through a digital-to-analog (D/A) converter (Data-translation DT2801A), low-pass filtered at 10 kHz for anti-aliasing, attenuated with a digitally programmable attenuator (Tucker-Davis Technologies, Model PA3), and amplified by a Crown D-75 audio amplifier (Crown Instruments). A loud speaker mounted 28 cm above the response panel transduced the stimuli. The timing of experimental events was done to 1 ms accuracy by resetting the interrupt rate on the IBM-AT during the experiment. Sound field calibration was accomplished by placing the microphone of a General Radio 1982 sound level meter in the position normally occupied by the bird's head during testing.

2.3 Software for Audiometric Tests

All experimental events are controlled by software. The programs used in this experiment were written in the popular language "C" but we have also had success with the threaded interpretive language Forth. The advantages of a software approach to stimulus generation, testing, data collection, and analysis are primarily flexibility and efficiency: The stimuli used for testing are no longer constrained by hardware considerations, experimental variables and paradigms can be easily modified requiring no additional hardware, and preliminary data analyses can be immediately available during or at the conclusion of individual test sessions.

The core of our testing software can be thought of as a timing loop which continuously plays out a digital sound file (background or target) at a pre-selected rate (typically 1/500 ms). The bird is trained to discriminate the acoustic change which occurs when one stimulus file (the target) is alternately substituted for another (the background). In a discrimination experiment, the bird is trained to discriminate the acoustic difference between two sound files when the background and the target are both audible and are alternately output through the D/A converter. This paradigm is used to measure difference limens. In a detection experiment, the background sound file consists only of zeros - in effect no sound is presented to the bird. In this case, when a trial is run, the bird's task is essentially to detect the presentation of a target sound file in the quiet. This paradigm is used to measure absolute thresholds as described below. The testing software for measuring difference thresholds and absolute thresholds is in most respects identical.

Before the experiment is begun, stimuli are synthesized and stored on disk. A text file called a parameter file is generated. This parameter file includes a list of stimulus files to be tested, values for all of the testing parameters in the experiment, the name of second file holding the data, and a random ordering, in blocks of ten trials, of the sequence of trials for the entire experiment. At the conclusion of a daily test session, the parameter file is updated showing which trials have been run and which remain to be run and the separate data file is updated with the results of the current session. At the end of the experiment, all of the data for each bird are stored in disk files. These data can be accessed for later analysis by custom written software or imported into commercially available spreadsheet software.

2.4 Training Procedures

Operant conditioning procedures for birds are readily available in the literature and are not reviewed here. We use one of two training strategies and sometimes a combination of both. Birds can be trained manually, by manipulating LED illumination and food reinforcement, in which case the experimenter monitors the bird's behavior and delivers reinforcement as behavior successively approximates the desired goal (ie. until the bird is pecking the LED with enough force to activate the food hopper). The birds can also be trained by an automatic process. A software program was developed in which specific behaviors are reinforced at different stages of the program ultimately resulting in the desired behavioral sequence in which the bird pecks an observing key to start a trial and then pecks the report key if a 2.0 kHz, 70 dB pure tone is heard (see, Okanoya, this volume).

2.5 Stimuli and Testing Procedures

Once birds reliably pecked the observing key to initiate a trial and pecked the report key when a 2.0 kHz, 70 dB tone burst is presented at a performance level of 90% or better, training in the testing procedure commences. Pure tone stimuli at nine frequencies from 0.5 kHz to 8.0 kHz in half-octave steps were generated in software at a 20 kHz sampling rate and stored on disk. All stimuli were 400 ms in duration with 5 ms rise and fall times. Seven intensities (ie. attenuator levels of the PA3) at each test frequency were selected which covered a 60 dB range and bracketed the bird's suspected absolute threshold. In practice, once the location of a bird's threshold was roughly determined using a 10 dB step size, the range of intensities was often further adjusted so that it covered a 30 dB range in 5 dB steps with the added constraint that one stimulus intensity was clearly below threshold, the next was at or slightly above threshold, and the remaining five were well above threshold.

The sequence of trial events were as follows. At the beginning of testing, both the observing and the report LEDs were illuminated. The bird's first peck on the observing key started a random variable interval timer which ranged from 1 to 10 seconds. Pecks on the observing key during this interval had no consequence. Pecks on the report key, however, resulted in a 5-10 s blackout period during which lights in the test chamber were extinguished. The first response on the observing key following the variable interval resulted in the presentation of the tone stimulus at the rate of 1/430 ms at one of seven intensities. If the bird pecked the report key within 2 s of this alternating pattern (Hit), it was reinforced (ie. the food hopper was raised for 2 s) with an 80% probability. Failure to respond (Miss) led to a 1s inter-trial-interval followed by a new trial sequence which was signaled by the illumination of both the observing and report LEDs.

Generally, at the beginning of a test session, the bird was tested on 5-10 warm up trials. These warm up trials involved the presentation of the loudest stimulus of the set and the results of these trials were not saved but rather served to remind the bird of the task at the beginning of a daily test session.

The birds were tested in 1-2 daily test sessions of 100 trials each. Within each block of ten trials, the seven stimulus intensities were each presented once in random order along with three sham trials during which no stimulus was presented. Responses during sham trials provide a measure of false alarm rate. A response on the report key during a sham trial (Incorrect Response) was punished with a timeout period during which the lights in the chamber were extinguished. The length of the food reinforcement interval, the length of the timeout period, and reinforcement probability in response to a hit were all manipulated to control the false alarm rate of the bird. In the experiment reported here, the length of the timeout interval within a session was automatically adjusted between 5 and 15 seconds depending on the bird's false alarm rate. The birds were tested repeatedly at each frequency until threshold values across sessions showed no further improvement. Any number of "rules" may be used to determine when a bird has been tested sufficiently on a given condition and that no further testing is necessary. To some extent this rule may depend on the stimuli under test. The important point is that this rule be decided on before the experiment begins and that it be applied to all conditions. Though it depends somewhat on the acoustic stimuli under test, a typical rule which works well is three successive sessions in which threshold is within +/- about 1/3 the step size. Under this rule, a final threshold estimate could then be defined as the mean or median threshold estimate from the last three test sessions. In the present experiment, test frequencies were tested in random order and a different order was used for each bird. As described above, each bird was tested repeatedly at each frequency until no further improvement in threshold was observed. Final threshold estimates were then taken from the last three blocks of 100 test trials. It should be mentioned that there is nothing to prevent the experimenter from retesting birds on particular conditions. It may be, for instance, that when examining the final audibility curves for four animals, one animal shows unusually high thresholds at a particular frequency compared to the other three birds. Retesting the bird with high threshold would be warranted to ensure that the original threshold estimates are accurate.

3 Results

Figure 2 shows a psychometric function for one quail tested at 2860 Hz with a 10 dB step size. This psychometric function represents somewhat the ideal in several ways. Five stimuli that are well above threshold are responded to 100% of the time while the lowest stimulus value is clearly below threshold. The bird responded to this stimulus at about the same rate as it did to sham trials (ie. the false alarm rate). The bird's false alarm rate is relatively low at only 8%. Each data point represents the percent correct out of 30 trials spread over the last three days (sessions) of testing. Threshold is defined here as 6.2 dB SPL which is the sound pressure level which corresponds to a hit rate of 50%. This level of performance corresponds to a d' of about 1.56. A slightly more conservative estimate of threshold (about 7.5 dB SPL) is obtained when the bird's hit rate is corrected for guessing (pc*) according to high threshold theory (Gescheider, 1985).

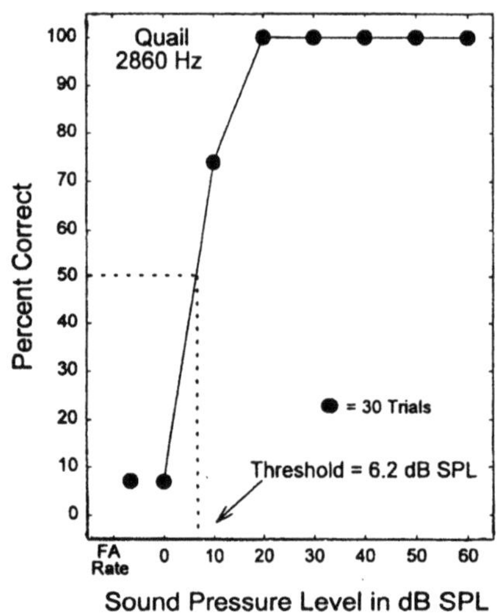

Figure 2. Psychometric function for one quail tested in a detection experiment with tones of 2860 Hz presented with a 10 dB step size.

All together, four quail were tested at nine frequencies throughout their range of hearing from 500 Hz to 8.0 kHz at half-octave steps as described above. The average audibility curve for the four quail is shown in Figure 3. As with many small birds, quail hear best in the frequency region of about 2-4 kHz with sensitivity declining steeply at frequencies above 4 kHz and somewhat more gradually at frequencies below about 1.0 kHz. Thresholds at high frequencies near 8.0 kHz and beyond are greater than 60 dB SPL which is also fairly typical of non songbirds as is the fact that non songbirds generally show lower thresholds than songbirds at low frequencies (Dooling, 1982).

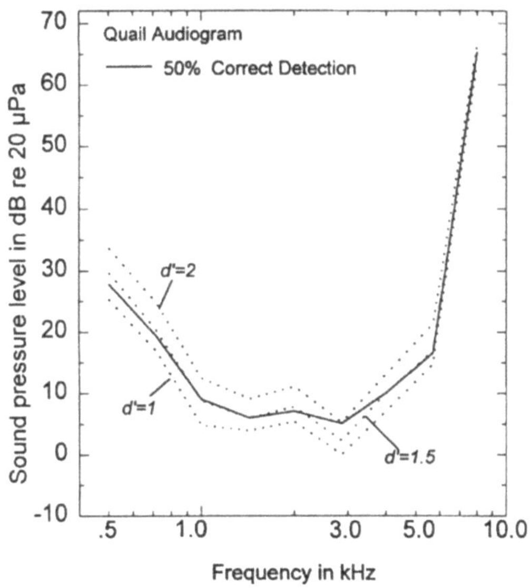

Figure 3. Audibility curves for four quail. The SPL corresponding to a hit rate of 50% is given by the solid line. Dashed lines show thresholds for a three different levels of d'.

The method of constant stimuli as applied here results in a threshold estimate that is always a combination of the birds' sensitivity and its response proclivity. It is difficult, if not impossible, to know precisely the influence of response bias (Penner, this volume). In this regard, it is important to realize that the false alarm rates for all birds at all frequencies were in the range of 7-12%. It has been our experience that well-trained birds generally perform in these tasks with a false alarm rate of around 8-10%. The consistency in false alarm rate is, in part, due to the strategy of discarding test sessions with false alarm rates higher than 20%. With these false alarm rates and with threshold defined as the SPL corresponding to a hit rate of 50%, threshold generally tracks a d' of about 1.5. To make this point, we also defined threshold by Signal Detection Theory. In other words, for each bird's false alarm rate at each test frequency, we calculated the hit rate corresponding to a given level of d'. Threshold was then defined as the SPL corresponding to this new hit rate. The average audibility curves for the four quail performing at a d' of 1.0, 1.5, and 2.0 are also shown in Figure 3 as dashed lines. The differences among these curves is not large. It can be seen that when threshold is defined as the SPL corresponding to a 50% correct detection rate, threshold tracks a d' of about 1.5.

We describe an application of the method of constant stimuli in a detection task to measuring absolute auditory sensitivity in the quail. This procedure works as well in a discrimination task where the bird is required to discriminate a change in a repeating background of sound. Okanoya and Dooling (1990a,b) determined masked thresholds and gap detection thresholds in budgerigars and quail using this same apparatus and procedure. In principle, these procedures should work well with a variety of avian species.

4 Acknowledgements

We thank Tracy Freeman and Michael Dent for the preparation of figures and Alison Linzenbold for data collection. This work has been supported by NIH grants NIH DC-00198 and MH-00982 to RJD.

5 Suggested Reading

Gescheider, G.A. (1985). *Psychophysics: Method, Theory, and Application.* New York: Lawrence Erlbaum & Assoc.

Stebbins, W.C. (1970). *Animal psychophysics: the design and conduct of sensory experiments.* New York: Appleton-Century-Crofts.

6 References

Dooling, R.J. (1982). Auditory Perception in Birds. In D.. Kroodsma and E.H. Miller (Eds.). *Acoustic Communication in Birds* (Vol. 2). New York: Academic Press.

Linzenbold, A., Dooling, R.J., and Ryals, B. (In Preparation). Auditory sensitivity of the japanese quail (*Coturnix coturnix japonica*).

Moody, D.B., Beecher, M.D., and Stebbins, W.C. (1976). Behavioral methods in auditory research. In C.A. Smith, J.A. Vernon (Eds) *Handbook of Auditory and Vestibular Research Methods* Springfield, IL: Charles C. Thomas, pp. 439-497.

Okanoya, K. and Dooling, R.J. (1990a). Detection of auditory sinusoids of fixed and uncertain frequency by budgerigars (*Melopsittacus undulatus*) and zebra finches (*Poephila guttata*). *Hearing Research, 50*, 175-184.

Okanoya, K. and Dooling, R.J. (1990b). Detection of gaps in noise by budgerigars (*Melopsittacus undulatus*) and zebra finches (*Poephila guttata*). *Hearing Research, 50*, 185-192.

Park, T.J., Okanoya, K., and Dooling, R.J. (1985). Operant conditioning of small birds for acoustic discrimination. *Journal of Ethology, 3*, 5-9.

G. M. Klump, R. J. Dooling, R. R. Fay & W. C. Stebbins (eds)

Sound Localization Studies in Non-specialized Birds

Georg M. Klump

Institut für Zoologie, Technische Universität München
Lichtenbergstraße 4, 85748 Garching, Germany

Summary. The chapter describes classical and operant conditioning paradigms applied in the study of sound localization in bird species with a small interaural distance (< 20 mm). In these species, the wavelengths of the sounds used in communicating often are up to an order of magnitude larger than the species' head dimensions. This has implications for the possible mechanisms used in locating the source of a sound.

1 Introduction

Many small bird species use communication sounds containing spectral components with wavelengths that are much larger than the diameter of their heads. For such sounds the bird's head is no obstacle, and it provides no significant sound shadow that would lead to intensity differences between the bird's ears which could be used to accurately determine the angle of sound incidence (at least in the horizontal plane, i.e., in azimuth). Therefore, a number of researchers have suggested that an acoustical coupling of the bird's ears by air spaces that connect the middle-ear cavities results in a pressure-gradient system which would generate additional interaural time and intensity differences between the signal at the two ears. Thus, this pressure-gradient system might enhance the locatability of signals (e.g., see Calford and Piddington, 1988). Results from other studies suggest, however, that the connection between the middle ears is probably not very effective in generating interaural cues for localization, since many sounds in the range of the frequencies that are evaluated by the birds are attenuated considerably when traveling through the tympanum and the interaural pathway to the other ear (e.g., see Rosowski and Saunders, 1980, Moiseff and Konishi, 1981, Klump and Larsen, 1992).

Stimulated by the discussion on the mechanisms involved in sound localization in birds, a number of behavioral studies were conducted to determine the accuracy of localization in different bird species for a range of frequencies (Klump, Windt and Curio, 1986; Lewald, 1987a; Park and Dooling, 1991). Earlier studies had already indicated a good sound-localization ability in small birds (e.g., see Granit, 1941; Schwartzkopff, 1950), but more quantitative data were needed to elucidate sound-localization mechanisms. Three methods have been applied in the study of sound localization. Studies on sound localization in the barn owl (*Tyto alba*) took advantage of natural orienting responses (see Wagner, this volume). Studies in the pigeon (*Columba livia*) used a classical conditioning paradigm (see Bräucker, 1986; Lewald 1987a, 1989), and studies in other small birds have applied operant-conditioning methods to study the accuracy of sound localization (e.g., see Schwartzkopff, 1950; Klump et al., 1986; Park and Dooling, 1991). The two conditioning methods are described in this chapter.

Two alternate methods of sound presentation have been used in studies of birds with small interaural distances (this classification follows Park and Dooling, 1991): In studies using the first method, the signal is presented from one of a number of sound sources located in different directions from the bird. The subject has to report from which of the possible sources the signal was broadcast. This task has been named an "absolute localization task". In experiments using the second method the subjects have to report the change in the direction from which a sound is received. First, one source broadcasts the signals, and then during the ongoing trial the first source is switched off and an alternate source located in a different direction is switched on and continues broadcasting the signal. To avoid audible transients, the switching is done during a silent interval between sounds. This task has been named a "relative localization task".

2 Classical Conditioning Methods

Lewald (1987a, 1989) studied the relative localization acuity in pigeons (*Columba livia*) using classical heart-rate conditioning. Pigeons were implanted with four stainless-steel electrodes, one being attached to each scapula and one to the left and right os pubis (see Bräucker, 1986). An electrocardiogram (ECG) of the bird was recorded using two of the electrodes (e.g. the two scapula electrodes). The other two electrodes served as stimulating electrodes that delivered a weak electric shock to the subject. In response to an electric shock, the rate of heart beats increased as an unconditioned response, which was measured using an analog circuit that triggered a microcomputer at the time of occurrence of the R-wave in the ECG. As a measure of heart rate, the computer calculated the mean interval between subsequent triggers during a certain time period (e.g., 3.5 s). A response was defined as a statistically significant change in the time interval between R-waves during the stimulus presentations (one-tailed $p < 0.025$, Student's t-test) compared to the R-wave interval determined during reference periods.

By presenting first a sound signal (3.5 s duration, 10 ms rise/fall) and then immediately at the offset of this signal a weak electric shock, the birds were conditioned to respond by increasing their heart rate starting at the onset of the signal. In the relative sound localization study by Lewald (1987a, 1989), The signal that was not paired with a shock (and thus should fail to elicit a conditioned response) was presented from a speaker at one location (e.g., located to the left of the body axis of the bird) whereas the same signal, paired with a shock was presented from an alternate speaker (e.g., located to the right of the body axis of the bird; the electrical signals driving the speakers were identical). After presenting a random number of reference sound signals (range 5 to 11, the presentations were repeated at a rate of 1 signal per 5 s) that were not followed by an electric shock from one speaker, the next signal in the series was presented from a different location and a shock followed immediately after it ended. Usually, data from 12 such series were pooled for the analysis (this pooled data set was called a group, no parameters of the stimulation changed during the group data collection). Including the interval of approximately 1 min. between consecutive series, data collection for a group can be calculated to last about 20 min. A session with a duration of up to 6 hours consisted of a number of groups. Each session started with a simple discrimination task (e.g., a large angle between the speakers as seen from the position of the bird) enabling the bird to learn the discrimination. When the subject showed significant discrimination (sometimes a number of groups were necessary to achieve a constant level of performance), the bird was presented with increasingly more difficult discriminations (i.e., smaller angles between the speakers) until it consistently failed to differentiate between the speakers. An easier task was then presented to the birds to enable them to relearn their response before the limits of discrimination were explored again. Initially, the angle between the sound sources as seen from the position of the bird were reduced in steps of 30 degrees. The minimum stepsize was 2 degrees in the measurement of small minimum resolvable angles. The individual subjects were tested every 2 or 3 days.

Lewald (1987a, 1989) defined the minimum resolvable angle as the smallest angle found in all sessions in which the same stimulus was presented. This definition of the discrimination limen is very different from the customary psychophysical criteria (e.g., a 50 % response probability or a d' of 1), which makes it difficult to quantitatively compare the data obtained by this method with results from other studies. Furthermore, the procedure involves a large number of statistical tests. The significance criterion used by Lewald (one-tailed $p < 0.025$) implies that on average 1 out of 40 tests by chance will result in a false positive response. This may lead to an unrealistically low estimate of the threshold, since the best discrimination performance is reported as the minimum resolvable angle. Lewald (1987a, 1989) found a considerable scatter in the results obtained with heart-rate conditioning. The minimum resolvable angles in 5 pigeons ranged from 6 to 50 degrees and from 4 to 40 degrees for 1 and 2 kHz-tones presented at 70 dB SPL, respectively. It is unclear whether this scatter is due to inter-individual variation between the different pigeons (e.g., as a consequence of domestication), or whether it results from the methods used in the sound localization experiments with pigeons.

The advantage of using the heart-rate conditioning in the pigeon is the fast learning of the discrimination task. In a simple task, i.e., a task in which the speakers were separated by 120 degrees and the sound used was a white noise that is easy to localize, one subject learned to discriminate between the speakers even within the first group (Lewald 1989). Considerable shaping of the response was generally not necessary. If the heart-rate conditioning procedure is applied however, it would pay off to use more customary criteria for defining the discrimination threshold. Heart-rate conditioning seems to be less suitable for studies in small birds. Bräucker (1986) reported considerable difficulties in heart-rate conditioning with European starlings (*Sturnus vulgaris*) and quail (*Coturnix c. japonica*) in comparison to pigeons. Whereas in pigeons the heart-rate decreased during the initial part of a session reaching a plateau, in the starlings the heart rate increased during the session making it more difficult to detect an additional rise. This continuous rise may indicate that small birds are stressed more than the pigeons by this classical conditioning procedure involving restraint. Bräucker (1986) reported, that in the European starling classical conditioning sessions were only possible up to a duration of 1.5 hours. Beyond this, the heart-rate increased so much that no further discrimination measurements were possible. Thus, operant procedures with positive reinforcement seem to be a more appropriate method of studying sound localization in small birds (see below).

3 Operant Methods

In the study of absolute sound localization in birds with a small interaural distance, a two-alternative forced-choice paradigm has been most frequently applied as the operant procedure (e.g., see Gatehouse and Shelton, 1978; Klump et al., 1986; Park and Dooling,, 1991). Here, the procedures used in sound-localization experiments with great tits (*Parus major*) are given in more detail. The birds were tested in a setup with three perches. The first perch (observation perch) was positioned at the place at which the bird initiated a trial. The subject was trained to wait on this perch until a sound was broadcast from one of two speakers that were located left and right of the body axis. After a randomized pre-signal waiting time of between 3 and 8 s, a single signal was presented from one of the two speakers. The bird had to indicate the source of the signal by flying to one of two response perches within 2 s of the onset of the sound. These two response perches were also positioned left and right from the body axis of the bird on the waiting perch and at a distance of 60 cm. The subject was rewarded at the response perch with a piece of mealworm delivered by a computer-controlled feeder if it made a correct choice, i.e., if it flew to the left perch when the left speaker had broadcast the signal and to the right perch if the right speaker had broadcast the signal. If the wrong response perch was chosen, no reward was given. A new trial was initiated when the subject returned to the observation perch. If the bird left the observation perch during the pre-signal waiting time, the trial was canceled and a new trial started when it returned to this perch.

To prevent the subjects from "homing in" on the correct speaker while en route to the response perch, the signal was gated off by a digital control circuit with a ramp duration of 5 ms when the bird left the response perch. This gating circuit, which also shaped the onset of the signal with a linear ramp of 5 ms, was under the control of a microcomputer (for brand-names and model numbers of the equipment used, the reader should refer to Klump et al., 1986; an account of up to date equipment is given by Tucker and Varosi, this volume) which also determined other aspects of the signal generation (e.g., such as the choice of the speaker presenting the next sound). When needed, the computer could generate signals via an A/D-converter. Furthermore, it determined the operant contingencies and protocolled the bird's behavior during the session by monitoring all perches via microswitches which were attached to the perches.

A session commenced until up to about 40 trials were collected or one of the two feeders, that could deliver a maximum of 12 rewards, was empty (the session ended when one of the two criteria was met). During each session, the angular separation between the speakers and the signal was kept constant.

The application of the two-alternative forced-choice procedure in the great tit was not without problems. Most importantly, with simply rewarding of every correct response it was difficult to keep the choice of the perch under the control of the stimulus. During the ongoing experiments with the 2-AFC paradigm the great tits developed a number of behaviors that, if not eliminated, would have acted against the collection of reliable discrimination-threshold data. The probability of getting a reward in this procedure just by chance (50 %) seemed to be high enough to let the birds use alternate strategies in obtaining the food rather than matching the choice of the perch to the side from which the sound was presented. It was easy to detect the alternate strategy of a side-constant response (e.g., the bird always chooses the right response perch irrespective of the side from which the stimulus was presented) and eliminate it by preferentially giving sounds from the less preferred side. To inhibit the birds from generating a side preference for the response, no more than three consecutive trials presented the stimulus from the same side (within this limitation the sequence of presentations from the left and right side were randomized). We had some indication, that especially in sessions with speaker separations that were close to the discrimination threshold, the birds were prone to just choose the response perch at random. This resulted in an unwarrantable variation in the discrimination threshold, since in some sessions the birds seemed to be well able to separate the sound sources whereas in other sessions they were not and the bird's performance was right at chance level. Although the 2-AFC procedure is insensitive to a change of the criterion of the subjects (see Penner, this volume), it is sensitive to a violation of the assumption that the subject's responses are a true indication of their discriminatory ability. To eliminate the strategy of the birds choosing a perch at random, the computer was programmed to reward the great tits based on the history of responses during sessions in which the angular separation of the speakers was well above the discrimination threshold. The subject had to make two correct choices in succession to be rewarded. To produce this pattern, the computer always looked back at the response history of the bird during the session (only during the first three trials of a session every correct response was rewarded). If the previous response in the session had also been correct, a reward was given following the current correct choice; if the previous response had been incorrect, no reward was given - even if the current response was correct. By following this reinforcement schedule, the probability of obtaining a reward by chance was reduced to 0.25, which was small enough to deter the subjects from using this strategy. In other bird species, however, these difficulties were not encountered and the 2-AFC procedure seemed to work well (e.g., see Park and Dooling, 1991; and see Hulse, this volume). In these studies, however, incorrect choices resulted in all lights in the experimental chamber being extinguished which deterred the subjects from using response strategies that were not related to acoustic stimulation. Park and Dooling (1991) also rewarded the subjects upon returning to the observation position.

The response of the birds in the 2-AFC procedure was shaped in several steps. First, the birds were rewarded when they sat on the observation perch, and one 2 kHz-tone (300 ms duration) was given when the feeder delivered a reward. In a second step, the waiting time on this perch were increased before a tone and a reward were given until the birds learned to sit and wait for a maximum of 30 s. During these two steps the left and right feeders delivered the rewards with equal probability, and the left speaker broadcast the tone when the left feeder was active, and vice versa. The angle between the speakers as seen from the position of the bird was 75 degrees. The third step in shaping the response began by only rewarding the subject when it flew to the left feeder when the left speaker was active and to the right feeder when the right speaker was active. If the bird did not land on the correct feeder perch during a 20 s time period, it was not rewarded. Finally, the time period for making the choice was reduced to 2 s forcing the birds to fly directly to the correct feeder perch when a tone was given. In the final training, the waiting times were reduced to intervals of random duration of between 3 to 8 s. Despite considerable training, the birds did not perform better than 75 % correct in this task when every correct response was rewarded. Only after the introduction of the sequencing (see above) did the great tits achieve a level of between 80 and 85 % correct responses on easy localization tasks (i.e., a large separation between the speakers).

The procedures described above were used to measure the bird's performance in an absolute sound localization task. The operant procedures applied by Park (1989) in a relative sound localization task in canaries, zebra finches and budgerigars were similar to those used in the alternating sound procedure (see Dooling and Okanoya, this volume). In relative sound localization experiments, the background stimulus was a sound presented from a speaker at one location whereas the test stimulus was the same sound presented from a speaker at a different location. The bird was trained to report the change in the location of the sound source.

4 The Minimum Resolvable Angle for Sounds in Small Birds: An Example for an Absolute Localization Task

Different threshold criteria have been used in studies of absolute sound-localization abilities in small birds. Either the threshold was defined as a statistically significant deviation from the performance of the subject expected by chance (e.g., 65 % correct responses in 60 2-AFC trials which is different from a 50 % correct performance expected by chance at the significance level of $p < 0.05$), or the threshold was defined as 75 % correct responses in a 2-AFC procedure (e.g., see Park and Dooling 1991). Figure 1 compares localization thresholds in the great tit that were calculated using both criteria (data from Windt, 1985, and Klump et al., 1986). On average, localization thresholds for tones determined in the great tit using the criterion of 75 % correct responses were 7 degrees larger than localization thresholds determined with the criterion of 65 % correct responses (the former criterion corresponds to a d' of 0.95, the latter to a d' of 0.54). The shape of the function describing the relationship between the minimum resolvable angle and the test-tone frequency is not different when determined with the two criteria. The two subjects of this study showed very similar minimum resolvable angles; they differed on average by 8 degrees (range 2 to 16 degrees) for the 75 % correct criterion and by 5 degrees (range 2 to 15 degrees) for the 65 % correct criterion.

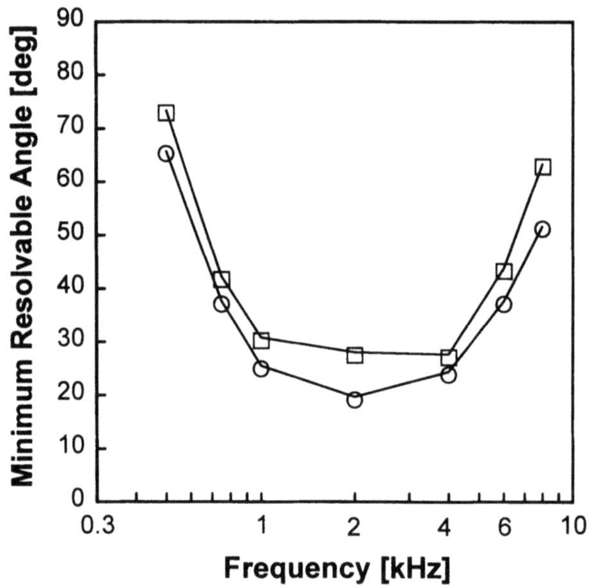

Figure 1. The minimum resolvable angle in azimuth in the great tit (n = 2 individuals) in relation to the threshold criterion: circles and squares show data for a threshold criterion of 65 % correct responses (d' = 0.54) and for 75 % correct responses (d' = 0.95, extrapolated for one of the subjects at 0.5 and 1 kHz, and for both subjects at 8 kHz), respectively.

Besides the investigation of the minimum resolvable angle in the great tit, absolute sound localization has been studied in detail in three other bird species with 2-AFC procedures: the budgerigar (*Melopsittacus undulatus*), the canary (*Serinus canarius*), and the zebra finch (*Poephila guttata*, Park and Dooling, 1991). Figure 2 shows data for the localization of pure-tone signals in these four species. In general, the worst locatability for pure tones (i.e., the lowest minimum resolvable angle) is found for low frequencies (1 kHz in the zebra finch, 0.5 kHz in all other species). With increasing frequency, the minimum resolvable angle decreases. In some species, this decrease is followed by an increase at the highest frequencies that were tested (e.g., in the great tit and in the canary). Table 1 presents minimum resolvable angles determined with wide-band noise signals in an absolute localization task. In the great tit, noise pulses of 300 ms duration and with a sound-pressure level of 50 dB SPL were used; in the other three species the noise pulses had a duration of 200 ms and a sound-pressure level of 63 dBA. In three of the four species, the minimum resolvable angles for wide-band noise is similar to the best minimum resolvable angles for pure-tone signals. Only in the zebra finch did the presentation of the noise pulse result in larger minimum resolvable angles than the

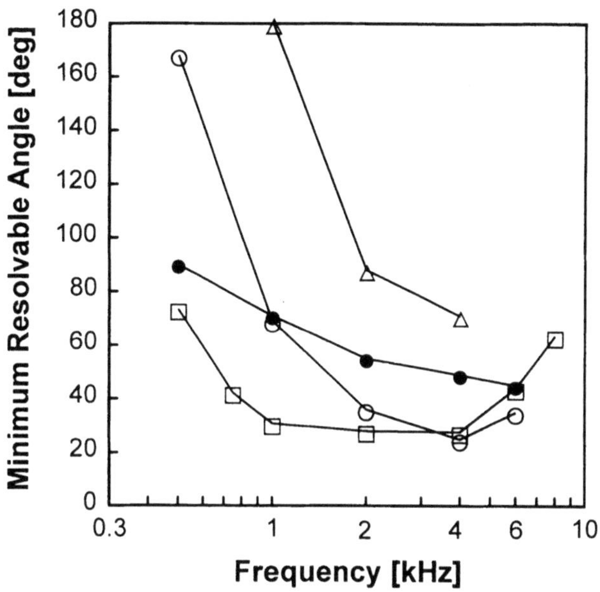

Figure 2. The minimum resolvable angle in azimuth for pure tones in different species of small birds for a threshold criterion of 75 % correct responses (d' = 0.95): O canary, ● budgerigar, △ zebra finch (Park and Dooling, 1991), and □ great tit (Klump et al. 1986, data from fig. 1).

smallest found for the pure tones. This suggests that a combination of the information from different frequency channels does not greatly enhance locatability of sounds in these bird species.

The data in the great tit are consistent with the hypothesis that these birds rely only on interaural time differences in the absolute localization of sound in azimuth. The best minimum detectable time differences that can be calculated from the minimum resolvable angles are in the order of 28 μs (these calculations are based on the minimum resolvable angle resulting in 75 % correct responses). In the great tit, the minimum resolvable phase differences that correspond to the minimum resolvable angles are constant up to 1 kHz and increase monotonically above that frequency (see Klump et al., 1986). The data available for the other three species (see Park and Dooling 1991) also suggest a monotonic increase of the minimum resolvable phase with increasing frequency. This supports the hypothesis that the ability of a small bird's auditory-nerve fibers to code the phase of the signal may be limit its performance in the absolute localization of sounds in azimuth.

Table 1. Size of the minimum resolvable angle (MRA) for different bird species stimulated with wide-band noise signals (thresholds for 75 % correct detection in a 2-AFC procedure, based on data from Park and Dooling, 1991, and Windt 1985)

Species	MRA (degrees)
Zebra finch	101
Canary	29
Great tit	23*
Budgerigar	27

* data for one of the two individuals were extrapolated, since the threshold criterion in this study was 65 % correct responses

Interaural intensity differences may additionally enhance the interaural time differences (for a more detailed discussion see Klump et al., 1986, and Klump and Larsen, 1992).

5 Concluding Remarks

The two methods that have been applied in sound localization studies in non-specialized birds, i.e., classical heart-rate conditioning and operant conditioning using positive reinforcement, have different advantages and disadvantages. In the heart-rate conditioning experiments the bird is restrained in its movements which offers the possibility of independent stimulation of the two ears with earphones (see Lewald 1987b, 1989). Such a stimulation is not possible in operant-conditioning studies where the bird is not restrained, and the sound is presented in the free field. On the other hand, restraint and the shock applied in heart-rate conditioning constitute stress to the experimental subjects that is not desirable and may even impair such experiments in small birds (see above). Operant-conditioning procedures with positive reinforcement seem to be much more suitable to study sound localization in small birds. Especially in these agile subjects, an adapted 2-AFC procedure allows a fast acquisition of the data with individual trials lasting in the order of 30 s. In contrast to the heart-rate conditioning method, the birds can be tested every day which further increases the speed of data collection. Provided the customary practise in controlling the motivation of the experimental subjects is used, the variation in the minimum resolvable angle is relatively small when operant procedures are used.

The two localization tasks, i.e., the absolute and the relative localization task, give very different results. The birds' minimum resolvable angles seem to be much smaller in the relative localization task than in the absolute localization task (e.g., see Park 1989, who compared results for the two tasks measured in the same bird species). In the relative localization task, the subject has the opportunity to make direct comparisons of the sounds presented in sequence and separated only by a brief time interval of up to a few seconds. This makes it easy for the bird to distinguish between the sounds by using any change in its characteristics that is correlated with a change of the speaker from which the sound is radiated. The absolute localization task may be more representative of the common sound-localization problem faced by the birds in nature. Based on sounds originating from one source, they have to identify the direction in which the sender can be found.

6 Acknowledgements

During the preparation of this chapter the author was funded by a grant from the Deutsche Forschungsgemeinschaft within the SFB 204 "Gehör". The comments of M. Dyson on a previous version of the manuscript are gratefully acknowledged.

7 Suggested Reading

Klump, G.M., Windt, W. and Curio, E. (1986) The great tit's (*Parus major*) auditory resolution in azimuth. *Journal of Comparative Physiology* A 158: 383-390.

Park, T.J. and Dooling, R.J. (1991) Sound localization in small birds: absolute localization in azimuth. *Journal of Comparative Phsychology* 105: 125-133.

Yost, W.A. and Gourewitch, G. (1987) *Directional hearing*. Springer, New York, Berlin, Heidelberg.

8 References

Bräucker, R. (1986) Die Herzfrequenzkonditionierung als Methode zur Messung psychophysischer Wahrnehmungs- und Unterscheidungsschwellen an Vögeln. Doctoral Dissertation, Ruhr-Universität Bochum.

Calford, M.B. and Piddington, R.W. (1988) Avian interaural canal enhances interaural delay. *Journal of Comparative Physiology* A 162: 503-510.

Gatehouse, R.W. and Shelton, B.R. (1978) Sound localization in the bobwhite quail (*Colinus virginianus*). *Behavioral Biology* 22: 533-540.

Granit, O. (1941) Beiträge zur Kenntnis des Gehörsinns der Vögel. *Ornis Fennica* 18: 49-71.

Klump, G.M. and Larsen, O.N. (1992) Azimuth sound localization in the European starling (*Sturnus vulgaris*): Physical binaural cues. *Journal of Comparative Physiology* A 170: 243 - 251

Klump, G.M., Windt, W. and Curio, E. (1986) The great tit's (*Parus major*) auditory resolution in azimuth. *Journal of Comparative Physiology* A 158: 383-390.

Lewald, J. (1987a) The acuity of sound localization in the pigeon (*Columba livia*). *Naturwissenschaften* 74: 296-297.

Lewald, J. (1987b) Interaural time and intensity difference thresholds of the pigeon (*Columba livia*). *Naturwissenschaften* 74: 449-451.

Lewald, J. (1989) Verhaltensphysiologische und neurophysiologische Untersuchungen zum Richtungshören der Taube. Doctoral Dissertation, Ruhr-Universität Bochum.

Moiseff, A. and Konishi, M. (1981) The owl's interaural pathway is not involved in sound localization. *Journal of Comparative Physiology* A 144: 299-305.

Park, T.J. and Dooling, R.J. (1991) Sound localization in small birds: absolute localization in azimuth. *Journal of Comparative Psychology* 105: 125-133.

Park, T.J. (1988) *Sound localization in small birds*. Doctoral Dissertation, University of Maryland.

Rosowski, J.J. and Saunders, J.C. (1980) Sound transmission through the avian interaural pathways. *Journal of Comparative Physiology* A 136: 183-190.

Schwartzkopff, J. (1950) Beitrag zum Problem des Richtungshörens bei Vögeln. *Zeitschrift für Vergleichende Physiologie* 32: 319-327.

Windt, W. (1985) Lokalisation von Kunst- und Naturlauten durch Kohlmeisen (*Parus major*). Diplom Thesis, Ruhr-Universität Bochum.

G. M. Klump, R. J. Dooling, R. R. Fay & W. C. Stebbins (eds)

Sound-localization Experiments in Owls

Hermann Wagner

Institut für Zoologie, Technische Universität München
Lichtenbergstraße 4, D-85747 Garching, Germany

Summary. The sound-localization capability of owls can be analyzed at many different levels of investigation, from observations in the natural environment to controlled studies in the sound-proof chamber. The virtues of the approaches at the different levels are reviewed in this paper with an emphasis on laboratory studies. The most widely utilized sound-localization behavior measured in the laboratory are movements of the head toward a sound source. A method is described for sound presentation, data recording and analysis.

1 Introduction

Barn owls are keen hunters of mice. The owl's exquisite sense of hearing provides the main source of information during the hunt. This was shown indirectly by Payne (1971) and Konishi (1973). The latter tied a piece of paper via a thread to a mouse's tail. When the mouse moved around, the paper made more noise than the mouse, and the owl was striking at the paper, not at the mouse. This elegant experiment ruled out vision, infrared sensitivity, and chemical senses as the cue used by the owl. Since echolocation doesn't play a role either (s. Payne 1971, p. 536), the owl must have utilized the rustling noises for the localization of the target.

Studies on the owls' ability to locate prey can be conducted at many different levels. There is, however, a trade-off between the richness of observable behaviors and the possibilities of controlling the stimulus. Thus, while - on a first level - observations in nature might reveal the complete behavioral repertoire, they are often not precise enough for drawing conclusions about the mechanisms underlying a particular behavior. On a second level, in a restricted environment, flight studies can be conducted in the laboratory (Payne 1962, 1971; Konishi 1973; Kolb 1988). The experiment mentioned above constitutes a nice example of the opportunities offered by such studies. During flight experiments in a sound-proof chamber it was also observed that owls would turn the head on the first sound they notice to fixate a potential prey, and that they will leave the perch for an attack only when further sounds are heard (Konishi 1983). This allowed experiments at a third level of investigation with an even better control of the stimuli. To measure head movements of an owl sitting on a perch, Knudsen and coworkers (1979) adapted the search-coil technique that had been introduced by Robinson (1963) for the measurement of eye movements in humans. The search-coil technique can be used with free-field stimuli (for example Knudsen and Konishi 1979; Knudsen and

Knudsen 1986; Wagner 1993) as well as with a pseudo closed-field system. By inserting earphones in the ear canals, the latter method allows for the independent control of the stimuli to the left and right ears (Moiseff and Konishi 1981; Moiseff 1989; Wagner 1991). In this way it was found that the main stimulus parameter used for azimuthal sound localization is the interaural time difference and that the prime cue used for elevational sound localization is the interaural level difference.

In the following I shall first present a summary of each methodological step, and then pick out several experiments to demonstrate the power of the methods. I will not present a thorough review of the subject, because such reviews are available (for example Knudsen 1981; Konishi 1993; Volman 1994 (which also contains an excellent list of important references)).

2 Methods

Studies in the field require good ornithological skills, but no special treatment of the animals. In this respect they stand apart, because at all other levels special experimental treatments like owl handling and training or stimulus control are necessary. Field methods are not described furtheron, and the methods used at the other levels are described en bloc in the following.

2.1 Owl Keeping and Owl Handling

So far, behavioral experiments have been conducted with barn owls (*Tyto alba*), great-horned owls (*Bubo virigianus*), and Tengmalm's owls (*Aegolius funereus*). Owls can easily be kept in aviaries. Since owls breed well in captivity, supply of birds is not a problem.

Tame birds are of great advantage when behavioral experiments are to be conducted with the larger owls, while this is less of a concern with small owls. Thus, barn owls have usually been raised by hand from around post-natal day 10, just before the animals open their eyes. Such birds may become very tame, but I always tried to avoid imprinting on humans. A critical phase in the taming process seems to be around 4 weeks of age, when the birds need constant attention in order to remain friendly to humans. Once owls got accustomed to handling, they will accept humans even if they are not touched and handled for some time. Handling of the owls is necessary for placing them in the experimental set-up, for putting on and removing search-coils and ear-rings and for inspection of the ear canals for ear wax. In delicate experiments, controls for ear wax should not only include the owl but also the earphones.

2.2 Anesthesia and Surgery

Some experiments require a metal post on the head and in some experiments ear-rings are used. The metal post is cemented on the skull, while the ear-rings are sutured to the skin near the ear opening. For these procedures surgery and anesthesia are necessary. Anesthesia in owls is easy. Ketamine (4-20 mg/kg/h) and valium (1 mg/kg/h) work very well and are very safe. As is done before human anesthesia, birds should not be fed on the day before the surgery. They should cast a pellet during the night. We use valium as pre-medication to calm the birds down, and we use a single dosis of atropine sulfate (0.05 mg/kg) to prevent salivation.

2.3 Owl Training

Owls cooperate well in behavioral experiments. As mentioned above, head turning is a natural reaction in the sequence of the behaviors occurring before an attack. To maintain a constant performance in head turning, positive reinforcement by feeding small pieces of meat in conjunction with food deprivation is used in most experiments. Without a reinforcement owls soon become uninterested in the task and it is very important to provide the reinforcement in a most stringent way. Food is provided from a semi-automatic feeder. The experimenter is watching the owl from outside via an infrared sensitive video camera, and decides when the owl has to be fed. We have used feeders of varying sizes, dependent on the needs of the birds. During experiments we routinely use a feeder with 12 sectors that we refill in the course of an experiment consisting of 30 to 60 trials. The refilling does not distract the owl.

The body weight of the birds is controlled every day and is kept 10-20% below the normal body weight in captivity which is probably higher than the natural weight. We have seen seasonal changes in body weight within this range, and in sick owls it was observed that only a loss of body weight of some 40% is dangerous. Exact control of body weight is the second most important factor - apart from the stringency of the experimental protocol - for a successful experiment. Variations of some 2% of the body weight might noticeably influence the owl's motivation.

Training in the head-turning experiments consists of several steps. First, owls have to learn to eat from the feeder. During this period the birds should be food deprived and fed only while sitting on the perch and only from the feeder. First small dead mice, and later pieces of meat of decreasing size are offered to the birds, initially from the top of the feeder and finally from the sectors. Some birds have problems in eating from the feeder because meat sometimes becomes stuck to the upper part of the beak. These birds will then use their feet to help with feeding, and this may distract them. In such birds varying the size of the pieces of meat might be helpful. In any case, birds have to be able to eat reliably from the feeder before the next step of training is initiated.

The second step of training consists of the reinforcement of head turning after a sound stimulus. The reinforcement is provided when the owl turns toward the sound within one second regardless of how precise the localization is. Additionally, the owl should focus on the target for at least a second. Negative reinforcement is not used. Most owls respond on 100% of the trials. Unreliable behavior on the part of the owls is most often due to inconsistent criteria on the part of the experimenter. Tests can be started when owls reach nearly 100% of performance on this simple head turning task. But, one problem with using only a test stimulus without a zero stimulus (see below) is that there is no internal control of the owl's alertness: it is the experimenter that decides, by watching the bird, when a new trial should be started. Another problem is that in order to begin a new trial, the owl must be looking straight ahead without moving. This might take considerable time. Therefore, in some experiments a third step of training has been used before starting an experiment. In these cases, a zero speaker is placed directly in front of the owl in addition to the test speaker which is positioned at different angles along a semicircular track (Knudsen et al. 1979). The owl has to learn to first react to the zero speaker, and then to fixate on the test speaker. This paradigm offers an internal control of the owl's alertness and helps to orient the owl within the external coordinate system. Since the first sound seems to dominate the head turning behavior measured in the laboratory at least in some owls (see also **Results and Discussion**), we abandoned this method.

Owls wearing earphones turn the head towards the side defined by the sign of the interaural time difference (for example to the right when the sign - as is our convention - is positive). Thus, in experiments using time-shifted, but correlated signals in the two ears, additional training is not necessary. When stimulated with a combination of correlated and uncorrelated noises, the owls had to learn to react to this dichotic stimulus in a correct way (Wagner 1991).

Since all steps during an experiment follow a clear logic, the paradigm can be implemented on a computer as well. Allowing the owl to initiate a trial might minimize the problems associated with varying attention.

2.4 Stimuli

Mice make broad-band noises when walking around, and they may squeak at high frequencies. That broad-band noise from a loudspeaker is indeed a good stimulus in the laboratory (Konishi 1973) suggests that owls do not identify the source before striking. Owls react with less enthusiasm to pure tones. The birds like faint sounds better than loud sounds and they also attack a prey if the sound wanes before the impact (open-loop behavior). Things may be more complex in nature though. Some owls may become specialized for hunting certain prey. For instance, some owls have been observed to preferentially capture shrews which make sounds at a frequency around 5 kHz.

It is necessary to use an open-loop paradigm for measuring the localization precision. Since the shortest latency observed in head turning has been 50 ms, sounds of 50 ms or shorter sounds can be regarded as an open-loop stimulus because the stimulus is over before the owl starts to respsond. Broad-band loudspeakers as used in cars (Mac Audio, ML103E, 10-cm speakers, bandwidth +/-5dB from 0.8 to 10 kHz) have worked well for us. Miniature earphones were obtained from Knowles (for example the model 1759). When earphones are used, a time shifter that should work in a range up to +/-200 μs and have a resolution of at least 2 μs should be available.

2.5 Data Recording

Barn owls are active during the dawn and dusk and during the night and they sleep during the day. Not surprisingly, we have found it difficult to conduct experiments during the late morning and early afternoon, in terms of the owl's behavior, and much easier to conduct experiments after 5 p.m..

a) Sound-localization Correlates in Flying Birds

Payne (1971) determined localization precision in flying birds with a simple method. The talons form a rectangle, and the distance of the center of the rectangle to the target was the localization precision. Konishi (1973) used an improved method with masonite plates having microswitches connected to them. With these methods it was found that the owl can locate prey in total darkness with an accuracy of at least one degree in both the vertical and horizontal planes (Payne 1971).

b) Head Turns Towards a Sound Source as a Correlate for Sound Localization

Head turns may be characterized by their amplitude, their latency and their maximal velocity. For the measurements of head position two methods have been used. In the simpler one, a marker is fixed on the owl's head, and the owl's head movements are video-taped and analyzed offline with an accuracy of +/- 2.5 degrees (Moiseff and Konishi 1981; Wagner 1991). A much better accuracy is provided by the modified search-coil technique (Knudsen et al. 1979). With this method accuracy can be well below one degree, and we have digitized data at 0.5 degrees. The search-coil technique takes advantage of Faraday's law, i.e. that a changing magnetic field induces a current in a loop of wire. A magnetic field is generated by large induction coils surrounding the bird. This field induces a current in the small search coils fixed to the owl's head (Fig. 1). When the phase of the induced current instead of its amplitude is

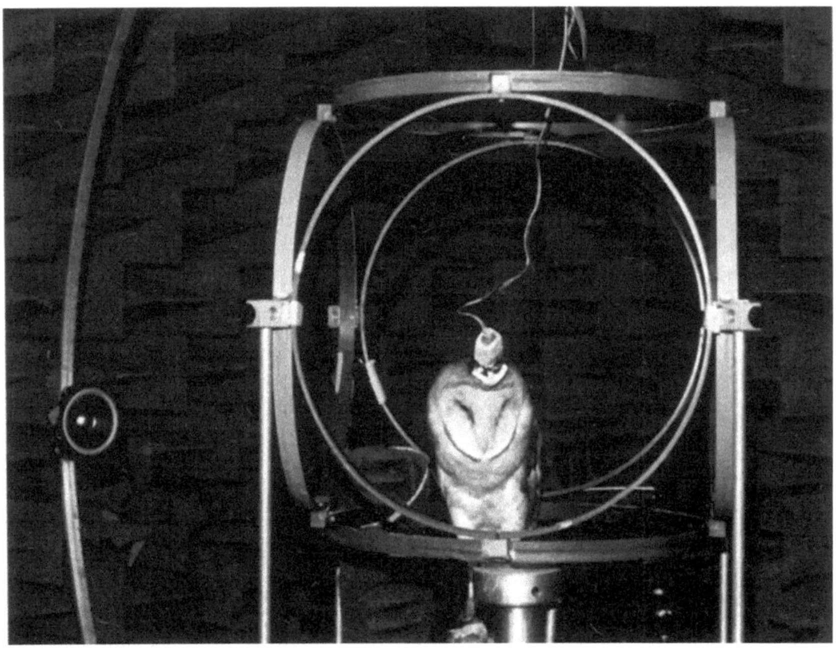

Figure 1. The search-coil set-up with an owl. On the left the semicircular track with the loudspeaker (test speaker) can be seen. The 3 pairs of induction coils are positioned around the owl that wears 3 search coils and the cables attached to them. In this scene, the owl fixates the feeder directly in front of it.

used for the measurement, a precise calculation is possible over almost the entire frontal hemisphere. To allow for measurements when the bird is stationary, the magnetic field is rotating. Such a rotating magnetic field is generated in each spatial plane by feeding periodic signals of one frequency in each of two orthogonally oriented pairs of induction coils. One pair of coils receives a sine and the other a cosine. If this is done in all spatial planes, rotations about all three spatial axes can, in principle, be analyzed. We have measured rotations about the vertical and the transverse axes and have used frequencies of 11 kHz in the horizontal plane and 25 kHz in the vertical plane, three pairs of induction coils (one of which is shared in the two planes) and three search coils with orthogonal orientation (Fig. 1). The induced currents are first converted to voltages, then separated by band-pass filters, sent to a phase-locked loop that calculates the position of the head, and finally fed into a computer. A sampling interval of 5 ms is sufficient.

So far, the size of the induction coils that have been used in owl experiments has not been larger than 40 cm. Larger induction coils would provide advantages. The size of the induction coil should be such that the owls can sit on the perch without any body parts touching the induction coils. This requires an induction coil with a diameter of about 70 cm. The search coils used are about 3 cm in diameter and have a thickness below 0.3 cm. These coils weigh some 25 g. Owls tolerate search coils without any indication of discomfort. Search coils connect to the electronic via cables. These cables are also well tolerated by the birds as long as they do not touch the head or the wings. This can be avoided by tying the wires to a stretchable telephone cable.

2.6 Data Analysis

The discussion of this point will be restricted to the coordinate system in which the data should be analyzed. Sound localization in three-dimensional space requires the determination of three variables. The most obvious coordinate system is a spherical coordinate system which consists of two angles, azimuth and elevation, and the distance between the owl and its target (Fig. 2a, b). Knudsen and coworkers (1979) used a double-polar coordinate system (Fig 2c) and listed several advantages of it over other possible coordinate systems. The equator of the coordinate system runs through the interocular axis of the owl when the animal stands upright. The owl's midsagittal plane defines the vertical plane of the coordinate system (Fig. 2d).

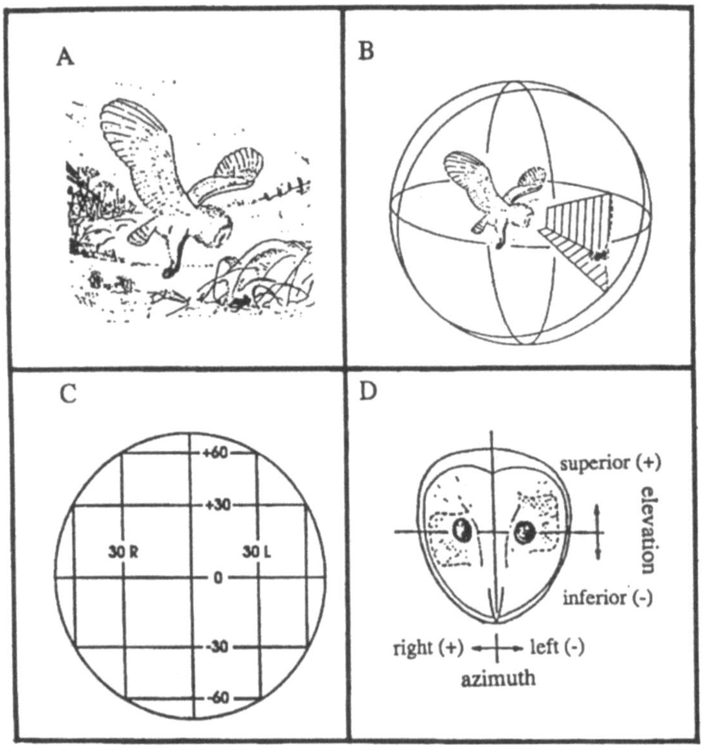

Figure 2. The framework for the analysis of sound localization. Please refer to the text for a description of this framework.

3 Results and Discussion

When we want to learn something about the auditory capabilities of animals we have to find an appropriate motor reaction through which the animal can tell us, unambiguously, that it senses a stimulus or "understands" a task. In owls, good motor responses are flying towards a post, turning of the head in a given direction, pecking food from a feeder and selecting food from one of two feeders. These different reactions have been used for different purposes. In the following I shall present examples investigated in some detail in the last 25 years.

3.1 Experiments with Flying Owls

Owls can be trained to fly toward a perch to pick up food. This method has been used to study the barn owl's absolute frequency-discrimination capabilities (Quine and Konishi 1974) and to study the discrimination of noise spectra by memory (Konishi and Kenuk 1975). In free-flight experiments more related to sound localization, Konishi demonstrated that the owls were able to make mid-flight turns, if the sounds were switched from one loudspeaker to the next (Konishi 1973; see also Payne 1971). It was also found in addition that both barn owls (Konishi 1973) and Tengmalm's owls (Kolb 1988) fly at slower velocities and at a more curved path when it is dark than when lights are on.

3.2 Head-turning Experiments with Free-field Sources

With the introduction of the head-coil system, most experiments have been conducted with owls sitting on a perch and turning their heads in response to a free-field sound. If the owls are trained well and have the correct weight, they act very reliably in this paradigm. There is, however, a considerable variation in turning amplitude (Fig. 3), especially at larger stimulus angles. Since these experiments were conducted under open-loop conditions, one source of variability comes from the decrease in localization precision for stimulus angles larger than 30 degrees in azimuth (Knudsen et al. 1979). Owls tend to localize targets in azimuth and elevation with similar precision. Stimuli positioned at 10 degrees of azimuth or elevation can be localized with a precision of about 1 deg (Knudsen et al. 1979). At very small stimulus angles, a further source of variability might be the lack of responses. I have also observed that an owl responded almost perfectly to stimuli covering a wide range of azimuths on one day, but was more sloppy on the next day. Thus, the number of trials or some unknown internal state of the birds may play a role in the variability as well.

By capitalizing on the observation that the owls are able to align their talons along the axis of motion of the prey independent of their own flight direction (Payne 1971), we were able to study the influence of acoustic motion on the owls' head turning. Two loudspeakers were positioned in front of the owl and 30 degrees apart. When the stimulus in one speaker was ramped on while the stimulus in the other speaker was ramped off at the same time, the owl nevertheless turned toward the speaker that emitted a loud sound first, and not in the direction of motion, which would have been away from the direction of the speaker emitting the loud sound at the onset. The response to a moving stimulus was not different from the response when the stimulus came only from the location of the first speaker (Wagner, unpublished). This demonstrates a dominating influence on the owl's response of the first sound it hears. The implications of this dominance in the natural situation are not well understood, because it seems - at least at first hand - to contradict the observation of mid-flight corrections cited above.

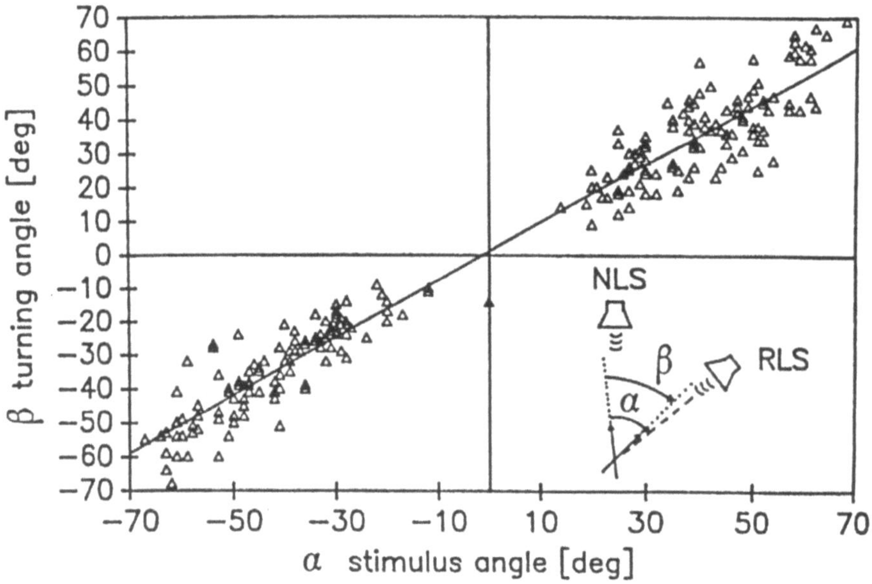

Figure 3. Horizontal component of head-turning (turning angle) of an owl in response to a free-field stimulus positioned at different azimuths (stimulus angle). The owl first received a zero signal from the zero speaker (NLS). When it reacted with a short-latency, high-velocity turn, a second sound was presented from the test speaker (RLS).

3.3 Head-turning Experiments with Earphone Stimulation

Moiseff and Konishi (1981) were the first to use the earphone system. They found a linear relation of the turning amplitude with interaural time difference (see also Moiseff 1989). Together with their demonstration that neurons in the owl's midbrain are tuned to the interaural time difference of sinusoidal signals having frequencies up to some 8 to 9 kHz, this finding lay ground to the notion that barn owls have extended the range over which they can use interaural time difference to higher frequencies than any other animal investigated so far (see also Konishi et al. 1988). With earphone stimulation it could be shown in a similar way that the owl uses interaural level difference as the main cue for elevational sound localization (Moiseff 1989). If stimuli having an interaural time difference or interaural level difference were repeated in a burst, the owls would repeatedly turn their heads in fast sequence (Konishi, pers. communication). In my experiments, owls clearly observed the sign of interaural time difference (Fig. 4), but turning amplitude showed a high variability - higher than in free-field studies - and it seemed that the birds reacted more in a flip-flop-like fashion. It may be that training for localization precision would help here to decrease variability.

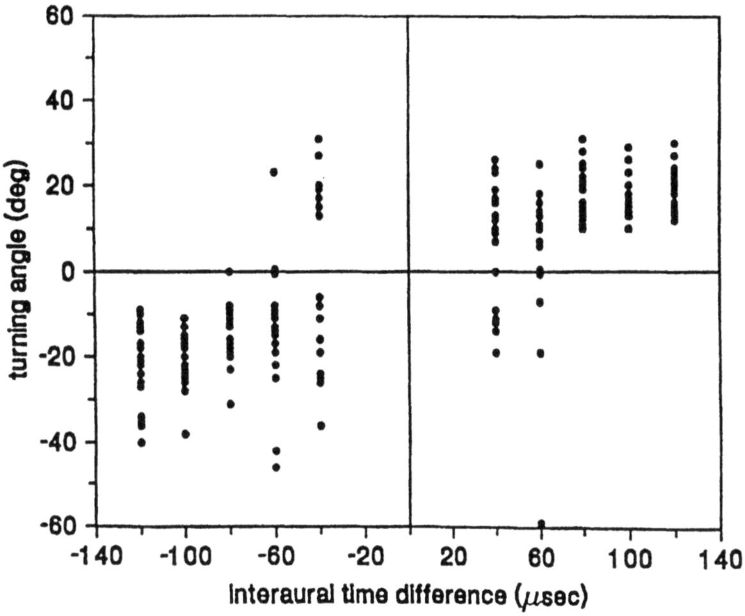

Figure 4. Responses of an owl to sounds presented via earphones. When the interaural time difference (ITD) was varied, the owl turned toward the left (negative turning angles) when negative ITDs were presented, and it turned to the right (positive turning angles) when the sound had a positive ITD.

4 Acknowledgements

I thank Henning Blanke, Dirk Kautz and Georg Klump for critical comments. This work was supported by a fellowship from the Deutsche Forschungsgemeinschaft (Heisenbergstipendium).

5 Suggested Reading

Knudsen, E.I., Blasdel, G.G. & Konishi, M. (1979) Sound localization by the barn owl (*Tyto alba*) measured with the search coil technique. *Journal of Comparative Physiolology, 133,* 1-11.

Konishi, M. (1973) How the owl tracks its prey. *American Scientist, 61*: 414-424.

Konishi, M. (1993) Listening with two ears. *Scientific American, 268,* 66-73.

Moiseff, A. & Konishi, M. (1981) Neuronal and behavioral sensitivity to binaural time differences in the owl. *Journal of Neuroscience, 1,* 40-48.

Payne, R.S. (1971) Acoustic location of prey by barn owls (*Tyto alba*). *Journal of Experimental Biology, 54,* 535-573.

6 References

Knudsen, E.I. (1981) The hearing of the barn owl. *Scientific American, 245,* 113-125.

Knudsen, E.I. & Knudsen, P.F. (1986) The sensitive period for auditory localization in the barn owl is limited by age, not by experience. *Journal of Neuroscience, 6,* 1918-1924.

Knudsen, E.I. & Konishi, M. (1979) Mechanisms of sound localization in the barn owl (*Tyto alba*). *Journal of Comparative Physiology A, 133,* 13-21.

Kolb, K.H. (1988) Das Verhalten des Rauhfußkauzes (*Aegolius funereus*) beim Beutefang. Diplomthesis, Tübingen, Germany: Universitiy of Tübingen.

Konishi, M. (1983) Neuroethology of acoustic prey localization in the barn owl. In: Huber, F. & Markl, H. (eds.) *Neuroethology and behavioral physiology*, Springer Verlag , Berlin, Heidelberg, pp. 303-317.

Konishi, M. & Kenuk, A.S. (1975) Discrimination of noise spectra by memory in the barn owl. *Journal of Comparative Physiology A, 97,* 55-58.

Konishi, M., Takahashi, T.T., Wagner, H., Sullivan, W.E. & Carr, C.E. (1988) Neurophysiological and anatomical substrates of sound localization in the owl. In: Edelman, G.M., Gall, W.E. & Cowan, W.M. (eds) *Auditory Function - Neurobiological bases of hearing*, John Wiley & Sons, Inc., New York, pp. 721-745.

Moiseff, A. (1989) Bi-coordinate sound localization by the barn owl. *Journal of Comparative Physiology A, 164,* 637-644.

Payne, R.S. (1962) How the barn owl locates prey by hearing. *The Living Bird, First Annual of the Cornel Laboratory of Ornithology*: 151-159.

Quine, D.B. & Konishi, M. (1974) Absolute frequency discrimination in the barn owl. *Journal of Comparative Physiology A, 93,* 347-360.

Robinson, D.A. (1963) A method of measuring eye movement using a scleral search coil in a magnetic field. *IEEE Transactions in Biomedical Electronics, 10,* 137-145.

Volman, S.F. (1994) Directional hearing in owls: neurobiology, behavior, and evolution. In: Davies, M.N.O. & Green, P.R. (eds) *Perception and motor control in birds*, Springer, Berlin, pp. 292-314.

Wagner, H. (1991) A temporal window for lateralization of interaural time difference by barn owls. *Journal of Comparative Physiology A, 169,* 281-289.

Wagner, H. (1993) Sound-localization deficits induced by lesions in the barn owl's space map. *Journal of Neuroscience, 13,* 371-386.

Amphibia

Traditionally, the study of hearing in amphibia derives from neuroethological approaches to the coevolution of sensory and communication systems. Most frogs have a simple vocal repertoire with a small number of species-specific, stereotyped calls. This stereotypicity allowed the evolution of physiological mechanisms that appear "matched" to call characteristics in order to aid in the recognition of species-specific signals. Simple temporal and spectral characteristics of the frogs' vocalizations are evaluated by its auditory system to recognize conspecific calls. The mechanisms involved in this analysis are probably very general to all land vertebrates, since the amphibia stand at the evolutionary roots of the land-dwelling vertebrates. On the other hand, amphibia show some specializations in their auditory system (e.g., two separate auditory papillae that encode different ranges of frequencies) that make them interesting subjects for comparative studies of neural integration.

Unlike in psychoacoustic studies in birds and mammals, only rarely have operant conditioning methods been used in the study of hearing in frogs. The reflex modification technique, described in detail in the chapter by Simmons and Moss, is a variation on a classical conditioning method which has provided us with absolute and masked auditory thresholds in frogs. In most studies, however, researchers typically exploit the natural behaviour of the frogs to investigate the function of their auditory system. When ready to mate, female frogs approach males guided by the male's advertizement calls. How this phonotactic behavior can be exploited to study the sensory analysis in the frog's auditory system is described in the chapter by Gerhardt. Methods to study sound localization in frogs are described in the chapter by Klump.

G. M. Klump, R. J. Dooling, R. R. Fay & W. C. Stebbins (eds)

Reflex Modification: A Tool for Assessing Basic Auditory Function in Anuran Amphibians

Andrea Megela Simmons[1] and Cynthia F. Moss[2]

[1]*Brown University, Departments of Psychology and Neuroscience, Box 1853, Providence RI 02912 USA*
[2]*Harvard University, Department of Psychology and Program in Neuroscience, 33 Kirkland Street, Cambridge MA 02138 USA*

Summary. Reflex modification is a psychophysical technique based on the phenomenon that the response to a reflex-eliciting stimulus is changed in amplitude by the prior presentation of a neutral prestimulus. This technique provides reliable data on hearing thresholds both in quiet and in noise for anuran amphibians. Potential difficulties with the reflex modification technique include variability in the effectiveness of the reflex-eliciting stimulus, individual variability between animals, variability in the slopes of psychometric functions, and the small size of reflex modification effects in anurans.

1 Introduction

1.1 Conditioning Techniques

Anuran amphibians (frogs and toads) are highly vocal animals that rely on acoustic communication to guide social and reproductive behaviors. Historically, attempts to study sensory function in anurans with traditional conditioning techniques have not met with great success (but see Elepfandt et al., 1985). Ambiguous or negative results have been reported with escape and avoidance conditioning (Boice, 1970; Crawford & Langdon, 1966; Hoyer, Shafer, Mauldin, & Corbett, 1971; McGill, 1960; Yaremko, Boice, & Thompson, 1969; Yaremko, Jette, & Utter, 1974), conditioning of appetitive instrumental responses (van Bergeijk, 1967), and classical conditioning of heart rate and eyeblink responses (Capranica, 1965; Goldstein, Spies, & Sepinwall, 1964). Studies using the unconditioned galvanic skin response to measure hearing sensitivity (Brzoska, 1980; Brzoska, Walkowiak, & Schneider, 1977; Strother, 1962; Weiss & Strother, 1965) found considerable variability in responses, and behavioral threshold curves derived from this technique do not well match threshold curves derived from neural responses (Brzoska, Walkowiak, & Schneider, 1977). These difficulties motivated the use of reflex modification as a psychophysical technique reported here.

1.2 Reflex Modification

Yerkes (1905) first introduced a technique for studying auditory function in frogs based on inhibition of reflex responses to sounds. In his experiments, the frog was restrained in a saddle-like apparatus. A rubber hammer tap on the frog's head elicited a reflex movement of the hind leg. The magnitude of this leg movement decreased when the head tap was preceded by prior sounds (whistles, tuning fork vibrations, bells). These effects disappeared when the eighth cranial nerve was cut. Yerkes concluded that frogs tested in this way could detect frequencies up to 10 kHz, but absolute detection thresholds were not measured.

Modern studies of reflex modification are based on the observation that a reflex elicited by one stimulus is changed when that stimulus is preceded by a "neutral" prestimulus that itself does not elicit the reflex response. Modification of the reflex may take the form of a shift in reflex latency, or a reduction in its amplitude, depending on the time interval between the neutral prestimulus and the reflex-eliciting stimulus (see Figure 1). At short time intervals, reflex latency is reduced and at long time intervals, reflex amplitude is reduced. The amount of reflex modification depends on the intensity of the prestimulus; inhibition can occur even to weak prestimuli that are near the detection threshold (Reiter & Ison, 1977; Young & Fechter, 1983). This means that reflex modification is not an all-or-none phenomenon but rather a graded, increasing function of prestimulus intensity. This relation underlies the potential usefulness of reflex modification as a psychophysical technique: The magnitude of the reflex inhibition effect should vary consistently with the amplitude of the prestimulus. Sensitivity functions obtained by both reflex modification and classical conditioning techniques in the same species are similar (Young & Fechter, 1983). Reflex modification has advantages over conditioning techniques in that the effects are seen on the first trial and require no instrumental response. It has produced reliable data in a number of vertebrate species, using visual, auditory, and cutaneous prestimuli, and a variety of reflex behaviors (Hoffman & Ison, 1980).

2 Methods

2.1 Apparatus

Our initial attempts to test the usefulness of reflex modification as a psychophysical technique involved use of an apparatus similar to that employed by Yerkes (1905). This was quickly abandoned, however, as the animals did not react well to direct restraint. We then designed an apparatus in which controlled levels of reflex-eliciting stimuli could be delivered without the need to directly restrain the animal. This was a small round plastic dish, 1-2 cm in diameter larger than the animal itself (approximately 5 cm diameter for small green treefrogs and 20 cm diameter for large bullfrogs). The small size of the dish constrained the amount of spontaneous movement the animal could make, but without direct restraint. Typically, once acclimated to the dish, the frog settled in one position throughout an experimental session. Two separate copper strips were cemented to the sides of the dish. These were connected to a Grass SD5 stimulator which delivered a mild electrical stimulus (shock). The bottom of the dish contained water to a depth of 1-2 mm. Fine nylon mesh was stretched across the top of the dish to further restrict the frog's movements. The dish was placed on top of a custom-built stabilimeter to detect any movements of the animal. Electrical stimuli delivered through the metal strips caused reflex movements of the animal's limbs. These movements vibrated the dish and produced a voltage change in the electrical signal from the stabilimeter. This voltage change was amplified and transmitted to a computer for display and processing. Prestimuli were generated by audio oscillators, gated, attenuated, amplified, and presented by a loudspeaker mounted 1 m above the animal in its dish. Delivery of both prestimuli and the reflex-eliciting stimuli were under computer control. A schematic of the apparatus is shown in Figure 1.

2.2 Experimental Procedure

Experiments were run in blocks of trials. Each block consisted of 6-8 test (prestimulus + reflex-eliciting stimulus) trials and 2 control (reflex-eliciting stimulus alone) trials. The order of presentation of these different trials was random within each block (method of constant stimuli). Within an individual block, intervals between trials varied from 30 to 60 sec. Intervals between blocks varied from 3 to 5 minutes. Both of these time intervals were allowed to vary across individual animals, and even within the same experimental session. At the beginning of the experiment, baseline "responses," that is, spontaneous movements in the absence of any stimulation, were recorded. This was to provide an estimate of the noise level in the apparatus. About 10 trial blocks were run per day for an individual animal.

Two different kinds of experiments were conducted. One kind of experiment measured the optimal time interval for producing reliable reflex inhibition across individual animals. The amplitude of the prestimulus was fixed at a level well above threshold, and only the time interval -- the lead time -- between the onset of the prestimulus and the reflex-eliciting stimulus

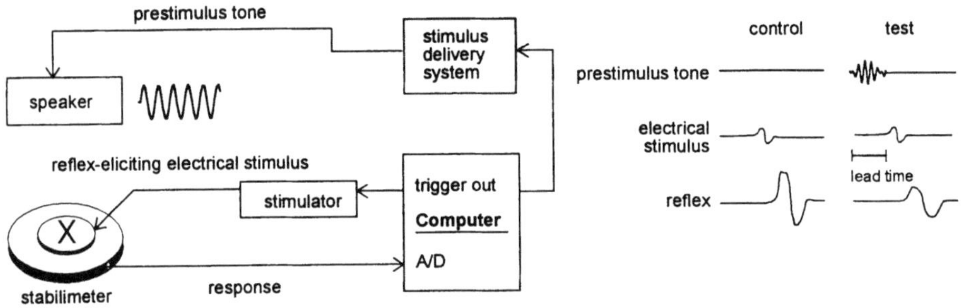

Figure 1. (Left) Schematic of the apparatus. (Right) Diagram of expected changes in reflex response in control and test trials.

varied. The offset of the prestimulus occurred simultaneously with the onset of the reflex-eliciting stimulus, and therefore the duration of the prestimulus equaled the lead time (see Figure 1).

The second experiment estimated thresholds to sounds. The time interval between the prestimulus and the reflex-eliciting stimulus was fixed at the appropriate lead time, and the amplitude of the prestimulus was varied in 10 dB steps, from a level well below to a level well above threshold. Complete data were collected for one frequency before testing the next frequency. The order of presentation of different frequencies varied across individual animals.

2.3 Data Analysis

The animals' responses to the reflex-eliciting stimulus were digitized and displayed as changes in amplitude over time. Sampling began immediately after onset of the reflex-eliciting stimulus and lasted for a duration of 512 msec. The reflex response itself was complete within 100 msec. Analysis included peak amplitude, trough amplitude, peak-to-trough amplitude, latency to peak, latency to trough, and integrated area under the response. The measure providing the most reliable data across individual animals was the combined peak-trough amplitude; changes in integrated area were correlated with changes in peak-trough amplitude but were somewhat more variable. We analyzed changes in the largest peak-trough amplitude within the entire 100 msec window, regardless of latency within that window. No responses to the prestimulus alone were ever observed.

Within each block of trials, responses to the test trials were normalized with respect to responses to the mean of the two control trials. These normalized responses were then averaged across all blocks, and expressed as % reflex inhibition in the test trials relative to the reflex magnitude in the control trials.

In experiments estimating lead times, data were plotted as % inhibition against the time interval between the prestimulus tone and the reflex-eliciting stimulus, and the time interval producing the most consistent inhibition was chosen from these plots. To estimate the animals' thresholds to various sounds, data were summarized in the form of psychometric functions plotting % reflex inhibition against prestimulus amplitude. The data points were fit by linear regression. The prestimulus amplitude producing a level of 10% inhibition on the linear regression was used as an estimate of threshold. Estimates of threshold across different frequencies were graphed together to produce audiograms.

2.4 Factors Influencing Reflex Modification

One of the most important variables affecting the success of the experiments was the animals' sensitivity to the reflex-eliciting stimulus itself. Shock level needed to be set carefully, and its effectiveness continually monitored throughout the experiment. The goal was to find a shock level that would reliably elicit a reflex, but that was not so strong as to serve as a punishing stimulus. There was great individual variability among animals in the shock level required. Sensitivity to shock could vary across experimental days in the same animal, or even within the same experimental session. For this reason, at the beginning of each experimental session, several trials were run to test the animal's responsiveness to shock. In these trials, shock level was varied until an appropriate level could be ascertained. There was some unavoidable subjectivity involved in finding this level. Once shock level was set, the animals were given a period (5-10 minutes) without stimulation before the experiment began.

During an experiment, the animal's responses to the shock sometimes changed. Responses to shock could either systematically increase over time (an apparent sensitization effect), or systematically decrease over time (an apparent habituation effect). Examples of these different effects are shown in Figure 2 (data from Simmons, 1988a). In the former case, a shock level that at the beginning of the experiment seemed to be well-tolerated and produced only a slight flexion of the animal's hind legs could over time appear to cause discomfort or large voluntary

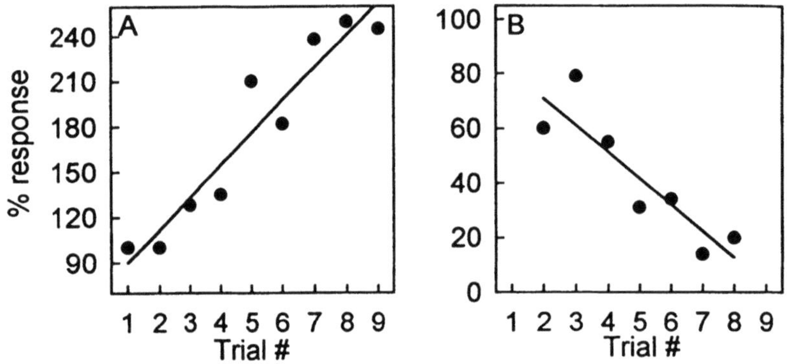

Figure 2. Sensitization (A) and habituation (B) of responses to the reflex-eliciting stimulus over trials.

movements. If such effects occurred, the shock level would be decreased from its initial setting for the next block of trials. Responses to the shock itself could also decrease, sometimes as much as 90% over successive trials, regardless of the order of the prestimulus tones (Figure 2). If this occurred, shock level was increased over its initial setting for the next block of trials. Any blocks showing significant and systematic increases or decreases in response amplitude regardless of the order of stimulus presentation were eliminated from the analysis. The number of trial blocks eliminated because of these effects varied but was relatively small; for example, in an experiment analyzing sensitivity to harmonic cues in complex sounds (Simmons, 1988a), 3% of all blocks were removed for these reasons.Responses to the electrical stimulus could also vary rather non-systematically in an experiment. At the end of an experiment, the magnitude of responses to the two control trials in each block were examined. If they differed from each other by 50% or more, that block was removed from analysis. In other cases, the animal apparently ceased all responding to the reflex-eliciting stimulus. Here, response magnitude to the shock did not vary at all within a stimulus block and did not differ from the baseline level of "response". Such effects usually occurred near the end of the experimental session, and blocks of trials showing these effects were also eliminated from analysis.

Part of the variability in shock responses was related to the amount of water placed in the animal's dish. It was important to maintain enough water to make good contact with the stimulating electrodes, but the water level had to be low enough to avoid covering the animal's entire body. Movements of the animal in the dish could cause water to splash out. For this reason, the water level was checked periodically during an experiment. The animals did not tolerate long experimental sessions; both intertrial intervals within a block and interblock intervals within one experimental session needed to be relatively long, and were allowed to vary between individuals. We found that apparent habituation or sensitization effects became more common near the end as compared to near the beginning of an experimental session.

3 Results and Discussion

3.1 Psychometric Functions

An example of a psychometric function of the responses of one green treefrog to a prestimulus tone of 2700 Hz at different prestimulus amplitudes is shown in Figure 3 (data from Megela-Simmons, Moss & Daniel, 1985; Moss, 1985). Several features of this psychometric function are important. First, the magnitude of the reflex modification effect for this animal is at maximum about a 25% reduction in reflex amplitude. The magnitude of the reflex modification effects varied across individual animals, and over all animals tested, ranged in size from about a 20% reduction in reflex response to at most a 50% reduction (see Simmons, 1988a). These are significantly smaller than the magnitude of reflex modification effects reported in mammals (Hoffman & Ison, 1980; Young & Fechter, 1983). Occasionally, some animals showed either no effect, or a very unreliable effect (large effects on one

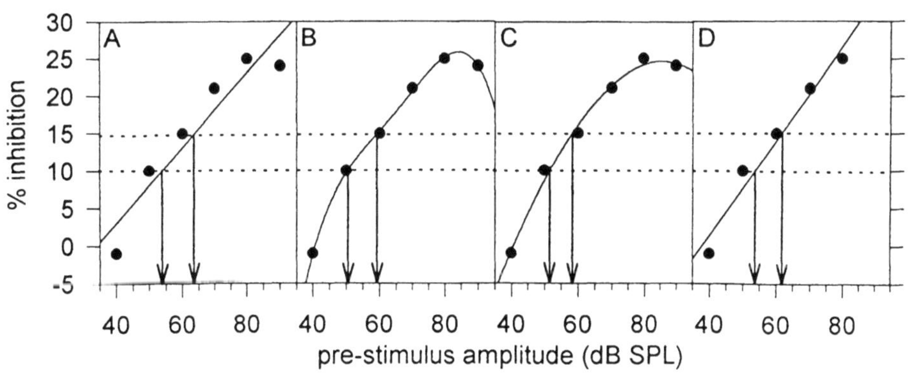

Figure 3. Psychometric functions of responses of one frog to a prestimulus of 2700 Hz under quiet (ambient) noise conditions, showing different fits to the data points and different criteria for thresholds. A. Linear fit. B. Quadratic fit. C. Cubic fit. D. Linear fit excluding the point at 90 dB SPL.

experimental session, and no effect on the next). Whether or not an animal showed any reliable or consistent reflex modification effects did not seem to depend on season of the year.

Another important feature of these data concerns the goodness-of-fit of linear regression to the psychometric functions. Linear regressions were used under the assumption that the magnitude of the reflex modification effect would increase systematically with the level of the prestimulus sound (Hoffman & Ison, 1980). This assumption did not always hold. This was typically due to responses at the least intense and/or the most intense prestimuli either remaining constant, or varying in the opposite way than predicted, indicating possible "floor" and "ceiling" (saturation) effects (compare the % inhibition produced by prestimulus amplitudes of 80 and 90 dB SPL in Figure 3). If such data points were eliminated from the linear regression (Figure 3D), the estimate of threshold varied little from that estimated when all data points were used (Figure 3A). As shown in Figures 3B and 3C, quadratic and cubic functions could also provide good fits to these data. Thresholds estimated from these different fits were relatively consistent, typically being within 5 dB of each other. Overall, linear regressions provided good fits to the data, accounting for 85 - 99% of the variance in different experiments, and the most consistent fits across individual frogs.

Figure 4 plots the slopes of psychometric functions against prestimulus frequency for 7 individual animals (data from Megela-Simmons, Moss, & Daniel, 1985; Moss, 1985).

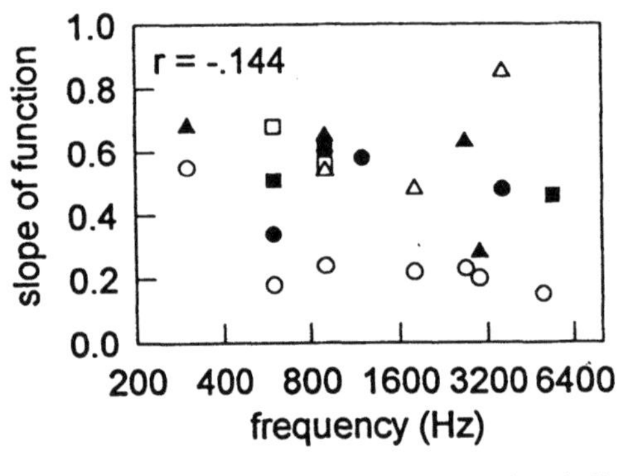

Figure 4. Slope of psychometric functions against prestimulus frequency for 7 individual frogs. Different symbols indicate data from different animals.

The slopes of the linear regressions fit to the data points on the psychometric functions varied, from about .18 to about .9 for this particular data set. This variability was due primarily to individual differences among animals and was not related to the particular prestimulus frequency tested. For the data shown in this figure, the correlation between slope and prestimulus frequency was not statistically significant (r = -.144). Slopes of psychometric functions generated for sound frequencies outside of the animal's range of hearing was close to 0 (.02 - .09) and were excluded from this analysis.

3.2 Criteria for Threshold

The criterion used to estimate threshold from the psychometric functions needs to be carefully chosen. The level of 10% inhibition used in our experiments was above the baseline variability in the data. Any experiment using the reflex modification technique should independently collect such baseline data for use in choosing a reliable threshold estimate. The sometimes shallow slopes of the psychometric functions could pose difficulties in making estimates of thresholds from these functions.

Different criteria for threshold could produce different estimates of auditory sensitivity. Figure 3 gives examples of thresholds obtained using two different criteria, 10% and 15% inhibition. The 15% inhibition criterion raises the threshold estimated from the 10% inhibition criterion by 5-10 dB. The choice of threshold criterion clearly influences the audiograms compiled from the psychometric functions. Figure 5 shows group audiograms from seven different green treefrogs (data from Megela-Simmons, Moss, & Daniel, 1985; Moss, 1985) compiled using these 2 threshold criteria. To compile these figures, estimates of threshold at each prestimulus frequency were made separately for each individual animal, and then averaged. The choice of criterion between 10% and 15% inhibition influences the absolute

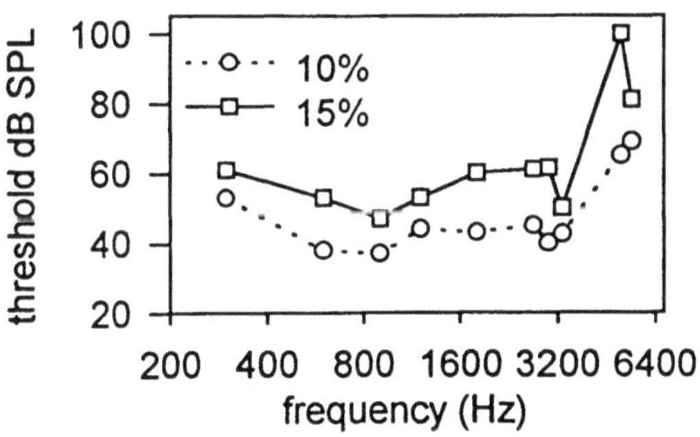

Figure 5. Audiograms for 7 individual green treefrogs estimated from psychometric functions using threshold criteria of both 10% and 15% inhibition.

threshold values; however, the shapes of the two audiograms are almost identical. The only marked difference appears at 5000 Hz, a frequency which produced very shallow psychometric functions and therefore large differences in threshold from the different criteria.

Two other criteria for threshold were also examined. These are the percent reflex inhibition that is twice the standard error of the mean (2xSEM) and the percent reflex inhibition that is 50% of the maximum inhibition value (50% MAX). These threshold criteria depend on the characteristics of the psychometric function itself, while the first two criteria (10 and 15% inhibition) are fixed. The 2xSEM criterion adjusted for the frog's response variability under particular stimulus conditions, and the 50% MAX criterion adjusted for the magnitude of the reflex modification effect that an individual animal showed. The 10% reflex inhibition and the 2xSEM criterion produced similar estimates (within 4 dB). The 50% MAX criterion was the most variable across the data, and its corresponding threshold estimates were sometimes above and sometimes below those determined by the other three criteria.

3.3 Comparison of Reflex Modification with Other Behavioral Techniques

Absolute threshold data collected using the reflex modification technique for both the bullfrog and the green treefrog (Megela-Simmons, Moss, & Daniel, 1985; Moss, 1985; Moss & Simmons, 1986) are similar in sensitivity to neural threshold data collected in the same species. They provide more sensitive estimates of absolute hearing sensitivity than estimates based on evoked vocalization and selective phonotaxis techniques (see Gerhardt, this volume). Masked threshold data using these different behavioral techniques provide much more similar estimates of sensitivity (Moss & Simmons, 1986; Simmons, 1988b). Experiments testing discrimination of complex, multiple harmonic sounds have, however, yielded conflicting data (Gerhardt, Allan, & Schwartz, 1990; Simmons, 1988a; Simmons, Buxbaum, & Mirin, 1993). The reflex modification technique described here may itself need to be modified to analyze discrimination as opposed to detection of sounds.

4 Acknowledgements

Work described in this chapter was supported by grants from NIH (NS 21911) and the Deafness Research Foundation to A.M.S. and a NSF Graduate Fellowship to C.F.M. We thank Mark I. Sanderson for preparing the figures.

5 Suggested Reading

Hoffman, H.S. & Ison, J.R. (1980). Reflex modification in the domain of startle: I. Some empirical findings and their implications for how the nervous system processes sensory input. *Psychological Review, 87,* 175-189.

Megela-Simmons, A., Moss, C.F. & Daniel, K.M. (1985). Behavioral audiograms of the bullfrog *(Rana catesbeiana)* and the green tree frog *(Hyla cinerea). Journal of the Acoustical Society of America, 78,* 1236-1244.

Moss, C.F. & Simmons, A.M. (1986). Frequency selectivity of hearing in the green treefrog, *Hyla cinerea. Journal of Comparative Physiology, 159,* 257-266.

Reiter, L.A. & Ison, J.R. (1977). Inhibition of the human eyeblink reflex: An evaluation of the sensitivity of Wendt-Yerkes method for threshold detection. *Journal of Experimental Psychology: Human Perception and Performance, 3,* 325-336.

6 References

Boice, R. (1970). Avoidance learning in active and passive frogs and toads. *Journal of Comparative and Physiological Psychology, 70,* 154-156.

Brzoska, J. (1980). Quantitative studies on the elicitation of the electrodermal response by calls and synthetic acoustical stimuli in *Rana lessona* Camerano, *Rana r. ridibunda* Pallas and the hybrid *Rana esculenta L. (Anura, Amphibia). Behavioral Processes, 5,* 113-141.

Brzoska, J., Walkowiak, W., & Schneider, H. (1977). Acoustic communication in the grassfrog *(Rana t. temporaria L)*: Calls, auditory thresholds and behavioral responses. *Journal of Comparative Physiology, 118,* 173-186.

Capranica, R. R. (1965). *The evoked vocal response of the bullfrog: A study in communication by sound.* Cambridge: MIT Press.

Crawford, F.T. & Langdon, J.W. (1966). Escape and avoidance responding in the toad. *Psychonomic Science, 6,* 115-116.

Elepfandt, A., Seiler, B. & Aicher, B. (1985) Water wave frequency discrimination in the clawed frog, *Xenopus laevis. Journal of Comparative Physiology, 157,* 255-261.

Gerhardt, H.C., Allan, S. & Schwartz, J.J. (1990). Female green treefrogs (*Hyla cinerea*) do not selectively respond to signals with a harmonic structure in noise. *Journal of Comparative Physiology, 166,* 791-794.

Goldstein, A.C., Spies, G. & Sepinwall, J. (1964). Conditioning of the nictitating membrane in the frog, *Rana p. pipiens. Journal of Comparative and Physiological Psychology, 57,* 456-458

Hoffman, H.S. & Ison, J.R. (1980). Reflex modification in the domain of startle: I. Some empirical findings and their implications for how the nervous system processes sensory input. *Psychological Review, 87,* 175-189.

Hoyer, W.J., Shafer, J.N., Mauldin, T.E. & Corbett, H.T. (1971). Discriminated avoidance and escape conditioning with the tadpole (*Rana pipiens*). *Psychonomic Science, 24,* 247-248.

McGill, T.E. (1960). Response of the leopard frog to electric shock in an escape learning situation. *Journal of Comparative and Physiological Psychology, 53,* 231-238.

Megela-Simmons, A., Moss, C.F. & Daniel, K.M. (1985). Behavioral audiograms of the bullfrog *(Rana catesbeiana)* and the green tree frog *(Hyla cinerea). Journal of the Acoustical Society of America, 78,* 1236-1244.

Moss, C.F. (1985). *Hearing sensitivity and frequency selectivity in the green treefrog, Hyla cinerea.* Unpublished doctoral dissertation, Brown University.

Moss, C.F. & Simmons, A.M. (1986). Frequency selectivity of hearing in the green treefrog, *Hyla cinerea. Journal of Comparative Physiology, 159,* 257-266.

Reiter, L.A. & Ison, J.R. (1977). Inhibition of the human eyeblink reflex: An evaluation of the sensitivity of Wendt-Yerkes method for threshold detection. *Journal of Experimental Psychology: Human Perception and Performance, 3,* 325-336.

Simmons, A.M. (1988a). Selectivity for harmonic structure in complex sounds by the green treefrog (*Hyla cinerea*). *Journal of Comparative Physiology, 162,* 397-403.

Simmons, A.M. (1988b). Masking patterns in the bullfrog (*Rana catesbeiana*). I: Behavioral effects. *Journal of the Acoustical Society of America, 83,* 1087-1092.

Simmons, A.M., Buxbaum, R.C. & Mirin, M. (1993). Perception of complex sounds by the green treefrog, *Hyla cinerea*: envelope and fine-structure cues. *Journal of Comparative Physiology, 173,* 321-327.

Strother, W.F. (1962). Hearing in frogs. *Journal of Auditory Research, 2,* 279-286.

van Bergeijk, W.A. (1967). Anticipatory feeding behaviour in the bullfrog (*Rana catesbeiana*). *Animal Behaviour, 15,*231-238.

Weiss, B.A. & Strother, W.F. (1965). Hearing in the green treefrog (*Hyla cinerea cinerea*). *Journal of Auditory Research, 5,* 297-305.

Yaremko, R.M., Boice, R. & Thompson, R.W. (1969). Classical and avoidance conditioning of the nictitating membrane in frogs (*Rana pipiens*) and toads (*Bufo americanus*). *Psychonomic Science, 16,* 162-164.

Yaremko, R.M., Jette, J. & Utter, W. (1974). Further study of avoidance conditioning in toads. *Bulletin of the Psychonomic Society, 3,* 340-342.

Yerkes, R.M. (1905). The sense of hearing in frogs. *Journal of Comparative Neurology and Psychology, 13,* 124-137.

Young, J.S. & Fechter, L.D. (1983). Reflex inhibition procedures for animal audiometry: A technique for assessing ototoxicity. *Journal of the Acoustical Society of America, 73,* 1686-1693.

Methods in Comparative Psychoacoustics
G. M. Klump, R. J. Dooling, R. R. Fay & W. C. Stebbins (eds)

Phonotaxis in Female Frogs and Toads: Execution and Design of Experiments

H. Carl Gerhardt

University of Missouri, Division of Biological Sciences
215 Tucker Hall, Columbia, MO 65211, USA

Summary. This chapter discusses methods for testing phonotactically active female frogs and toads in playback experiments designed to assess the accuracy of sound localization or the selectivity for particular sound patterns. Procedures are suggested for inducing phonotaxis in captive females, for observing phonotactic behavior, establishing response criteria, and devising controls for motivational state and the effects of visual stimuli. Experimental design is discussed both in terms of arrangements of loudspeakers and collection and analysis of data required to address specific hypotheses.

1 Introduction

In many species of frogs and toads, the vocalizations of males elicit selective phonotaxis by females. Gravid females approach loudspeakers that emit appropriate sounds in the same manner as they approach calling males in nature; females choose between different acoustic stimuli that simulate choices that they face in nature. Acoustic differences representative of conspecific vocalizations and those of other, sympatric species typically mediate strong, intensity-independent preferences; female selectivity is also sufficiently well developed to serve as a source of sexual selection on the calls of conspecific males (e.g., Gerhardt, 1991).

The relative importance of selective phonotaxis in mate choice is influenced by breeding biology. In some species, female phonotaxis will be relatively unimportant because male frogs fight among themselves for oviposition sites or retreats used by females, and thus females may assess the quality of territories rather than, or in addition to, male vocalizations. In other species, some males actively search for females or dislodge other males that are clasping a female, thus negating any choice the female may have made based on call qualities. Moreover, the ability of females to make fine distinctions among males based on call variation may be limited by masking interference generated by the dense choruses typical of many species. Thus, even though females may show impressive selectivity in laboratory situations, predictions about the relative mating success of males whose calls have particular qualities may be compromised by the acoustic complexity of the environment. If the central question to be addressed by studies of female phonotaxis involves mate choice, then these biological factors should be taken into consideration in choosing a study species and designing experiments.

If an investigator is mainly interested in mechanisms of sound pattern discrimination, then the choice of a species will simply be based on the reliability with which female phonotaxis can be observed in the laboratory in simple experimental designs. However, phonotaxis in anurans is an unconditioned behavior that also depends on the reproductive state of the female and test conditions. Thus, a failure to observe phonotaxis in a playback experiment does not prove that the frog cannot hear the sound nor does it rule out the possibility that the stimulus could be effective under other conditions. Similarly, if females do not show a preference for one sound pattern over another pattern, we can not conclude anything about their ability to distinguish between the two patterns. In other words, until methods of conditioning frogs to show phonotaxis are developed, we can safely characterize minimum differences that result in selective phonotaxis as "just meaningful differences" rather than "just noticeable differences"(Nelson and Marler, 1990).

In this chapter I discuss methods for studying phonotaxis in females, with an emphasis on the execution of experiments. I briefly discuss experimental questions and the analysis and interpretation of data. I do not consider procedures for generating and varying the properties of synthetic vocalizations nor I provide details about stimulus control. However, the reader should be aware that varying one acoustic property by necessity often causes variation in one or more additional properties and that the acoustics of the testing environment are as important as the playback system. These issues are discussed in Gerhardt (1992).

2 Sources and Treatment of Female Frogs and Toads

2.1 Collecting Females in the Field

In many species of frogs and toads females ovulate and oviposit on the same night. Once ovulation is well underway, the female will lay her eggs even in the unlikely event that she does not find a conspecific male. The most reliable way to obtain phonotactically active females is to collect them as a member of a mating (amplexing) pair in the field. Ovulation usually will have begun in most females, which are likely to remain responsive to playbacks until they have oviposited. If the frogs cannot be tested conveniently on the night of capture, refrigeration at about 4°C will inhibit oviposition for one to three days. However, the longer the delay between the time of capture and the time of testing, the lower the probability that a female will show phonotactic behavior. Gravid females collected before mating sometimes respond phonotactically. One important question, which has not yet been formally addressed, is whether gravid females in the early stages of ovulation and phonotactic responsiveness are more selective than females that have completed ovulation and must soon lay eggs.

In some species, especially those in the genus *Rana*, oviposition is not tightly coupled to ovulation, and these frogs appear to be more sensitive than other frogs to handling and visual stimuli. Thus, not every species is amenable to the study of female phonotaxis, at least in the laboratory.

2.2 Inducing Phonotaxis in Frogs and Toads

Procedures that cause ovulation in gravid females may also produce a state of phonotactic responsiveness. These methods include injection of anuran pituitaries (Rugh, 1948) or human chorionic gonadotropin and progesterone (Schmidt, 1969). Ovulation, which is typically indicated by the observation of abdominal contractions, occurs between 9 to 24 hr after injection of the hormones, depending on the time of year and temperature (see Schmidt, 1969 for details for *Hyla cinerea* and *Bufo americanus*). Females must have ripe eggs in their ovaries for these procedures to be effective. Animals that have recently ovulated and oviposited should be fed heavily for one to three months before attempting to induce ovulation and phonotaxis. The accumulation of ripe ova can be observed through the skin of the inguinal area in some species of treefrogs.

Schmidt (1985) describes a protocol for inducing phonotactic behavior in non-gravid anurans. In the toad, *B. americanus*, the most successful treatment consisted of injections of arginine vasotocin and prostaglandin F_2 about 19-22 hours after a priming dose of progesterone. The combination of progesterone and prostaglandin was effective in inducing phonotactic responses in gray treefrogs (*Hyla chrysoscelis* and *H. versicolor*), and no difference was detected in the selectivity of females collected in the field and tested prior to oviposition and that of females induced to respond with progesterone and prostaglandin (Gerhardt, 1994). Although no formal study has been conducted, this method appears to be effective from late Spring through late summer. In both toads and treefrogs, females began responding to playbacks within a few minutes to about an hour after the injection of prostaglandin.

3 Experimental Methods and Procedures

3.1 Test Procedures, Response Criteria, and Measurements of Phonotactic Accuracy

Females are usually placed by hand in an acoustically transparent container, such as a hardware cloth cage, until the acoustic stimuli have been started. Because they often orient and move rapidly to the first attractive sounds they hear, females should be restrained for several repetitions of alternative stimuli before being released. An arrangement as simple as a string attached to the top of the releasing cage and run through a pulley can allow the researcher to free the female without moving near her. In two-speaker tests, females should be placed so that they face neither speaker.

Although most researchers define a response as an approach to within some predetermined distance of a loudspeaker, I recommend that phonotactic orientation movements also be used as a response criterion. Such movements are obvious: the female turns her head, her whole body, or both during or just after the emission of an acoustic stimulus. Females usually take a zig-zag path toward a loudspeaker (e. g., Rheinlaender et al., 1979) and typically make relatively small hops. Some females crawl during part of their approach. Long, fast hops or seemingly random wandering are typical of non-responsive females and yet, by chance, may bring the frog near a speaker.

Frogs are very sensitive to visual stimuli over a wide range of wavelengths (Backwell & Passmore, 1990; Buchanan, 1993). Individuals that are not highly responsive to sounds often show obvious phototactic (positive or negative) orientation movements, and they may crawl or hop around an object such as a speaker that is not emitting appropriate sounds (e.g., Gerhardt et al., 1994). Nevertheless, females usually show reliable phonotactic behavior while illuminated by a dim flashlight or other light sources. It is important to place the source of illumination so that it does not initially guide the female towards or away from a loudspeaker. In multiple-speaker designs, overhead lighting of speakers of nearly identical appearance is preferable. In two-speaker experiments in the field, the observer's position and source of illumination should be at right angles to an imaginary line connecting the loudspeakers and passing through the release point. Similarly, the loudspeakers and release point should be oriented so this imaginary line is also perpendicular to any other extraneous light sources (e.g., moonlight) in the natural environment. Females of some species (e.g., spring peepers, *Pseudacris crucifer*) are much more likely to respond phonotactically under infrared rather than visible light. Indeed, using infrared illumination to observe phonotactic behavior is a good idea for observing even species that respond well in visible light because there may always be some subtle, undetected bias in lighting that could influence female choices in tests where preferences are weak.

If the paths taken by females and the extent of orientation movements such as head and body scanning are of interest, then a grid of squares can be drawn on the substratum. The positions of the frog relative to an active loudspeaker can be sketched on a data sheet, videotaped, preferably using an infrared-sensitive camera, or entered as coordinates in a data acquisition software program (e.g., Jørgensen & Gerhardt, 1991). If infrared viewing or recording equipment is unavailable, then a single loudspeaker should be hidden behind some kind of acoustically transparent material or an additional, inactive loudspeaker of nearly identical appearance should be placed in an equally visible position in the arena. This procedure provides some control for the possibility that females merely orient and move toward a visually detectable object whenever attractive sounds are played back from anywhere in or near the testing arena. However, it does not eliminate the possibility that the combination of sounds plus a visual target at the same location will be more attractive or more accurately located than an invisible sound source. That females of arboreal species use visual cues in phonotactic approaches was obvious in experiments in which the frogs were observed moving within a 3-D grid of sticks or elevated perches (e.g., Backwell & Passmore, 1990; Jørgensen & Gerhardt, 1991).

In two-speaker designs, side biases may also be introduced in experiments conducted outdoors by slopes or wind within the playback area. Asymmetries that are difficult to detect or control may also occur indoors, and some females may have an inherent tendency to move in one direction or another. If a preference is strong, then side biases and other extraneous factors will have little influence on the results of an experiment. But if the preference is subtle, and the bias strong, then females will continue to move to the same speaker position even when alternative stimuli are switched between speakers. In other words, the bias will be obvious, but a preference cannot be convincingly demonstrated.

3.2 Body Temperature

Anurans are poikilothermous, and temporal properties of male calls are usually correlated with the male's body temperature. Female preferences for particular values of such properties may shift with temperature in a roughly parallel fashion (Gerhardt, 1978). Even though changes in temperature usually cause only small percentage changes in spectral properties such as dominant frequency, female preferences for dominant frequency may be dramatically affected by temperature. For example, in the green treefrog (*H. cinerea*) cooling females to about 18°C reversed their preference at normal breeding temperatures (about 24°C) for a low-frequency spectral peak of 900 Hz (typical of conspecific males) over a low-frequency peak of 500 Hz (Gerhardt & Mudry 1980). Temperature had only a slight effect on preferences for different values of the high-frequency spectral peak (2700-3600 Hz). Stieber & Narins (1990) found neurophysiological correlates of these behavioral results: tuning properties of neurons innervating the frog's amphibian papilla (most sensitive to relatively low-frequency sounds) shifted strongly with changes in temperature; tuning properties of neurons innervating the basilar papilla (most sensitive to relatively high-frequency sounds) were only slightly affected by changes in temperature.

These results serve to emphasize the importance of knowing the range of temperatures over which mating takes place in the study species and controlling the temperature during tests of selective phonotaxis. I recommend acclimating females to the target temperature for at least thirty minutes before testing. If a temperature-regulated testing arena is available, adjusting it to about 2°C above the target temperature usually results in a body temperature that is very close to the target temperature, depending on the relative humidity. Frogs lose water through their skin and when the air is relatively dry, evaporative cooling can be substantial. In any event, water should be sprinkled at the release point and nearby areas to reduce dehydration of the subjects.

4 Experimental Designs: Number and Arrangement of Loudspeakers

4.1 Single-Stimulus Experiments

In addition to exploring mechanisms of sound localization by quantifying phonotactic accuracy (e.g., Klump & Gerhardt, 1989; Klump, this volume), single-speaker tests often are used to learn if some stimulus elicits phonotaxis at all or to estimate a threshold sound pressure level (SPL) for phonotaxis to a stimulus (e.g., Gerhardt, 1981; Gerhardt & Klump, 1989). Because, as mentioned above, an animal might fail to respond because it has laid most of its eggs rather than because a stimulus is unattractive, one control procedure is to follow any "no response" by the playback of a stimulus known to be highly attractive. If the female responds vigorously to this stimulus, then the failure to respond to previous stimulus is unlikely to have been caused by the frog's motivational state, and the researcher can confidently score the previous trial as a "no response" (e.g., Gerhardt & Klump, 1989; Ryan & Rand, 1993).

Because the probability of a response is only loosely correlated with the female's reproductive state, which is always changing, and the effects of handling are uncertain, response latencies are likely to vary considerably from animal to animal and between different tests of the same animal. This means that if two different stimuli are similar in their relative attractiveness, then many trials with a large number of females will be needed to show that the mean latency of response is different for the two alternatives. Two-speaker, simultaneous choice designs are much more efficient and powerful in detecting differences in relative attractiveness.

4.2 Multiple-Stimulus Experiments

a) Conventional design

In the most widely used design, females are released midway between two speakers, each of which broadcasts a different acoustic stimulus (Fig. 1a). The female chooses one of the sounds by the criteria discussed above. Females that fail to respond provide no information about the relative attractiveness of the stimuli. As a check for side biases, the alternative stimuli are switched between speakers so that about one-half of the females tested will experience each configuration. It is critical to record the acoustic stimuli at the release point of the females; analyses of these recordings will detect non-linearities in the playback system and arena acoustics and indicate how well the two speakers are matched. It is especially important that the two channels of the system have equivalent amounts of distortion, reverberation, background noise, and non-linearities in the frequency-response (see details and examples in Gerhardt, 1992).

The usual procedure is to adjust the sound pressure level (SPL) of alternative stimuli to the same value at the release point. Gerhardt (1992) provides some suggestions concerning the choice of a meter-time-constant in relation to the temporal structure of the acoustic stimuli. Particular care must be taken when the alternative stimuli differ significantly in their rate of amplitude modulation or duration. After an experiment has established a preference for one stimulus over another stimulus, the strength of the preference can be established by lowering the SPL of the preferred sound until the preference is abolished or reversed. Experiments with the green treefrog have established that preferences can be affected by the SPL to which the alternatives are equalized, the relative SPL of the two alternatives or both (Gerhardt, 1987).

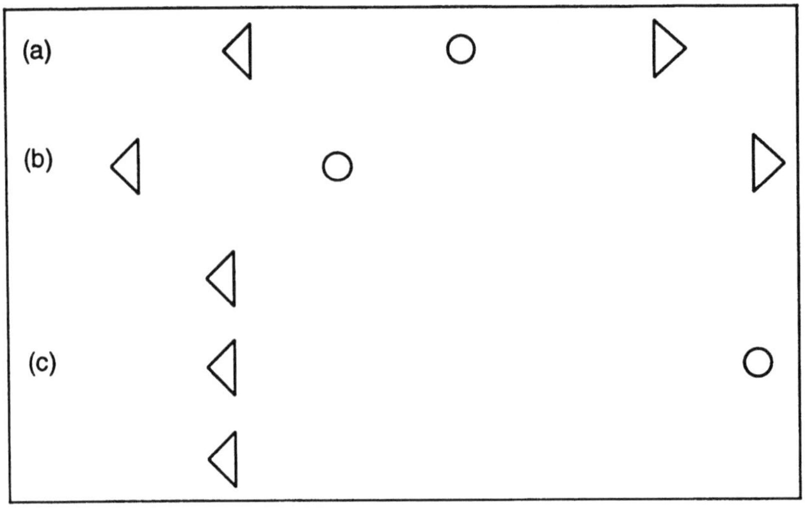

Figure 1. Arrangements of speakers (triangles) relative to release points of females (circles) in (a) conventional design; (b) design in which a difference in playback level is created by changing the release point; and (c) design which more realistically represents the spatial relationship between a female and calling male frogs in nature.

An example of the last effect was observed when females had a choice of sounds with low-frequency spectral peaks of 600 and 900 Hz; the SPL of the latter stimulus was 12 dB less than that of the 600 Hz-call. When the stimulus of 600 Hz was adjusted to 75 dB and that of the 900 Hz-stimulus, to 63 dB, most females chose the stimulus of 900 Hz. However, when the 600 Hz-call was at 85 dB, and the 900 Hz call at 73 dB, most females chose the stimulus of 600 Hz.

Differences in SPL can also be created by moving the release point closer to one of the speakers (Fig. 1b), each of which broadcasts a stimulus at the same SPL at a fixed distance. Experiments with gray treefrogs (*Hyla versicolor*) indicate that the effects of differences in SPL generated by distance are similar to those generated by differential amplification of the sounds when the release point is equi-distant from the two speakers.

b) Designs influenced by spatial distributions of males and females in nature

In early experiments, females were placed in the center of an array of four speakers, each separated by 90° with respect to the starting point; the selectivity of females of the green treefrog was reduced somewhat compared with the simpler two-speaker design (Gerhardt, 1987). This result led to the hypothesis that the more complex acoustic environment of a treefrog was reduced somewhat compared with the simpler two-speaker design (Gerhardt, 1987). This result led to the hypothesis that the more complex acoustic environment of a typical chorus reduced the ability of females to choose among conspecific calls that do not differ greatly in their acoustic characteristics. Field and further experimental studies have generally supported this idea (e.g., Dyson & Passmore, 1988), and females may even have difficulty in detecting a single calling male if the SPL of his calls do not at least equal that of the chorus background noise (Gerhardt & Klump, 1988). Phonotactic accuracy was unimpaired by acoustic complexity of the background in one species (Backwell & Passmore, 1991).

Most phonotactic approaches in nature do not start from a point between two calling males. Instead, the angle between two males initially will probably be much less than 180 , depending on the typical spacing between males around, say, the edges of a pond and the distance at which the female begins her assessment. Obviously the angle between males will increase as the female moves closer to the calling males. Designs in which the angle between a pair of speakers can be varied relative to the release point provide an opportunity not only to assess mate choice in a more natural spatial arrangement, but also to explore the potential interplay between mechanisms of discrimination and sound localization (Fig. 1c). Directional hearing can improve the detectability of signals in broadband noise (Schwartz & Gerhardt, 1989) or, more significantly, reduce the masking interference of fine-temporal properties that occurs when the calls of neighboring frogs overlap (Schwartz & Gerhardt, in preparation).

5 Experimental Designs: Hypotheses, Data Analysis, and Interpretation

If the goal of a single-stimulus experiment is to estimate phonotactic thresholds, then the data consist of the proportions of females that respond to a given stimulus as a function of playback SPL. Again, it is important to test the motivational state of any female that does not respond to the test stimulus. Ryan and Rand (1993) conducted a simple statistical test of phonotactic potency by comparing the proportion of females that responded to the stimulus of interest with the proportion that responded to a non-specific, presumably unattractive, acoustic stimulus such as white noise.

In most studies, females have been tested only once with the same pair of stimuli, a procedure that guarantees statistical independence of the data. The results can be tested with a binomial or sign test, usually against a null hypothesis of $p = q = 0.5$. Such tests are so conservative that rejection of the null hypothesis at the 0.05-level with the relatively small sample sizes usually available (N = 10-20) represents strong evidence that the females have shown a preference. However, if the results are not so clear cut, additional animals should be tested in order to obtain an estimate of the proportions (and confidence limits) of females preferring one, or the other stimulus. Indeed, a series of such estimates for pairs of stimuli that differ in a single variable, whether or not they are statistically significant in a binomial test, are often useful in generating functions such as "behavioral tuning curves" (e.g., Gerhardt, 1987). The recommendation that a researcher stop testing additional females when a preference has obviously been demonstrated (e.g., when all 10 females choose one stimulus over its alternative) and continue to test when a preference is not so well established runs counter to the usual dictate of classical statistics that the sample size be fixed beforehand. However, there is no such requirement if the researcher's aim is to estimate the magnitude of the preference, or even to test hypotheses about preference within the philosophical framework of Bayesian statistics. A full discussion and suggested readings are provided in Gerhardt (1992). The main conclusion is that when sample sizes are small, as they usually are in tests of selective phonotaxis in anurans, and no strong preference has been demonstrated, then there will be always be some degree of uncertainty about the existence of a preference (i.e., in terms of classical statistics, the probability of a Type II error is large). At the very least, a researcher should summarize the data in such a way that readers can drawn their own conclusions.

In many experiments, females do not all choose the same stimulus, thus suggesting that there is within-population variation in preference. However, this could also be caused by handling effects, side biases or other experimental errors. A more rigorous demonstration of phenotypic variability requires multiple tests of the same animal, with the goal of estimating the proportions of females that consistently choose each alternative (Gerhardt, 1991). One example of such a design would consist of ten tests of each female with the same pair of stimuli, thus generating a score of from 0 to 10; after an interval of several weeks, the females would again be tested ten times each with the same pair of sounds. The scores of a sample of females in their two sets of tests would generate data that could be analyzed with a nested analysis of variance, which partitions variance into within- and between-female components. The between-female component is an estimate of repeatability, which sets an upper bound on heritability. Butlin (1993) used another design, based on single-stimulus playbacks, to estimate variation in female preference in an insect species. Gerhardt (1981) and Butlin (1993) also discuss ways to estimate and minimize carry-over effects during a multiple-test session with an individual female.

Finally, it is always critical to choose the appropriate sampling unit for the hypothesis being tested. For example, if the vocalizations of a species show geographical variation, then to assess the behavioral significance of such variation, a researcher would need to sample females from a number of populations from each region of interest and to use a number of different exemplars (or synthetic calls based on natural geographical variation) in order to draw general conclusions. Stated another way, if fifty females from one population chose a representative call of a male from the same population over a representative call of male from another, distant population, the data would show only that the preference exists for those particular calls and for that particular population of females. With respect to the hypothesis of interest, the sample size would be one and not fifty. A full discussion of this and other problems involving pseudoreplication is found in McGregor (1992).

6 Acknowledgements

I thank S. D. Tanner, J. J. Schwartz, G. M. Klump, and G.F. Watson for helpful comments on the manuscript. My research has been funded generously by the National Science Foundation and National Institutes of Mental Health.

7 Suggested Reading

Fritzsch, B., Ryan, M.J., Wilczynski, W., Hetherington, T.E. & Walkowiak, W.(1988). *The evolution* of *the amphibian auditory system.* New York, Chichester, Brisbane, Toronto, Singapore: John Wiley & Sons.

Gerhardt, H.C. (1991). Female mate choice in treefrogs: static and dynamic acoustic criteria. *Animal Behaviour, 42*, 615-635.

McGregor, P.K. (1992). *Playback and studies of animal communication*. New York: Plenum Press.

8 References

Backwell, P.R.Y. & Passmore, N.I. (1990). Suitable perches affect female phonotaxis in an arboreal frog. *Herpetologica, 46*, 11-14.

Backwell, P.R.Y. & Passmore, N.J. (1991). Sonic complexity and mate localization in the leaf-folding frog, *Afrixalus delicatus. Herpetologica, 47*, 226-229.

Buchanan, B.W. (1993). Effects of enhanced lighting on the behaviour of nocturnal frogs. *Animal Behaviour, 45*, 891-899.

Butlin, R.K. (1993). The variability of mating signals and preferences in the brown planthopper, *Niaparva lugens* (Homoptera: Delphacidae). *Journal of Insect Behavior, 6*, 125-140.

Dyson, M.L. & Passmore, N.I. (1988). Two-choice phonotaxis in *Hyperolius marmoratus* (Anura: Hyperoliidae): the effect of temporal variation in presented stimuli. *Animal Behaviour, 36*, 648-652.

Gerhardt, H.C. (1978). Temperature coupling in the vocal communication system of the gray treefrog (*Hyla versicolor*). *Science, 199,* 992-994.

Gerhardt, H.C. (1981). Mating call recognition in the green treefrog (*Hyla cinerea*): importance of two frequency bands as a function of sound pressure level. *Journal of Comparative Physiology A, 144,* 9-16.

Gerhardt, H.C. (1987). Evolutionary and neurobiological implications of selective phonotaxis in the green treefrog, *Hyla cinerea. Animal Behaviour, 35,* 1479-1489.

Gerhardt, H.C. (1991). Female mate choice in treefrogs: static and dynamic acoustic criteria. *Animal Behaviour, 42,* 615-635.

Gerhardt, H.C. (1992). Conducting playback experiments and interpreting their results. In: P.K. McGregor (Ed.), *Playback and studies of animal communication.* New York: Plenum Press, pp. 59-77.

Gerhardt, H.C. (1994). Reproductive character displacement of female mate choice in the gray treefrog, *Hyla chrysoscelis. Animal Behaviour, 47,* 1490-1503.

Gerhardt, H.C., Dyson, M.L., Tanner, S.D. & Murphy, C.G. (1994). Female treefrogs do not avoid heterospecific calls as they approach conspecific calls: implications for mechanisms of mate choice. *Animal Behaviour, 47,* 1323-1332.

Gerhardt, H.C. & Klump, G.M. (1988). Masking of acoustic signals by the chorus background noise in the green treefrog: a limitation on mate choice. *Animal Behaviour, 36,* 1247-1249.

Gerhardt, H.C. & Klump, G.M. (1989). Phonotactic responses and selectivity of barking treefrogs (*Hyla gratiosa*) to chorus sounds. *Journal of Comparative Physiology A, 163,* 795-802.

Gerhardt, H.C. & Mudry, K.M. (1980). Temperature effects on frequency preferences and mating call frequencies in the green treefrog, *Hyla cinerea* (Anura: *Hylidae*). *Journal of Comparative Physiology, 137,* 1-6.

Jørgensen, M.B. & Gerhardt, H.C. (1991). Directional hearing in the gray treefrog *Hyla versicolor*: eardrum vibrations and phonotaxis. *Journal of Comparative Physiology A, 169,* 177-183.

Klump, G.M. & Gerhardt, H.C. (1989). Sound localization in the barking treefrog. *Naturwissenschaften, 76,* 35-37.

Nelson, D.A. & Marler, P. (1990). The perception of birdsong and an ecological concept of signal space. In: W.C. Stebbins & M.A. Berkley (Eds.), *Comparative perception. Vol. II: complex signals.* New York: John Wiley & Sons, pp. 443-478.

Rheinlaender, J., Gerhardt, H.C., Yager, D.D. & Capranica, R.R. (1979). Accuracy of phonotaxis by the green treefrog (*Hyla cinerea*). *Journal of Comparative Physiology, 133,* 247-255.

Rugh, R. (1948). *Experimental embryology.* Minneapolis: Burgess.

Ryan, M.J. & Rand, A.S. (1993). Species recognition and sexual selection as a unitary problem in animal communication. *Evolution, 47,* 647-657.

Schmidt, R.S. (1969). Preoptic activation of mating call orientation in female anurans. *Behaviour, 35,* 114-127.

Schmidt, R.S. (1985). Prostaglandin-induced mating call phonotaxis in female American toad: facilitation by progesterone and arginine vasotocin. *Journal of Comparative Physiology A, 156,* 823-829.

Schwartz, J.J. & Gerhardt, H.C. (1989). Spatially mediated release from auditory masking in an anuran amphibian. *Journal of Comparative Physiology A, 166,* 37-41.

Stieber, I.B. & Narins, P.M. (1990). Temperature-dependence of auditory nerve response properties in the frog. *Hearing Research, 46,* 63-82.

Methods in Comparative Psychoacoustics
G. M. Klump, R. J. Dooling, R. R. Fay & W. C. Stebbins (eds)

Studying Sound Localization in Frogs with Behavioral Methods

Georg M. Klump

Institut für Zoologie, Technische Universität München
Lichtenbergstrasse 4, 85748 Garching, Germany

Summary. The chapter describes how the natural phonotactic behavior of frogs and their orienting responses towards a sound source can be used to measure the acuity of sound localization. First, some basic considerations in designing sound-localization experiments in frogs are discussed. Two examples of behavioral sound localization experiments are then given that demonstrate the implementation of such studies.

1 Introduction

In the mating season, frogs of various species form large aggregations in which the males emit species-specific advertisement calls to attract females that are ready to mate. In the cacophony produced by a chorus of males, female frogs have to find potential mates guided only by the acoustic signals, since mating in many species takes place at night when visual cues are absent. Female frogs are well adapted to analyze the acoustic features of the males' vocalizations (e.g., to distinguish between males of different species or choose among conspecific males; see Gerhardt, this volume) and to locate calling conspecifics. This latter ability has attracted the attention of a number of researchers for the same reason that made the study of sound localization in small birds so interesting, the characteristics of the signals make it appear unlikely that the mechanisms used to locate the sound source are the same as those described in mammals (see Brown, this volume). Since frogs use signals with a wavelength that is frequently more than an order of magnitude larger than the distance between their ears (which also lack pinnae and have tympana that are flush with the body surface), the operation of additional mechanisms that enhance interaural cues used in localization has been postulated (e.g., pressure-gradient receiver mechanisms, see reviews by Eggermont, 1988, and by Jørgensen and Gerhardt, 1991). Behavioral studies have explored the acuity of sound localization in frogs to provide a reference for physical and physiological studies that concentrate on the mechanisms of localization (e.g., see review by Eggermont, 1988, and see Schmitz et al., 1992).

Sound-localization studies in frogs have exploited natural orienting responses towards a sound source and phonotaxis, i.e., the approach behavior of individuals to a sound source that is, for example, broadcasting the species' advertisement or territorial calls. Depending on the species, both sexes or only one sex may show phonotactic behavior. Male green treefrogs (*Hyla cinerea*), for example, approach other vocalizing males and silently sit close to these active males in an attempt to intercept incoming females (Perill, Gerhardt and Daniel, 1978). In some species, males are territorial and will approach another caller in their own territory to drive it away (e.g., in the dendrobatid frog *Colostethus nubicola*, see Gerhardt and Rheinlaender, 1980). Most commonly, however, phonotaxis has been observed in females that locate partners for mating in a chorus by using advertisement calls of conspecific males. In the green treefrog, for example, females that have ovulated and are ready to mate approach the male guided by his advertisement calls. In many frog species, phonotaxis by females can be readily elicited by collecting pairs in amplexus and then separating the female from the male (for this and other methods to elicit phonotaxis see the chapter by Gerhardt, this volume).

This chapter will describe phonotaxis experiments for studying the acuity of sound-localization in frogs. After discussing some basic questions of the experimental design of such studies, two examples will be given in more detail that elucidate different species' abilities to locate a sound source.

2 Basic Design Principles of Phonotactic Sound-localization Experiments in Frogs

Sound-localization experiments in frogs are generally conducted by observing the responses of a male or female to attractive sounds presented in an arena in which the acoustics and other environmental factors (e.g. lighting, temperature, etc.) can be controlled or, at least, can be well described. The requirements for this type of experiment are the same as for phonotactic experiments designed to study preferences for certain call parameters (e.g., those used in species recognition). The reader is referred to the chapter by Gerhardt (this volume) who discusses the general prerequisites for experiments involving phonotaxis. Here I will focus on topics that are especially important for sound-localization experiments.

2.1 Open-loop and Closed-loop Stimulation

The goal of a sound-localization experiment determines the appropriate experimental design. If the goal is simply to assess the animal's sensory capabilities when stimulated in a natural fashion, a sound-localization experiment with a closed-loop presentation of the signals is usually conducted. In a closed-loop presentation, the sound stimulus persists throughout the animal's orienting response. This enables the subject to use a number of strategies to localize the source that are indistinguishable to the experimenter. The animal could, for example, determine the angle of sound incidence instantaneously and then turn towards the sound source through the same angle. Alternatively, the animal could use a lateralization mechanism to determine the direction from which the sound signal originates; it could simply turn until the sound source is perceived as coming from the other side of the body axis. Despite the difficulties in interpreting the results, the closed-loop presentation of sound signals is commonly used because the repeated or ongoing stimulation more readily elicits a response from the experimental subject. If, however, the study of the mechanisms that underlie the ability to localize a sound source is the goal, a different experimental design with an open-loop presentation of the stimulus is more suitable. In the open-loop presentation, the sound is terminated before the subject starts an orienting response. This mode of presentation limits the strategies that are available to the subject to localize the sound source (e.g., the animal cannot simply turn until interaural differences are at a minimum). Thus, only experiments with an open-loop stimulation can unambiguously determine a subject's ability to perceive the angle of sound incidence directly.

2.2 Determining the Accuracy of Localization

In sound-localization experiments that exploit phonotaxis behavior, the animal is first positioned in an arena at a specific release point (in some species it is confined in a small release cage that can be opened by the experimenter from a remote position, e.g., with the help of a pulley). The release point can either be in the center or at one side of the arena. If phonotaxis is studied in three dimensions, the release point may be on a small platform elevated above the ground. A sound signal (e.g., an acoustic replica of a natural call or other signals that elicit phonotaxis) is then broadcast and the behavior of the animal is either observed directly by the experimenter, or is recorded with a video camera and analyzed afterwards using single video frames. To make a quantitative analysis of the frog's movements easier, the arena is usually marked with a regular grid that allows one to determine the position of the animal more accurately. If the accuracy of localization is only studied in the horizontal plane, this grid is marked out on the floor of the arena; if the localization is studied in three dimensions, the arena consists of an array of vertical sticks that are positioned at the intersections of a regular quadratic grid on a horizontal ground plane (e.g., formed by a plywood board).

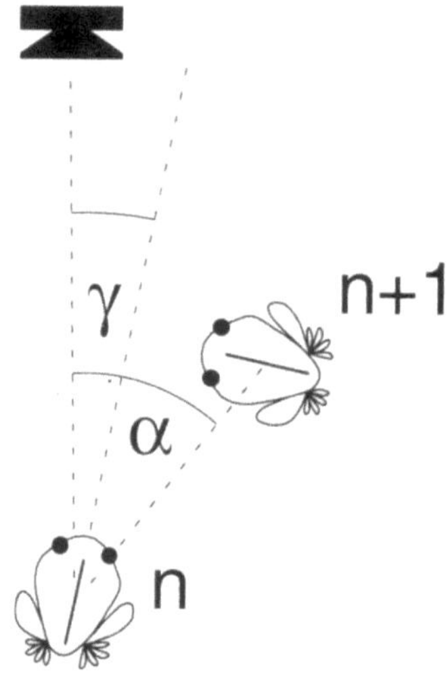

Figure 1. The definitions for the different error angles: (α) jump error angle when moving from position n to position n+1, (γ) orientation error angle.

A typical phonotactic approach consists of a series of head-scanning movements and other orienting responses that are observed before the frog leaves its position in the arena, and crawls and hops by which the frog moves from its current position in the arena to a new position. Three different measures have been used to describe the accuracy of localization in the phonotactic approaches. First, the angle between the longitudinal axis of the frog's head and the angle of sound incidence has been determined before every movement on the grid (this angle is termed the head orientation error angle or scanning error angle). Second, the angle between the longitudinal axis of the frog's body and the angle of sound incidence has been determined in species that orient with their whole body to the sound source (this angle is termed the orientation error angle, γ , see figure 1). Most frequently a third measure was used to determine the accuracy of localization: the angle between the vector that is defined by the frog's movement from the current to a the new position and the vector connecting the position of the frog at the beginning of its advance and the position of the speaker broadcasting the sound (this angle is termed the jump error angle, α , see figure 1). In a two-dimensional arena, the measures for the orientation error angle and the jump error angle are relatively easy to take; the head-orientation angle may be more difficult to observe. In three dimensions, only the jump error angle θ has been determined by measuring the distances between the starting and stopping position of the frog (d_1), between the starting point of the frog and the position of the sound source (d_2) and between the stopping position of the frog and the position of the sound source (d_3):

$$\theta = \arccos\left(\frac{d_1^2 + d_2^2 - d_3^2}{2d_1 d_2}\right) \qquad (1)$$

The distances can be calculated using the coordinates of the respective locations; e.g., the distance d_1 between the starting position $P_0(x_0, y_0, z_0)$ and the stopping position $P_1(x_1, y_1, z_1)$ is given by:

$$d_1 = \sqrt{(x_1 - x_0)^2 + (y_1 - y_0)^2 + (z_1 - z_0)^2} \qquad (2)$$

The other distances can be calculated accordingly. A more detailed derivation of the formula for the error angle is given in a publication by Passmore, Capranica, Telford and Bishop (1984). Horizontal and vertical jump error angles can be calculated separately by first projecting the frog's and the speaker's positions upon the appropriate two-dimensional plane and then applying the formula (1). The accuracy of sound localization is then described by the distribution of the error angles.

2.3 How to Determine the Frog's Motivation to Approach the Sound Source

The results of sound localization experiments in which phonotaxis of frogs is exploited depend critically upon the motivation of the experimental subjects to approach the sound source. If, for example, a sound signal is unattractive to a frog because it lacks species-specific features used in the recognition of potential mates, the experimental subject may not approach the sound source but move through the arena in search of a place to hide. If the locatability of different signals is compared, ideally they should be of similar attractiveness to the frogs. To assure that this is the case, the relative attractiveness of the signals could first be tested in a two-alternative choice experiment with two speakers (see chapter by Gerhardt) before their locatability is explored. If signals cannot be matched in relative attractiveness, their ranking in an attractiveness test should be compared with the their ranking in terms of locatability (e.g., see Jørgensen and Gerhardt, 1991).

In general, in the design of sound localization experiments the signals should either be of about equal sensation level (i.e., at a constant level above the auditory threshold for the respective signals), or the signal components should be at least far enough above the species' auditory thresholds so that the sensation level probably is of little importance in determining the localization performance. The sound-pressure level should be high enough to elicit a steady phonotactic approach. Usually this is the case if stimulation levels of about 85 dB are used that correspond to a calling male at a distance of about 1 to 4 m (for sound-pressure levels of calling treefrog males see Gerhardt, 1975).

It is a good practise in a sound-localization experiment involving phonotaxis to get some measure of the frogs interest in the sound source that either is independent of the localization accuracy or inversely related to it. In the latter case one can at least conclude that the frog was interested in the signal, but could not localize it accurately. The frequency of scanning or orienting responses before making a decision about an approach trajectory may be such a measure. Klump and Gerhardt (unpublished data) observed in green treefrogs, that differences between the locatability of signals of various spectral content were considerably reduced if only those runs of frogs were included in the analysis in which the experimental subjects showed a minimum of three scanning movements rather than leaving the arena without scanning. This condition insured that only subjects that were motivated to locate the source were included in the analysis.

3 An Example for a Closed-loop Localization Experiment

Behavioral studies of closed-loop sound localization have been conducted in a number of frog species both in two dimensions on a horizontal plane (see review by Rheinlaender and Klump, 1988) as well as in three-dimensional grids (see Gerhardt and Rheinlaender, 1982, Passmore et al., 1984,. and Jørgensen and Gerhardt, 1991). In such experiments the localization accuracy is, for example, determined from the distribution of jump error angles accumulated in a number of phonotactic approaches (for other measures see above). Figure 2 shows a distribution of jump error angles that were measured by observing the phonotactic approaches of gray treefrog females in a three-dimensional grid towards a speaker broadcasting synthetic advertisement calls (the data in this example stem from a study in *Hyla versicolor* by Jørgensen

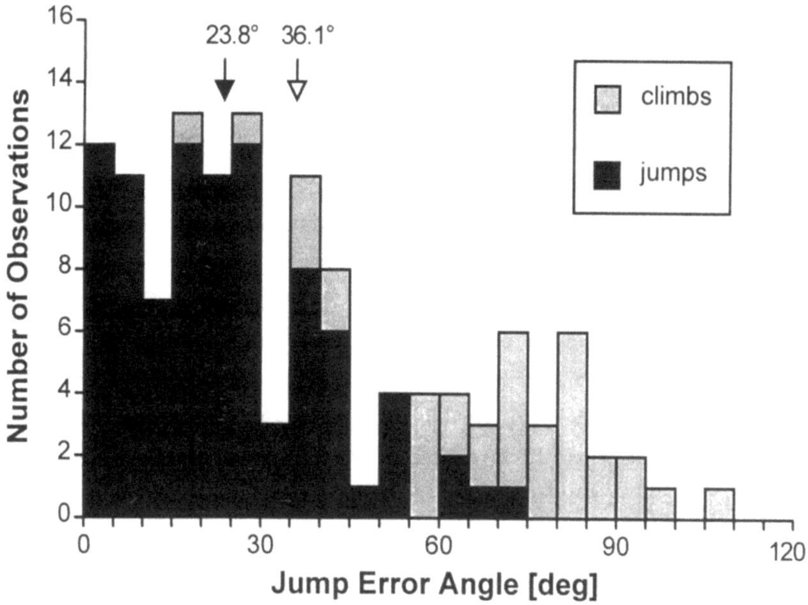

Figure 2. The distribution of of jump error angles determined when observing gray treefrogs performing phonotaxis in a three-dimensional grid (drawn after Jørgensen and Gerhardt, 1991). The arrows point at the mean jump error angles that were measured with the climbs included (open arrow) or excluded (filled arrow).

and Gerhardt, 1991). Jørgensen and Gerhardt constructed the grid out of 36 vertical wooden dowel sticks (6 rows of 6 each) that were 110 cm long and emerged from the intersections of a regular square pattern (square size 20 cm * 20 cm) on a quadratic plywood base plate. The sticks were marked in 10 cm intervals. The base plate was covered with sound-absorbing foam rubber (Sonex) to reduce echoes. The frogs started their phonotactic approach from a small release platform (10 cm * 5 cm) that was mounted 10 cm in front of the middle of one side of the grid at a height of 50 cm. The speaker was mounted 10 cm behind the opposite side of the grid at a height of 100 cm. It played the synthetic calls with an amplitude that produced a sound-pressure level of 85 dB at the position of the female which is similar to the level produced by a calling male at a distance of about 2 m (in most experiments Jørgensen and Gerhardt adjusted the sound pressure level after every change of the frog's position using a computer-controlled attenuator) Since in the gray treefrog advertising males often call from perches in trees or bushes, three-dimensional localization of sound sources is a natural task for females.

During the phonotactic approach, Jørgensen and Gerhardt (1991) recorded the position of the frogs after all jumps or climbs on the sticks in three-dimensional coordinates and calculated the jump error angles. During the recording of the female's position, the sound was turned off which reduced the likelihood that the frog would move to another position. The data shown in figure 2 summarize the phonotactic approaches when the females were stimulated with a synthetic call having spectral energy at 1.1 and 2.2 kHz which are the dominant frequencies in the natural advertisement calls of males. Two observations by Jørgensen and Gerhardt (1991) should be mentioned, since they show how the result of a phonotactic localization study may depend on details in conducting the experiment. First, they observed that if females were released on the base plate, they remained on the ground for much of their approach before leaping into the three-dimensional grid. Therefore, conducting the experiment with the release point placed on the ground would have resulted in apparently less accurate localization than was measured when the release point was placed above the ground. Second, when climbs were included in the analysis, much larger localization errors were found than if only jumps were analyzed. This may be the result of the vertical sticks offering fewer degrees of freedom for the climbs than for the jumps. It could also be the cause of the apparently larger localization errors when both data from climbs and jumps were pooled.

In any case, the measure of the accuracy of localization in a behavioral experiment in frogs using phonotaxis includes both a component that is based on the error made in the perception of the angle of sound incidence and a component that results from errors introduced by locomotion. Thus the localization error found in these experiments represents a conservative upper limit for the perceptual abilities of the animal. Despite these limitations, remarkably accurate sound localization has been found in the different species studied in closed-loop experiments (see Table 1). A comparison of different measures of the accuracy of localization has shown that perceptual errors may be much smaller than jump errors. A comparison of head orientation error with jump angle error in the green treefrog (*Hyla cinerea*, see Rheinlaender et al., 1979) has revealed for example, that the average perceptual localization error may be smaller than 8.4° (as can be inferred from the mean head orientation error of 8.4°) which is far less than the average jump angle error of 16.1° that is observed when data from all phonotactic approaches are summarized.

Table 1. Means and standard deviations of jump error angles reported in closed-loop localization experiments in different frog species when tested with natural signals or signals representing the most prominent spectral components. The three-dimensional jump error angles include errors from climbing movements on the vertical sticks in the grid.

Species	mean	±SD	Reference
Two-dimensional localization:			
Hyla cinerea	16.1°	±14.5°	Rheinlaender et al. 1979
Hyla versicolor *	19.4°	±17.2°	Jørgensen and Gerhardt 1991
Colostetus nubicola	23.2°	±17.4°	Gerhardt and Rheinlaender 1980
Hyperolius marmoratus	22.0°	±29.7°	Passmore et al. 1984
Three-dimensional localization:			
Hyperolius marmoratus	43.0°	±31.7°	Passmore et al. 1984
Hyla versicolor	36.1°	±26.8°	Jørgensen and Gerhardt 1991

* horizontal jump error angle determined in a three-dimensional grid

4 An Example for an Open-loop Localization Experiment

In frog species that respond to single brief signals, open-loop experiments are possible. The barking treefrog, *Hyla gratiosa*, is one such species. It is especially suitable for open-loop measurements of sound localization because when performing phonotaxis, females do not readily move before the playback of an attractive sound. This allows for easy measurement of orientation error angles and jump error angles. Klump and Gerhardt (1989) positioned female barking treefrogs in the center of an arena, aligning the frog's longitudinal body axis along a 0° reference line on the arena. Before positioning the frog, a speaker was placed at a randomly prechosen position at a distance of 1 m from the release point of the frog that resulted in angles of sound incidence of between -45° and 45° (stepsize 15°). A single advertisement call was then played back to the frog (one digitized exemplar prerecorded on a tape was used throughout the experiments, see Klump and Gerhardt, 1989), and two observers recorded its movements. If the frog did not move after the playback of the first call, up to nine additional calls were played at a rate of 1 call per second. When the frog oriented by turning after a call had been presented, or when it jumped after the playback of a call, no further signals were broadcast and the new position and orientation of the body axis were recorded by the observers (the orientation was measured in steps of 7.5° relative to the reference line at 0°). Before a new trial started, the frog was positioned back in the center of the arena and the procedure described above was repeated with a new choice of the speaker position. Trials in which the female moved prior to or during the playback of a call were excluded from the analysis. This insured that only trials with an open-loop stimulation were analyzed.

Figure 3 shows the results of the experiment. The dotted lines indicate the orientation and jump angles that would be expected if the frog perfectly oriented or jumped towards the position of the sound source. The mean orientation and jump angles follow this line. A statistical analysis of the data revealed that the orientation angles were significantly more negative if the speaker was located at -45° than if it was positioned at -30° or -15° (positive angles indicate speaker positions to the right of the body axis, negative angles positions to the left). Similarly, the orientation angles were significantly larger if the speaker was located at +45° than if it was positioned at +15°. The jump angles showed a similar relation to the speaker location as the orientation angles, but there was more scatter in the data. Thus, only at -45°, +30° and +45° were the distributions of the jump angles different from the distribution determined with a speaker location of 0°. Jumps were significantly less accurate than the corresponding orientation responses. The data from these open-loop experiments suggest that the barking treefrog has the ability to discriminate between different angles of sound incidence. Only this ability would allow the females to discriminate between sound-incidence angles of different sizes. If they only lateralize the sound source (i.e., determine whether it is located left or right of the longitudinal body axis), the orientation and jump angles should only depend on the sign of the angle of sound incidence and not on its size. Nevertheless, an analysis of the distribution of orientation and jump errors indicated that the frogs have a tendency to turn if the sound source is located between -15° and 15° (Klump and Gerhardt, 1989). This pattern may represent a behavioral strategy of the frogs to enhance binaural disparities if they are low.

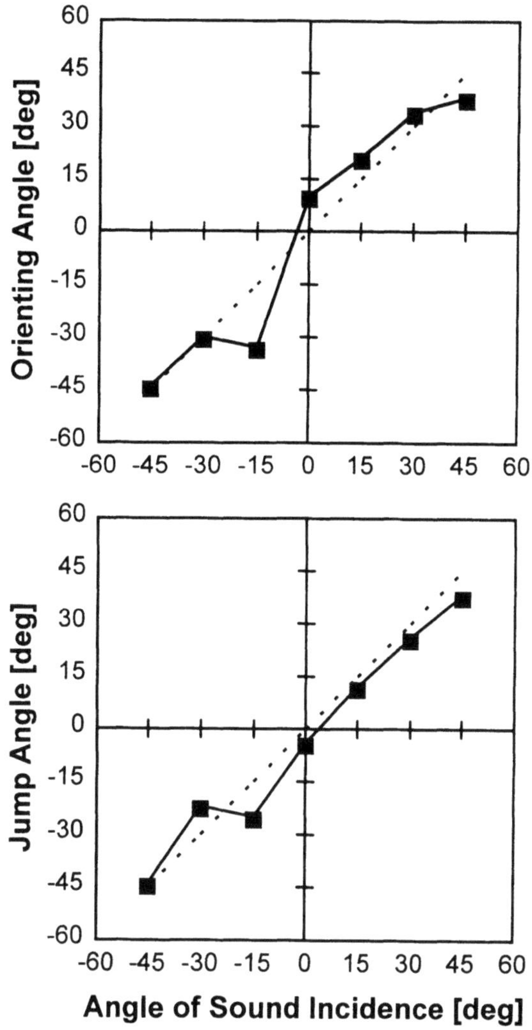

Figure 3. The relationship between the angle of sound incidence (speaker location) and the orienting (top) or jump angles (bottom) in the phonotaxis behavior of the barking treefrog (after Klump and Gerhardt, 1989).

5 Concluding Remarks

The behavioral experiments on sound localization in frogs described in this chapter have demonstrated a remarkable accuracy in perceiving the direction from which a sound comes. The study in the barking treefrog using open-loop stimulation has shown that, at least in this species, the mechanism of localization allows them to discriminate between different angles of sound incidence. How the mechanism works, however, is not yet understood (see the discussion in Jørgensen and Gerhardt, 1991). More experiments combining behavioral, physical and physiological methods are needed before the impressive ability of frogs to localize sound sources can be fully understood.

6 Acknowledgements

During the preparation of this chapter the author was funded by a grant from the Deutsche Forschungsgemeinschaft within the SFB 204 "Gehör". The comments of M. Dyson on a previous version of the manuscript are gratefully acknowledged.

7 Suggested Reading

Jørgensen, M.B. and Gerhardt, H.C. (1991) Directional hearing in the gray tree frog *Hyla versicolor*: Eardrum vibrations and phonotaxis. *Journal of Comparative Physiology* A 169: 177-183.

Klump, G.M. and Gerhardt, H.C. (1989) Sound localization in the barking treefrog. *Naturwissenschaften* 76: 35-37.

Rheinlaender, J and Klump, G.M.(1988) Behavioral aspects of sound localization. In: *The evolution of the amphibian auditory system* (Fritsch, B., Ryan, M.J., Wilczynski, W., Hetherington, T.E. and Walkowiak, W., eds.). John Wiley, New York, pp. 297-305.

8 References

Eggermont, J.J. (1988) Mechanisms of sound localization in anurans. In: *The evolution of the amphibian auditory system* (Fritsch, B., Ryan, M.J., Wilczynski, W., Hetherington, T.E. and Walkowiak, W., eds.). John Wiley, New York, pp. 307-336.

Feng, A.S., Gerhardt, H.C. and Capranica R.R. (1976) Sound localization behavior of the green treefrog (Hyla cinerea) and the barking treefrog (H. gratiosa). *Journal of Comparative Physiology* A 107: 241-252.

Gerhardt, H.C. (1975) Sound pressure levels and radiation patterns of the vocalizations of some north american frogs and toads. *Journal of Comparative Physiology* 102: 1-12.

Gerhardt, H.C. and Rheinlaender, J. (1980) Accuracy of sound localization in a miniature dendrobatid frog. *Naturwissenschaften* 67: 362-363.

Gerhardt, H.C. and Rheinlaender, J. (1982) Localization of an elevated sound source by the green tree frog. *Science* 217: 663-664.

Jørgensen, M.B. and Gerhardt, H.C. (1991) Directional hearing in the gray tree frog *Hyla versicolor*: Eardrum vibrations and phonotaxis. *Journal of Comparative Physiology* A 169: 177-183.

Klump, G.M. and Gerhardt, H.C. (1989) Sound localization in the barking treefrog. *Naturwissenschaften* 76: 35-37.

Michelsen, A., Jørgensen, M.B., Christensen-Dalsgaard, J. and Capranica R.R. (1986) Directional hearing of awake, unrestrained treefrogs. *Naturwissenschaften* 73: 682-683.

Perill, S.A., Gerhardt, H.C. and Daniel, R. (1978) Sexual parasitism in the green treefrog (*Hyla cinerea*). *Science* 209: 523-525.

Passmore, N.I., Capranica, R.R., Telford, S.R., and Bishop, P.J. (1984) Phonotaxis in the painted reed frog (*Hyperolius marmoratus*). *Journal of Comparative Physiology* A 154: 189-197.

Rheinlaender, J., Gerhardt, H.C., Yager, D.D. and Capranica R.R. (1979) Accuracy of phonotaxis by the green treefrog (*Hyla cinerea*). *Journal of Comparative Physiology* A 133: 247-255

Rheinlaender, J., Walkowiak, W. and Gerhardt, H.C (1981) Directional hearing in the green treefrog: A variable mechanism. *Naturwissenschaften* 68: 430-431.

Rheinlaender, J and Klump, G.M.(1988) Behavioral aspects of sound localization. In: *The evolution of the amphibian auditory system* (Fritsch, B., Ryan, M.J., Wilczynski, W., Hetherington, T.E. and Walkowiak, W., eds.). John Wiley, New York, pp. 297-305.

Schmitz, B., White, T.D., & Narins, P.M. (1992) Directionality of phase locking in auditory nerve fibres of the leopard frog Rana pipiens pipiens. *Journal of Comparative Physiology* A 170: 589-604.

Fish

Fish differ in a number of respects from the other vertebrates. Here just a few key points are given with respect to the detection and analysis of sound. First, the acoustic environment under water is quite different from that of a terrestrial vertebrate. Water is much less compressible and more dense than air. Since a fish's body is similar to water in acoustic impedance, sound waves easily penetrate the body and travel directly to the ears. Due to the relatively high impedance and wavelength of underwater sound, the near field is much more important for hearing in fish than in land-living vertebrates. These and other differences will effect the physical constraints under which sensory systems of water-living animals evolved. Second, the acoustic receptors of fishes are one or more of the otolith organs (saccule, utricle, and lagena) that have a vestibular function in most terrestrial animals. The otoliths have a higher density than the surrounding tissue and water, and appear to function as mass-loaded accelerometers in detecting acoustic particle motion as sound passes through the body. Some fish have evolved an additional path for stimulation of the ears by using the swim bladder or another gas sack as a pressure receiver. Sound pressure fluctuations cause the swim bladder to expand and contract, and in some species, these motions of the swim bladder's wall are efficiently transmitted to one or more of the otolith organs. Thus, some of the fishes are unique among vertebrates in that they simultaneously acquire information about sound pressure and acoustic particle motion. Thirdly, with their lateral line system fish have an additional sensory modality which is stimulated by low frequency, relative motions between the animal's skin and the local nearfield of underwater sound sources. However, despite these differences between fishes and other vertebrates, the hair-cell receptors of all lateral line, otolith, auditory papillae, and cochlear organs are essentially similar in their most basic structures and functions. The study of the sense of hearing among fish is interesting as an example of a functionally homologous auditory system making use of receptor organs that are only analogous to the auditory papillae and cochleae of other vertebrates.

Psychoacoustic studies in fish have utilized the range of behavioral paradigms applied in other vertebrates. In her chapter, Coombs gives an example how natural orienting responses can be used to study the function of the lateral-line system. The chapter by Fay demonstrates the success of classical conditioning in the study of a wide range of basic hearing capabilites in the goldfish. These include the analysis of the goldfish's judgement of the similarities among simple and complex sounds using a stimulus generalization paradigm. Finally, The chapter by Yan shows that fish can be successfully trained using operant techniques to study sensitivity and discrimination acuity.

Methods in Comparative Psychoacoustics
G. M. Klump, R. J. Dooling, R. R. Fay & W. C. Stebbins (eds)

Natural Orienting Behaviors For Measuring Lateral Line Function

Sheryl Coombs

Parmly Hearing Institute, Loyola University of Chicago
Chicago, Il. 60626 USA

Summary. The naturally-occurring orienting response found in some fish and aquatic amphibians has been the behavior of choice for investigating lateral line sensory capabilities. Although this unconditioned behavior is most useful in measuring the ability of animals to localize sources of water disturbance, it can also be used to measure threshold levels of detection. Furthermore, the orienting response can be operantly conditioned to shape discriminatory behavior. Potential problems associated with this natural behavior include (1) uncertainty as to the sensory components or range of stimuli that drive the behavior, (2) difficulty in maintaining high levels of motivation and response reliability, and (3) frequent spontaneous movements of animals. The major advantage of the orienting response and natural behaviors in general are that they give us insights on sensory performance and function in biologically relevant contexts.

1 Introduction

1.1 Natural Behaviors for Measuring Sensory Function

This chapter will examine the use of a naturally-occurring, unconditioned orienting response in measuring lateral line sensory function. The orienting response is part of an overall feeding behavior that is found in many surface-feeding fish (Schwartz 1967, Bleckmann 1989) and amphibians (Görner 1973) and in the benthic-feeding mottled sculpin, *Cottus bairdi* (Hoekstra & Janssen 1985, 1986). In the mottled sculpin, a fish routinely used in our lab for studies of lateral line function, this behavior consists of an initial orientation of the animal's body towards a vibrating source and a subsequent step by step movement towards the source resulting in an eventual strike (bite) at the source (Fig. 1). Orienting responses like this have been used to measure threshold levels of detection (Coombs & Janssen 1990, Bleckmann 1980, Bleckmann & Topp 1981) and how well animals can determine both the distance (Schwartz 1967, Bleckmann & Schwartz 1982, Hoin-Radkovsky et al. 1984) and the angular location (Elepfandt 1982, Müller & Schwartz 1982, Tittle 1988, Görner & Mohr 1989, Janssen 1990) of sources of water disturbance. In addition, operant conditioning of the orienting response has been employed to measure frequency (Bleckmann et al. 1981, Elepfandt et al. 1985) and intensity (Coombs & Fay 1993) discrimination abilities. Reaction time of the response has also been measured as a function of source intensity and distance (Bleckmann 1980), but its usefulness in measuring perceptual phenomena like loudness equivalency (e.g. Moody 1970) has not been explored. Finally, the orienting response has been used to measure the distance or working range of the lateral line system for both artificial (e.g. vibrating sphere, Coombs in press) and natural (e.g. live prey, Coombs & Janssen 1989) sources.

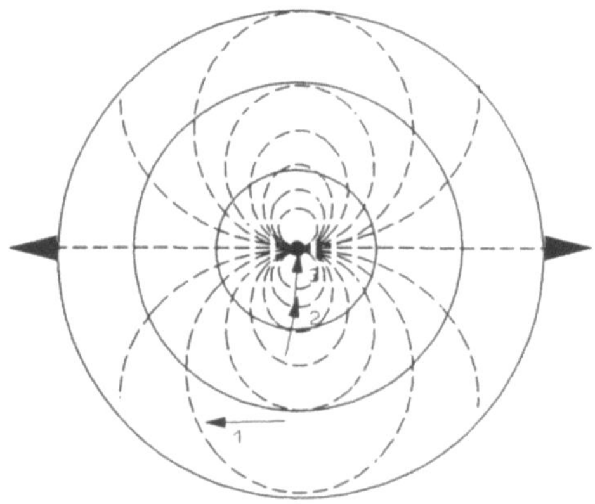

Figure 1. Schematic representation of a mottled sculpin approaching a sinusoidally-vibrating (50 Hz) sphere (filled circle in center of the field). Large arrow heads represent the axis of vibration and dashed lines represent the dipolar flow field about the source. Arrow #1 represents the position of the fish before the stimulus is turned on and arrow #3 represents the position of the fish as it strikes at the source. Arrows are drawn from the tip of the fish's snout (arrowhead) to the point of pectoral fin insertion. Circles around the source are references for fixed radial distances (see section 2.1).

Although animal psychophysics can be defined in the broadest sense as the behavioral analysis of sensory function (Stebbins 1970), it has relied heavily, if not exclusively, on behavioral responses that have been conditioned. The use of unconditioned, naturally-occurring behaviors to examine sensory function has largely been the domain of neuroethology. One of the often-cited advantages of using naturally-occurring behaviors is that they give us insights on sensory function in biologically-relevant contexts that we might not otherwise get from a strictly psychophysical approach. In other words, they tell us something about the role of natural selection in shaping the design of the sensory system. The elegant body of work on the electrosensory system of fish is an excellent example of how naturally-occurring behaviors, such as electrolocation (Bastian 1990) and the jamming avoidance response (Heiligenberg 1991), have been put to good use for understanding sensory function.

One of the major drawbacks of unconditioned behaviors is that they are less useful as a tool for systematically measuring the limits and dimensions of sensory function for the purpose of comparing across species (e.g. Fay 1988). For one thing, comparable behaviors that can be measured in a large number of diverse species are difficult to find, although orienting and escape behaviors are among the most common. For another, unconditioned behaviors may be controlled by (1) more than one sensory system, (2) subcomponents of one or more systems, or

(3) a small set of stimuli that don't represent the entire range of stimuli to which the sensory system under investigation is sensitive. C-start escape behavior in teleost fish, for example, is elicited by signals that signify imminent danger only and is mediated by a network of brainstem cells, including the Mauthner cell, which receives auditory, lateral line and visual input (Eaton et al. 1991).

Unconditioned behaviors are also subject to many of the same problems that plague conditioned behaviors, such as those associated with making sure that the behavior is under stimulus control and that non-stimulus variables, such as the animal's state of attention or motivation are properly controlled for. Although the field of psychophysics has focused considerable attention on non-stimulus variables that might affect the response (e.g. signal detection theory), less rigorous attention of this kind has been paid to naturally-occurring responses used in neuroethological approaches. Nevertheless, the unconditioned orienting response remains one of the most powerful behavioral tools we have for measuring sensory function of the lateral line system. The major goals of this chapter will be to describe different ways in which this behavior can be used to measure the sensory capabilities of this system and to point out some of the problems and limitations associated with this particular behavioral approach.

1.2 The Lateral Line and Psychoacoustics

There are probably two good reasons for including the lateral line system in a methods book on psychoacoustics. One is that the lateral line system has been historically treated as part of a single octavolateralis system - with some investigators having regarded it as an accessory auditory structure (see review by Sand 1981). Although recent neuroanatomical, physiological and behavioral lines of evidence confirm Dijkgraaf's (1962) original view of the lateral line system as a separate sensory system of "distance touch" (Coombs, Görner & Münz 1989), it is important to point out that there is still considerable overlap in responsiveness between the two systems (Sand 1981, Platt et al. 1989, Coombs et al. 1992). Any suprathreshold sound source or mechanical disturbance of the water, if it is close enough (within one or two body lengths away) or lower in frequency than about 200 Hz, can potentially stimulate both the lateral line and auditory system. Thus, the second reason for including the lateral line system in this book is that behavioral studies on underwater hearing in fish and amphibians need to be concerned about potential stimulation of the lateral line system (see section 2.5).

2 Methods

2.1 Behavioral Set-Up

The behavioral approaches described in this chapter take advantage of two naturally-occurring behaviors of the mottled sculpin - the orienting response of the fish and its benthic, sedentary behavior. Because these fish tend to sit on the bottom of tanks, they can be easily trained to rest motionless on a moveable platform (Fig. 2), separated from the bottom of the experimental tank by a sheet of plastic grating (egg crate). The platform can be moved to position the fish relative to the stimulus source, which in typical lateral line experiments has been a small (6 mm in diameter) plastic sphere mounted on the end of a cylindrical shaft (a 16 gauge stainless steel, blunt-tipped syringe needle). The height of the sphere above the fish can be changed, but for most experiments is around 12 mm above the platform (corresponding to the average position of the trunk lateral line canal on the fish). As landmarks for positioning the fish, there are a series of concentric circles drawn on the bottom of the tank (as illustrated in Fig. 2) indicating fixed radial distances (3.0, 6.0, and 9.0 cm) from source center. A video camera directly below the tank affords a ventral view of the fish, silhouetted against the concentric circles so that the fish can be accurately positioned relative to the source.

Figure 2. Schematic diagram of behavioral set-up showing how the fish is moved relative to the stimulus source, a small vibrating sphere (Adapted from Coombs & Janssen 1990).

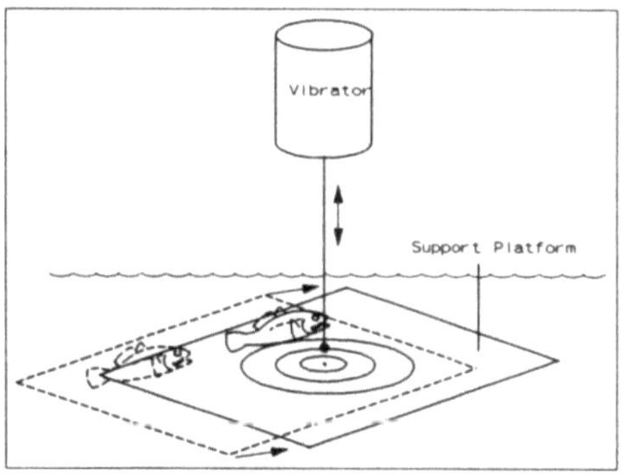

One of the advantages of this set-up and the natural behaviors of the mottled sculpin is that the stimulus conditions are very easy to replicate for physiological preparations for the purpose of comparing neural and behavioral responses (Fig. 3). The position of the stimulus relative to the animal can be controlled in both set-ups and the immobilization of the animal required for physiological preparations is not a severe departure from the animal's normal sedentary

behavior. Furthermore, additional apparati (e.g. bars for pressing etc.) are not required and thus, do not interfere with the stimulus field. An additional advantage is that this set-up can be used in different ways to measure different aspects of lateral line sensory function as described in sections 2.2 - 2.4.

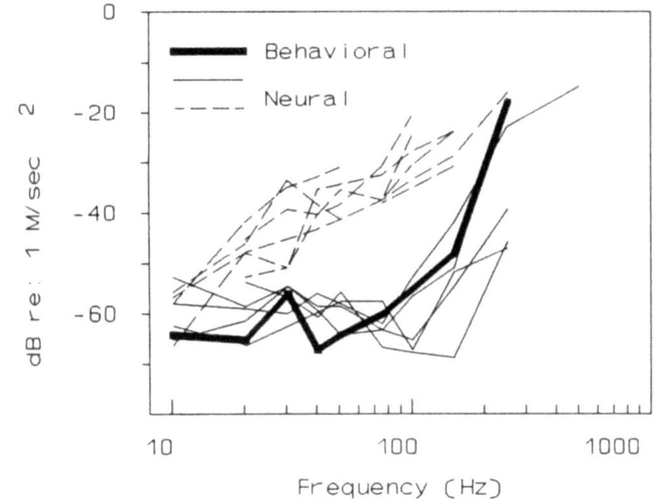

Figure 3. Mean behavioral threshold curve (heavy solid line) obtained from 4 fish when the source was 1.5 cm away from the trunk of the fish. Neural threshold curves were obtained under the same stimulus conditions from posterior lateral line nerve fibers innervating superficial (dashed thin lines) and canal (thin solid lines) neuromasts on the trunk (From Coombs & Janssen 1990).

2.2 Using the Unconditioned Orienting Response as a Detection Response

Once fish are positioned, the experimenter initiates a computer-controlled sequence that determines a random inter-trial interval (from 5 to 15 sec) followed by a 5-sec trial during which the source is vibrated. The sculpin's response is scored as either a detection (movement towards or strike (bite) at the source) or a non-detection (no movement or movement away) response; responses are keyed into the computer which uses this information to compute the stimulus level of the next trial (see adaptive tracking method below). Orienting responses are reinforced by feeding the fish with small pieces of squid (see section 2.4). If the fish moves before the trial begins, it is repositioned, and a new trial sequence is begun.

The animal's false alarm rate is monitored by randomly presenting between thirty to fifty percent of all trials as blank trials during which the source does not vibrate but the response is recorded. To help reduce observer bias, observers know only when a trial occurs and not whether it is a stimulus or blank trial. If detection responses to blank trials occur more than 20% of the time during any given experimental session, the orienting responses during that session are judged to be under poor stimulus control, and data from that session are excluded from the final data analysis. It is important to monitor false alarm rate because the natural behavior of these fish when searching for prey is saltatory -that is, fish rest motionless for a couple of seconds, presumably to "listen" for prey and then spontaneously move to a new position, where they stay for another couple of seconds before moving again. The wait times associated with the search behavior are generally less than 10 seconds, so spontaneous movements can be frequent. For this reason, inter-trial intervals are also kept relatively short.

For detection experiments we have used an adaptive tracking method in which the fish's response to a trial determines the amplitude of the subsequent trial, such that a detection response results in decreasing the amplitude of the next trial by 5 dB, and a non-detection response results in increasing the amplitude of the next trial by 5 dB. The signal level midway between that for consecutive yes and no responses is defined as a transition threshold. Approximately three to seven of these transition thresholds (ranging over 5 to 15 stimulus presentations) can be obtained for each experimental session, which generally lasts between 20 and 45 minutes. Mean or median transition thresholds can be obtained as a function of both vibration frequency and source location to yield behavioral threshold curves (Fig. 3; Coombs & Janssen 1990).

The adaptive tracking method has been chosen over other methods, like the method of constant stimuli, because it represents a very efficient way of measuring threshold (Stebbins 1970), given that the number of trials/experimental session and the number of sessions/week are severely constrained by motivational (i.e. hunger) factors (see section 2.6). In addition, this method minimizes the number of stimulus presentations that are way above threshold. Occasionally we have observed suprathreshold stimuli (> 50 dB SL) to elicit what look like escape responses, which by our criteria would be incorrectly scored as non-detection responses.

2.3 Conditioning the Orienting Response for Discrimination Tasks

Since the orienting response is based on a naturally occurring, unconditioned feeding response, conditioning the animal to orient to one of two stimuli for discrimination studies is necessary. In studies of intensity discrimination (Coombs & Fay 1993), for example, where the task for mottled sculpin was to orient towards the source only when they detected a change in source level, fish had to become habituated to the standard stimulus, which was on-going throughout the session, and reinforced for orienting only when they detected an increment in source level. In our experiments on intensity discrimination, it took fish from 7 - 13 experimental sessions before they were sufficiently habituated to the standard and the orienting response was judged to be under acceptable stimulus control (i.e. false alarm rates less than 20%).

2.4 Using the Orienting Response for Measuring Localization Abilities

Probably the most direct and powerful way that the unconditioned orienting response can be used is as a measure of the animal's ability to localize sources. In our lab, we have recently begun to videotape the pathways followed by blinded sculpin, starting just before the time of stimulus onset to the time that the fish strikes at the source. Since the fish approaches the source in discrete steps, each step can be captured as a frozen video frame and the position of the fish relative to the source can be specified with a vector, drawn from the tip of the snout to the fish's approximate center of mass. Vectors can then be mapped onto a computer-generated representation of the flow field (Fig. 1) in order to compute (1) the angle formed by the fish vector and the flow (velocity) vector and (2) the pressure gradient (the primary stimulus to the lateral line) across and along the fish's body. Additional measurements, such as the distance and angular location of the source with respect to the fish before and after the fish has changed position, can be obtained to assess the ability of fish to determine distance and angular location of the source, as has been done for surface feeding fish and amphibians (Bleckmann 1988, Elephandt 1982). This kind of approach will not only allow us to describe the distance determination and localization ability of fish but also to determine the behavioral strategies and sensory cues used by fish in finding vibratory sources.

2.5 Techniques for Ruling Out Other Sensory Systems

In all of our lateral line behavioral studies with the mottled sculpin, we have blinded animals to rule out the possibility that animals are using the visual system to orient towards vibrating sources. Since these fish normally feed at night and often on buried prey (Hoekstra & Janssen 1986), it is unlikely that vision plays a major role for them under natural conditions. Ruling out contributions from the auditory system is more difficult, however, because any technique used to completely knock out this system is likely to interfere with the equilibrium and overall behavior of the animal. Thus, we cannot completely rule out the possibility that the auditory system is responding to the vibratory sources used in these experiments. However, mechanical and pharmacological (e.g. immersion in CoCl2, Karlsen & Sand 1987) techniques for blocking the lateral line system without affecting the ear indicate that the lateral line system is necessary, if not sufficient, for the orienting response (Hoekstra & Janssen 1985; Janssen 1990). Behavioral threshold results using the orienting response provide a second line of evidence that this response is lateral-line mediated, as behavioral threshold curves agree quite well with threshold response curves measured physiologically from primary lateral line fibers under stimulus conditions nearly identical to those used for behavioral studies (Fig. 3; Coombs & Janssen 1990).

In this regard, however, it is interesting to point out that behaviorally-measured threshold curves can be predicted solely on the basis of input from fibers innervating canal neuromasts (Fig. 3), one of two subclasses of endorgans in the lateral line system. These results leave open the question of whether the orienting response is tapping into a particular subset of lateral line endorgans - a question that is difficult to answer experimentally because of the sheer numbers and dispersion of lateral line endorgans on the body of the fish.

Although questions remain about whether the orienting response may tap into one or more types of lateral line endorgans or whether the auditory system may somehow modulate the response, there is little doubt that the behavior is driven primarily by the lateral line and not the auditory system. In contrast, recent experiments in which conditioned suppression of respiration was used to measure detection thresholds in the goldfish to a small 50 Hz source placed near the trunk lateral line canal indicated that this behavior was driven primarily by the ear and not the lateral line (Coombs, in press). The reason for this may be simply that the exquisite sound pressure-sensitivity of the goldfish ear overides the hydrodynamic sensitivity of the lateral line. Nevertheless, this example serves to illustrate that without techniques for selectively eliminating the ear, behavioral measures of lateral line function must be interpreted with caution.

2.6 Techniques for Maintaining Good Stimulus Control

Although the unconditioned orienting response can be elicited from naive animals in the lab to novel, unnatural stimuli, such as a vibrating sphere, not all animals respond with equal reliability and probability. Over the course of any one experiment, many animals may be rejected because they simply fail to show the behavior consistently. Orienting responses seem to be particularly variable for the clawed frog, *Xenopus* (Elephandt 1982). Unfortunately, not all of the variables contributing to the lack of response are understood and seldom in these kinds of experiments is there any direct measure of response variability or a report of the percentage of unused animals. One factor that is understood, however, is that it is extremely important to maintain the behavior by keeping animals hungry and rewarding them with food during experimental sessions, even though food-reward itself is not necessary to condition the response.

Experiments in our lab have examined the effects of feeding schedules on the percentage of times that the orienting response was elicited by a suprathreshold stimulus (40 dB SL). When animals were fed (reinforced with food for correct responses) during the experimental session only, response rates to stimulus trials were generally greater than 80% and those to blank trails less than 20% (Fig. 4, panel A). Response rates fell outside of these limits for less than 10% of the total number of trials. When reinforcement was withheld during the experimental session, responses from all fish were slow to extinguish, usually taking 6 to 8 sessions before response rates dropped below 80% (Fig. 4, panel B). These results were quite similar to our experience with level discrimination studies in which it took several experimental sessions of unrewarded responses before habituation to an on-going source of vibration was established (see section

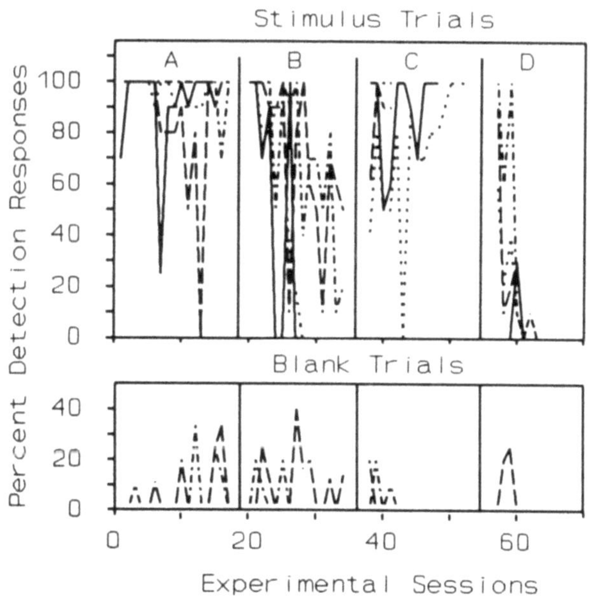

Figure 4. Correct response rates for four fish as a function of feeding regimen: (A) Fed during experiments only, (B) not fed during experiments, (C) same as A, (D) fed before experiments (Redrawn from Coombs & Janssen 1990).

2.3). Reinstatement of reinforcement resulted in an almost immediate return to response rates greater than 80% (Fig, 4, panel C). Feeding fish before experiments until they were satiated resulted in a dramatic and nearly immediate decline in response rate (Fig. 4, panel D), even though reinforcements were offered for correct responses. The various feeding regimens appeared to have no appreciable affect on false alarm rates, as shown by the bottom panel of Fig. 4.

The bottom line from these types of experiments is that (1) it is important to have some measure of response reliability both before and during experiments and (2) food deprivation and the association of food reward with the response is critical to the maintenance of reliable response levels. In order to maintain reliable response levels in these small fish, we have found it necessary to feed during experimental sessions only, to limit the number of reinforcements per daily session to less than 10 bite-size (550 mg) pieces of squid, and to run experiments every other day. Note, however, that even under conditions that yielded the highest and most reliable response rates, there are experimental sessions for almost all individuals when performance is inexplicably poor (Fig. 4, panel A & C).

3 Concluding Remarks

Behavioral studies of lateral line sensory capabilities are still at a very infant stage, especially with respect to subsurface water disturbances. Limiting factors besides those associated with behavioral approaches have included an inadequate understanding of the stimulus to the lateral line and an inability to easily measure and specify lateral line stimuli. As our ability in these areas improve, more creative and sophisticated behavioral techniques, such as those developed by Hassan (1985, 1986) for determining the ability of blind cavefish to use their lateral line in forming spatial images of the hydrodynamic environment, will undoubtedly ensue.

4 Acknowledgements

I wish to thank John Janssen for discovering the lateral-line mediated orienting response in the mottled sculpin and for his initial work in helping to develop the behavioral protocol. This work has been supported by grants from the National Institute for Deafness and Communicative Disorders and from the Max Goldenberg Foundation.

5 Suggested Reading

Bleckmann, H. (1993) Role of the lateral line in fish behavior. In T.J. Pitcher (ed) *Behavior of Teleost Fishes*, 2nd Edn. London: Chapman & Hall (pp. 177-202)

Bleckmann, H., Tittel, G. & Blubaum-Gronau. (1989). Lateral Line System of Surface-Feeding Fish: Anatomy, Physiology and Behavior. In S. Coombs, P. Gorner & H. Munz (Eds.). *The Mechanosensory Lateral Line: Neurobiology and Evolution* (pp. 501-526). New York; Berlin: Springer-Verlag

Coombs, S. & Janssen, J. (1990). Water flow detection by the mechanosensory lateral line. In W.C. Stebbins and M. Berkley (Eds.). *Comparative Perception: Complex Signals* Vol. 2 (pp. 89-124). New York: Springer-Verlag

Elephandt, A. (1989). Wave analysis by amphibians. In S. Coombs, P. Gorner & H. Munz (Eds.). *The Mechanosensory Lateral Line: Neurobiology and Evolution* (pp. 527-542). New York; Berlin: Springer-Verlag

Martin, P. & Bateson, P. (1993). *Measuring Behavior: An Introductory Guide* 2nd Ed. Cambridge University Press (222 p)

6 References

Bastian, J. (1990). Electroreception. In W.C. Stebbins and M. Berkley (Eds.). *Comparative Perception: Complex Signals* Vol. 2 (pp. 35-88). New York: Springer-Verlag.

Bleckmann, H. (1980). Reaction time and stimulus frequency in prey localization in the surface-feeding fish *Aplocheilus lineatus*. *Journal of Comparative Physiology A 140*, 163-172.

Bleckmann, H. & Schwartz, E. (1982). The functional significance of frequency modulation within a wave train for prey localization in the surface-feeding fish *Aplocheilus lineatus* (Cyprinodontidae). *Journal of Comparative Physiology A 165*, 331-339.

Bleckmann, H. & Topp, G. (1981). Surface wave sensitivity of the lateral lineorgans of the topminnow *Aplocheilus lineatus*. *Naturwissenschaften 68*, 624-625.

Bleckmann, H., Waldner, I., & Schwartz, E. (1981). Frequency discrimination of the surface-feeding fish *Aplochelius lineatus* - a prerequisite for prey localization?. *Journal of Comparative Physiology A 143*, 485-490.

Coombs, S. (In press). Nearfield detection of dipole sources by the goldfish (*Carassius auratus*) and the mottled sculpin (*Cottus bairdi*). *Journal of Experimental Biology*.

Coombs, S. & Fay, R.R. (1993). Source level discrimination by the lateral line system of the mottled sculpin. *Journal of the Acoustical Society of America 93*, 2116-2123.

Coombs, S., Gorner, P. & Munz, H. (1989). A Brief Overview of the Mechanosensory Lateral Line System and the Contributions to This Volume. In S. Coombs, P. Gorner & H. Munz (Eds.). *The Mechanosensory Lateral Line: Neurobiology and Evolution* (pp. 3-5). New York; Berlin: Springer-Verlag.

Coombs, S. & Janssen, J (1989). Peripheral processing by the lateral line system of the mottled sculpin (*Cottus bairdi*). In S. Coombs, P. Gorner & H. Munz (Eds.). *The Mechanosensory Lateral Line:Neurobiology and Evolution* (pp. 229-319). New York; Berlin: Springer-Verlag.

Coombs, S. & Janssen, J. (1990). Behavioral and neurophysiological assessment of lateral line sensitivity in the mottled sculpin, *Cottus bairdi*. *Journal of Comparative Physiology A 167*, 557-567.

Coombs, S., Janssen, J. & Montgomery, J. (1992). Functional and evolutionary implications of peripheral diversity in lateral line systems. In D. Webster, A.M. Popper & F.F. Ray (Eds.). *The Evolutionary Biology of Hearing*. New York: Springer-Verlag.

Dijkgraaf, S. (1963). The functioning and significance of the lateral-line organs. *Biological Review 38*, 51-105.

Eaton, R.C., DiDomenico, R. & Nissanov, J. (1991). Role of the Mauthner Cell in sensorimotor integration by the brainstem escape network. *Brain Behaviour and Evolution 37*, 272-285.

Elepfandt, A. (1982). Accuracy of taxis response to water waves in the clawed toad (*Xenopus laevis* Daudin) with intact or with lesioned lateral line system. *Journal of Comparative Physiology 148*, 535-545.

Elepfandt, A., Seiler, B., & Aicher, B. (1985). Water wave frequency discrimination in the clawed frog, *Xenopus laevis*. *Journal of Comparative Physiology A 157*, 255-261.

Fay, R.R. (1988) *Hearing in Vertebrates: A Psychophysics Databook* Winnetka, Il: Hill-Fay Associates (621 p).

Görner, P. (1973). The importance of the lateral line system for the perception of surface waves in the clawed toad, *Xenopus laevis* Daudin. *Experientia 29*, 295-296.

Görner,P. & C. Mohr (1989). Stimulus localization in *Xenopus*: Role of Directional sensitivity of lateral line stitches. In S. Coombs, P. Gorner & H. Munz (Eds.). *The Mechanosensory Lateral Line: Neurobiology and Evolution* (pp. 543-560). New York; Berlin: Springer-Verlag.

Hassan, E.S. (1985). Mathematical analysis of the stimulus for the lateral line organ. *Biol. Cybern. 52*, 23-36

Hassan, E.S. (1986). On the discrimination of spatial intervals by the blind cave fish (*Anoptichthys jordani*). *Journal of Comparative Physiology A 159*, 701-710.

Heiligenberg, W.F. (1991). *Neural Nets in Electric Fish*. Cambridge, Ma; MIT Press (179 p)

Hoekstra, D. & Janssen, J. (1985). Non-visual feeding behavior of the mottled sculpin, *Cottus bairdi*, in Lake Michigan. *Envir Biol Fish, 12*, 111-117.

Hoekstra, D. & Janssen, J. (1986). Lateral line receptivity in the mottled sculpin (*Cottus bairdi*). *Copeia 1986*, 91-96.

Hoin-Radkovsky, I., Bleckmann, H. & Schwartz, E. (1984). Determination of source distance in the surface-feeding fish *Pantodon buchholzi* (Pantodontidae). *Animal Behaviour 32*, 840-851.

Janssen, J. (1990). Localization of substrate vibrations by the mottled sculpin (*Cottus bairdi*). *Copeia 1990*, 349-355.

Karlsen, H.E. & Sand,O. (1987). Selective and reversible blocking of the lateral line in freshwater fish. *Journal of Experimental Biology 133*, 249-262.

Moody, D.B. (1970). Reaction time as an index of sensory function. In W.C. Stebbins (ed) *Animal Psychophysics: The Design and Conduct of Sensory Experiments*. New York: Appleton-Century-Crofts (pp. 277-302).

Platt, C., Popper, A.N. & Fay, R.R. (1989). The Ear as Part of the Octavolateralis System. In S. Coombs, P. Gorner & H. Munz (Eds.). *The Mechanosensory Lateral Line: Neurobiology and Evolution* (pp. 633-651). New York; Berlin: Springer-Verlag.

Sand, O. (1981). The lateral line and sound reception. In W.N. Tavolga, A.N. Popper & R.R. Fay (eds) *Hearing and Sound Communication in Fishes*. New York: Springer-Verlag (pp.459-480).

Schwartz, E. (1967). Analysis of surface-wave perception in some teleosts. In P.H. Cahn (Ed.). *Lateral Line Detectors* (pp. 123-134). Bloomington, IN: Indiana Univ Press.

Stebbins, W.C. (1970). Principles of Animal Psychophysics. In W.C. Stebbins (ed) *Animal Psychophysics: The Design and Conduct of Sensory Experiments*. New York: Appleton-Century-Crofts (pp. 1-19).

Tittel, G. (1988). *Untersuchungen zum Beutefangverhalten des Streifenhechtlings Aplocheilus lineatus (Cu. & Val.). Ein Modell zur Richtungsdetermination unter Berücksichtigung von ontogenetischen Prozessen des peripheren Seitenliniensystems* (Ph.D. Thesis) Giessen, W. Germany: Justus-Liebig-Universität Giessen.

G. M. Klump, R. J. Dooling, R. R. Fay & W. C. Stebbins (eds)

Psychoacoustical Studies of the Sense of Hearing in the Goldfish using Conditioned Respiratory Suppression

Richard R. Fay

Department of Psychology, Loyola University of Chicago
6525 N. Sheridan Rd., Chicago, IL 60626, USA

Summary. The methods used and some illustrative data from studies on hearing in the goldfish using classical conditioning of respiratory suppression are presented. Three psychophysical methods, and numerous psychoacoustical experimental designs are discussed. Presented in detail are: 1) A psychophysical study on level discrimination using the magnitude of respiratory suppression as if it were a confidence rating, and 2) A study of the perception of temporal acoustic patterns using a respiratory suppression in a stimulus generalization paradigm.

1 Introduction

This chapter discusses the methodologies used and data obtained in psychoacoustical studies on hearing in the goldfish (*Carassius auratus*) using classically conditioned respiratory suppression (Otis, et al., 1957). These methods have been used to describe the sense of hearing in the goldfish in more quantitative detail than for any other species of fish.

The rationale for developing these methods included the following considerations: 1) Studies were undertaken to help define the range of hearing phenomena among vertebrates and thus contribute to a greater understanding of the evolution of the sense of hearing. Fishes have the most simply organized vertebrate auditory systems, and systematic data on their hearing would provide significantly greater breadth to our understanding of hearing as a set of vertebrate characters. Goldfish were chosen because they are specialized to respond to sound pressure, are readily available and hardy, and had been studied anatomically and physiologically (Furukawa and Ishii, 1967). 2) A behavioral method was sought that could be used in restrained animals. This permits a precise estimate of the acoustic stimulus impinging on the animal, and permits neurophysiological experiments to be carried out with the fish in comparable sound fields. 3) A method was sought that leads to rapid acquisition of a conditioned response and that could be applied to the same individuals in many sessions without compromising the animals' health. The report by Otis et al. (1957) led to the first experiments (Fay, 1969a; Fay and MacKinnon, 1969).

2 Classical Respiratory Conditioning

An electric shock through the body (the unconditioned stimulus or UCS) causes a brief, unconditioned suppression of respiration (the unconditioned response, or UCR). A detectable acoustic signal of several seconds in duration that coterminates with the UCS becomes a conditioned stimulus (CS) after several conditioning trials, evoking a conditioned response (CR) of respiratory suppression. A typical conditioning trial is illustrated in Figure 1.

2.1 Measuring Respiration

A thermistor is positioned several mm in front of the animal's mouth. As the animal pumps water through the mouth and gills, heat is carried away from the thermistor. The temperature change is converted to a voltage by a bridge circuit. This is amplified, low-pass filtered at 6 Hz, and then digitally recorded. The signal's amplitude depends on the distance between the fish's mouth and the thermistor, and this distance may change as the animal changes its position in the restraining bag during a session. Thus, the signal's amplitude must be monitored, and the position of the thermistor adjusted when necessary.

Respiratory activity is quantified as the "length" of the respiratory waveform over several seconds, minus the length of the waveform expected if there were no respiration at all (a flat line). This line-length metric reflects both changes in respiration rate and amplitude. It is calculated as

$$\text{(1)} \quad \text{Sum over } i \text{ of } ((1 + v_i^2)^{0.5} - 1)$$

where v_i is the voltage at each sampling time *i*.

2.2 Measuring Respiratory Suppression

Suppression of respiration (SR) is quantified as

$$\text{(2)} \quad L_{cs} / (L_{cs} + L_{pcs})$$

where L_{cs} is the respiratory length during the CS, and L_{pcs} is the respiratory length for the same period just preceding the CS. SR thus tends to vary between 0 (complete suppression) and 0.5 (no suppression). SRs greater than 0.5 indicate accelerated respiration. The CS is six

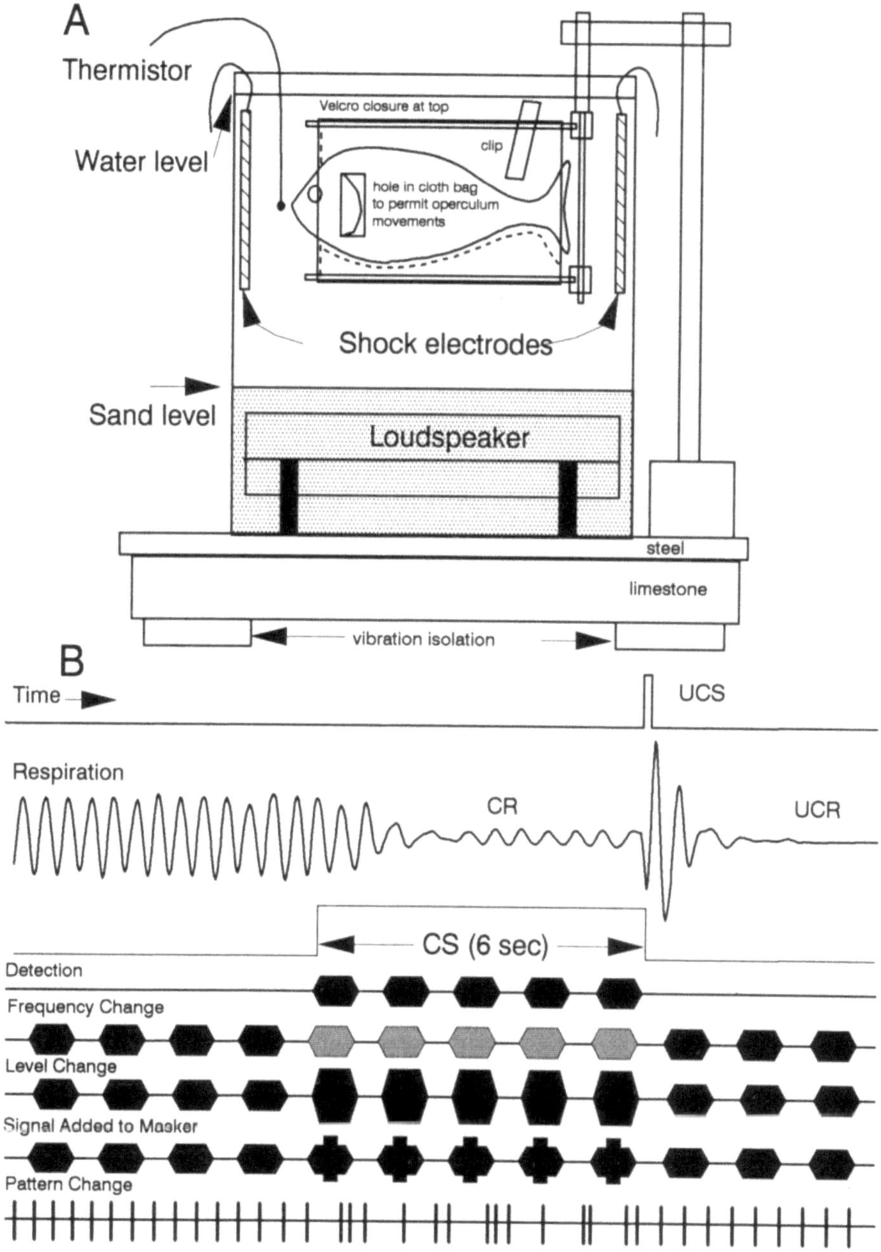

Figure. 1. A - a schematic diagram of the test tank and fish in its restrainer. B - A typical conditioning trial and a depiction of some of the sound stimulus paradigms used.

seconds in duration, but L_{CS} is measured only over the last four seconds where respiration tends to stabilize at its lowest values. L_{pcs} is also measured over four sec just preceding the CS (see Fig. 1).

2.3 The UCS

A 100 to 200 msec duration, 60 Hz ac shock (2-10 volts) from a variable transformer produces a robust UCR. Voltage is applied between steel hardware cloth grids located about 3 cm in front and behind the fish. Our impression is that head-tail voltages produce the most reliable response without causing tissue damage or responses that are incompatible with respiratory suppression (e.g., respiratory acceleration). However, shock across the animal's body has also been successfully used in respiratory (e.g. Fay, 1969a), and cardiac conditioning (e.g. Browman and Hawryshyn, 1992). The current passing through the animal depends on the resistances of the animal and the water. We use water from the animal's home aquarium.

2.4 The UCR

The unconditioned response is a suppression of respiration rate and amplitude lasting from one to 10 seconds. It is usually preceded by a transient caused by body movement. The voltage required for a robust UCR varies across individual animals, within a session, over days within an experiment, and appears to vary with the season of the year. Conditioning will not occur, and established Crs will extinguish if the UCR is not present, or is very weak.

It is sometimes difficult to evaluate the effectiveness of the UCS in trials having a robust CR; often the UCR appears attenuated or weak following a CR. Thus, in order to estimate the effectiveness of the UCS, it is necessary to observe the UCR in trials without a CS, or in trials not showing a CR. Successful conditioning and the maintenance of robust Crs requires a constant monitoring of UCS effectiveness, and adjustment of UCS voltage.

The UCR magnitude tends to decline during a test session. This appears to occur due to habituation and to factors that tend to accelerate respiration generally. Habituation can be controlled by appropriately long intertrial intervals; variable intervals ranging between two and five minutes seem to be effective in most cases. Respiration tends to accelerate over a test session (several hours). Unusually high respiration rates and amplitudes are difficult or impossible to suppress with a CS or UCS. In general, if the oxygen content becomes inadequate or if the water temperature rises, respiration magnitude will rise, and the UCS will become less effective. When animals show increasing respiration, conditioning or testing is suspended and the tank aerated for 10 to 20 minutes.

2.5 The CS

The conditioned stimulus is six or seven seconds in duration. The CS may be continuous or a train of bursts, and may be presented against a background of silence or of other sounds (see Fig. 1). The CS may consist of a change in some acoustic characteristic in an ongoing sound or train of bursts, such as level, frequency, or temporal pattern. Initial conditioning is usually carried out with the CS presented 30 dB or more above its estimated absolute threshold.

2.6 The CR

The conditioned response tends to appear in the first conditioning session after from five to 20 trials. The CR appears to be very similar to the UCR; it usually consists of a reduction of both the amplitude and rate of respiration. The CR may take two to five seconds to reach its maximum, particularly for near-threshold signals.

2.7 Restraining the Animal

Goldfish are gently restrained in a cloth bag (see Fig. 1). Loosely suspended between two horizontal rods, the bag opens at the top for introducing the animal, and is then closed over the top rod using Velcro strips. The bag is open in the back and has openings in front for the animal's snout and for the gill covers, and is contoured at the bottom to fit the fish's ventral profile. Once the animal is placed in the bag, the horizontal rods are separated just enough to restrain the animal without impeding respiratory mouth and gill movements. A spring clip is often used to hold the bag together at the rear and to prevent the animal from backing out. The restrainer is supported by clamps to a rod and magnetic base.

It is critical to have enough restraint to prevent the animal from escaping, but not enough to interfere with respiration. If an animal escapes once from the restrainer, it will struggle in the bag for the next several sessions. Animals that never have escaped seldom struggle. The positioning, sizes, and shapes of the mouth and gill cover holes are critical for successful restraint with unimpeded respiration and without struggling.

2.8 The Test Tank and Acoustics

The test tank is a plexiglass cylinder, 23 cm in diameter and 28 cm high, 6.2 mm wall thickness. The sound source is a University Sound UW-30 underwater loudspeaker, supported on legs, that projects upward from the bottom of the tank. All external chambers of the loudspeaker must be filled with water. The loudspeaker is buried in water-saturated sand with the speaker about 2 cm below the sand surface. The water-saturated sand tends to smooth out the frequency response characteristics of the tank system. The water saturating the sand is left in the tank indefinitely. When an animal is placed in the tank, a small plastic garbage bag is used to line the tank above the sand level, and is filled with water from the animal's home aquarium. In addition, the cloth bag restrainer is rinsed before each use. These two procedures help reduce stress possibly caused by secretions from the animal previously tested.

The restrainer is suspended in the center of the tank with the animal's head about 2 cm from the water surface. The water surface is 5 mm from the tank top. The tank is vibration-isolated within an Industrial Acoustics, Inc. single-walled audiometric booth.

Acoustic signals are measured by a calibrated hydrophone placed in the restrainer bag, replacing the fish. The hydrophone signal is amplified, bandpass filtered, and digitized at 5 kHz. The frequency response of this tank is uniform within 10 dB from 100 to 1000 Hz.

3 Psychophysical Paradigms

A wide variety of psychophysical paradigms and experimental designs have been used to study the sense of hearing in the goldfish with respiratory conditioning. In general, psychophysical paradigms include a modified method of limits, single staircase tracking, and the method of constant stimuli. Each may be appropriate for particular applications. For each paradigm, each trial is immediately preceded by a blank trial (without a CS and UCS). The purpose of the blank trials is to estimate the distribution of SRs occurring by chance, and to estimate false alarm rates in the staircase tracking and constant stimuli methods.

3.1 Modified Method of Limits

Stimulus values begin so as to be highly detectable, and are reduced on successive trials until the response magnitude or the stimulus value falls below some criterion. Then the series is repeated several times within a session. Suppression ratios (SR) are averaged and plotted as a function of stimulus value to give a performance function analogous to a psychometric function. An SR criterion is chosen as the definition of threshold, and thresholds determined as the interpolated stimulus value corresponding to the criterion SR. Usually three to four descending stimulus value runs are completed per session, for at least two sessions.

This method has been used in studies of sound detection (Fay, 1969a, 1969b), frequency discrimination (Fay, 1970a), critical masking ratios (Fay, 1974), psychophysical tuning curves (Fay et al., 1978), detection thresholds for direct head vibration (Fay and Patricoski, 1979), detection of sinusoidal amplitude modulation for tones and noise, and temporal modulation transfer functions (Fay, 1980), sound pulse interval discrimination with and without interval jitter (Fay and Passow, 1982), and amplitude modulation rate discrimination (Fay, 1982).

3.2 Single Staircase Tracking

Animals are initially conditioned using a clearly detectable stimulus value. After stable performance appears (often within the initial session) stimulus values are reduced following each response producing an SR below a criterion value (usually 0.4), and are otherwise increased. Usually eight to ten stimulus value reversals define a single track, and one to three tracks are run per session. At least two sessions are run per animal per stimulus to define a final threshold. Usually, four or more animals are run in each experiment. Note that unlike the method of limits and the method of constant stimuli, each individual response in the tracking method is classified as a "yes" or a "no" on the basis of the suppression ratio. A response larger than the criterion SR of 0.4 was shown to occur about 12% of the time by chance (Fay and Coombs, 1983).

This method has been successfully used in studies of temporal summation at threshold (Fay and Coombs, 1983), repetition noise delay discrimination (Fay et al., 1983), amplitude modulation detection (Coombs and Fay, 1985), simultaneous and forward masking (Fay and Coombs, 1988; Coombs and Fay, 1989), level discrimination (Fay, 1985; Fay, 1989a), frequency discrimination (Fay, 1989b), and gap detection (Fay, 1985).

3.3 Method of Constant Stimuli

In this method, only one stimulus value occurs per session of about 50 trials, and one to three sessions are run per day per animal. A threshold is defined after at least two days of testing per animal. This method has been applied in a study of psychometric functions and the effect of stimulus duration on level discrimination (Fay and Coombs, 1992). In this study, SRs were treated as confidence judgements that a level increment occurred (see below).

4 Detection and Discrimination Psychoacoustic Paradigms

A number of psychophysical studies on detection and discrimination have been carried out on the goldfish using respiratory conditioning. The following summarizes the stimulus variables manipulated.

Sound detection has been studied in the quiet and in the presence of masking noise for continuous (Fay, 1969a; 1969b; 1970a; Fay, et al., 1978) and pulsed tones (Fay and Coombs, 1983; 1988; Coombs and Fay, 1989) and noise (Fay, 1985).

Level and frequency discrimination have generally been studied by introducing a change (an increment or decrement) in an ongoing sound stream consisting of pulsed or continuous tones or noise (Fay, 1970b, 1985, 1989a, 1989b; Fay and Coombs, 1992).

Amplitude modulation detection (including the temporal modulation transfer function and gap detection) has been studied by introducing modulation to both constant and pulsed backgrounds (Fay, 1980, 1985; Coombs and Fay, 1985). Amplitude modulation rate discrimination has been studied using tone and various spectrally complex carriers, with and without jitter of the modulation period (Fay, 1982; Fay and Passow, 1982).

The perception of repetition noise ("rippled noise") has been studied systematically. Goldfish were conditioned to detect the substitution of a rippled spectrum for an otherwise flat or differently rippled noise spectrum (Fay et al., 1983).

5 Stimulus Generalization Paradigms

In a stimulus generalization paradigm, animals are conditioned to one or more stimuli and then tested (without the UCS) to novel stimuli. The magnitude of the response to the novel stimuli is a measure of the perceived similarity between the conditioning and testing stimuli. In our experiments, a naive animal is first conditioned (40 trials) and then tested for generalization (40 trials) in one session (Fay, 1969a, 1970b, 1972, 1992, 1994a, 1994b; Fay and Patterson, 1994). The test session consists of eight test stimuli presented four times each in random order without the UCS. On every fifth trial, the conditioning stimulus is presented with the UCS in order to maintain the conditioned response. Typically, conditioning and testing is suspended for 20-40 min in order to aerate the test tank water and maintain a slow, steady respiration rate.

A median suppression ratio (SR) is calculated for each of the stimuli presented during the testing session. The generalization metric is normalized with respect to the median SR for the conditioning stimulus measured during the test session. Percent generalization is defined as

$$(3) \quad ((0.5-med_T)/(0.5-med_C))*100$$

where med_T is the median SR to the test stimulus, and med_C is the median SR to the conditioning stimulus as determined during the test session. Percent generalization varies between 0 (no respiratory suppression) and 100 (suppression equal to that in response to the conditioning stimulus). Values above 100 occur when the suppression to a test stimulus is greater than that to the conditioning stimulus. Mean or median SRs are plotted as a function of the stimulus dimension(s) manipulated during the test phase to give a generalization gradient across the test stimuli for each animal. A session may be repeated for some individuals on different days.

Stimulus generalization has been used to study the similarity of tones of different frequency (Fay, 1969a, 1970b, 1992), the perception of amplitude-modulated tones (Fay, 1972; Fay and Patterson, 1994), pitch-like perceptions (Fay, 1972; Fay, 1994b), the ability to hear out sinusoidal components in multi-component stimuli (Fay, 1992), and the perception of click trains differing in temporal pattern (Fay, 1994a).

6 The SR as a Confidence Rating

In Fay and Coombs (1992), the method of constant stimuli was used to generate psychometric functions in a study of level discrimination. Psychometric functions were required so that neurophysiological predictions of behavioral performance could be quantitatively evaluated. Animals were conditioned to detect a change in the level of ongoing tone bursts.

A session consisted of 50 trials of a fixed signal. Performance was quantified as a percentage analogous to percent correct from a confidence-rating experiment (e.g., Watson et al., 1964) (see Fig. 2). On the left of Fig. 2, suppression ratios (SR) are shown for 50 test trials at four level differences. Solid lines connect SR values for signal trials, and dotted lines connect SR values for blank trials. Frequency distributions of SRs for signal and blank trials are shown in the middle panel of Fig. 2. SRs can be viewed as confidence ratings that the signal is present.

The differences between the signal and blank trial SR distributions are summarized by empirical receiver operating characteristics (ROC) (Fig. 2, right). Empirical ROCs are a plot of the cumulative distribution of SRs for the signal trials versus that for the blank trials. The area under the ROC (P(A)) is an estimate of the proportion correct for an unbiased observer in a two-interval, forced-choice experiment (Macmillan and Creelman, 1991). P(A) plotted as a function of signal level defines a psychometric function, and the signal level at P(A)=0.76 is defined as the level discrimination threshold. Three- to five-point psychometric functions were fit with Weibull functions (Relkin and Pelli, 1987) from which thresholds and slopes were measured.

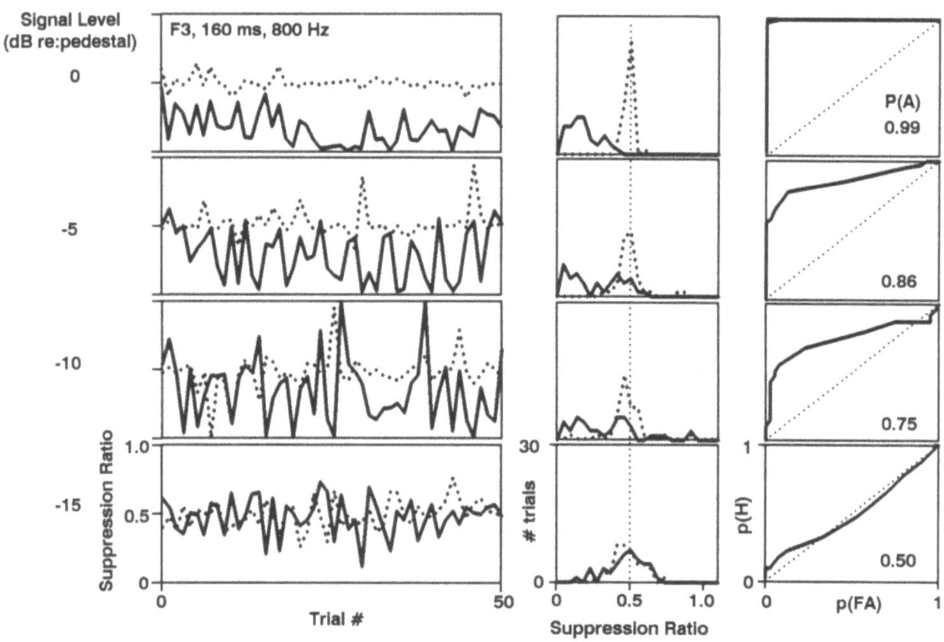

Figure 2. The method of ROC analysis for obtaining psychometric functions from SR data treated as confidence ratings. Left panels show SRs for 50 signal trials (solid lines) and 50 nonsignal trials (dotted lines). Middle panels show the corresponding frequency distributions of SRs from the last 40 of 50 signal (solid lines) and nonsignal (dotted lines) trials. The right panels show ROCs. P(A) is the area under the ROC (from Fay and Coombs, 1992).

Psychometric functions for one goldfish are shown in Fig. 3A. Weibull function slopes varied between 0.5 and 1.2. A typical psychometric function for one goldfish is compared with psychometric functions for several other vertebrates in Figure 3B. The close correspondence between psychometric functions for goldfish and other vertebrate species investigated argue for the general validity of animal psychophysics, and for the usefulness of this method.

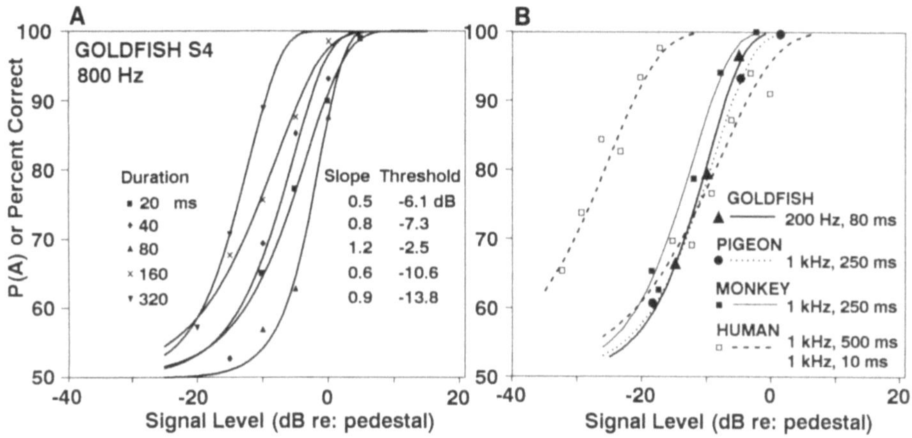

Figure 3. A - Psychometric functions for level discrimination in one animal tested at five durations for 800 Hz signals. Smooth lines are Weibull functions (see text) which best fit the plotted P(A) values. B - Psychometric functions for sound level discrimination for several vertebrate species compared (From Fay and Coombs, 1992). Functions for two signal durations are shown for the human (left function 500 ms, right function 10 ms).

7 Stimulus Generalization and the Perception of Temporal Patterns

These experiments (Fay, 1994a) focus on the responses to filtered clicks repeated with both periodic and complex temporal patterns. Previous studies showed that goldfish conditioned to pure tones produce symmetrical and monotonic gradients of generalization as a function of the frequency separation between conditioning and test tones (Fay, 1969a, 1970b; 1992). These data suggest that goldfish have a perceptual dimension that is a monotonic function of stimulus frequency and is possibly analogous to pitch perception in humans. The present study investigated whether goldfish may have a similar perceptual dimension determined by waveform periodicity that may be analogous to periodicity pitch in humans (e.g., de Boer, 1976). Stimuli were 6 sec trains of a pulse repeated periodically at different repetition rates. Five groups of four animals were conditioned to different pulse rates and then tested to the same set of eight novel rates.

Figure 4 shows mean generalization gradients. For all groups, generalization gradients are substantially symmetric and monotonic with the difference between the conditioning and test pulse rates. If generalization can be taken as a measure of perceptual similarity, periodic pulse trains are judged similar to one another to the extent that they have similar repetition rates. These data are consistent with the notion that there is a perceptual continuum, perhaps similar to periodicity pitch in humans, that is monotonic with stimulus pulse repetition rate between 40 and 251 pps. Other experiments using these stimuli (Fay, 1994a) suggest that this perceptual dimension arises from estimates of interpulse time rather than pulse count, and does not depend on the existence of spectral lines in the stimuli. However, it is not yet clear whether the dimension might best be termed "pitch-like" or "roughness-like."

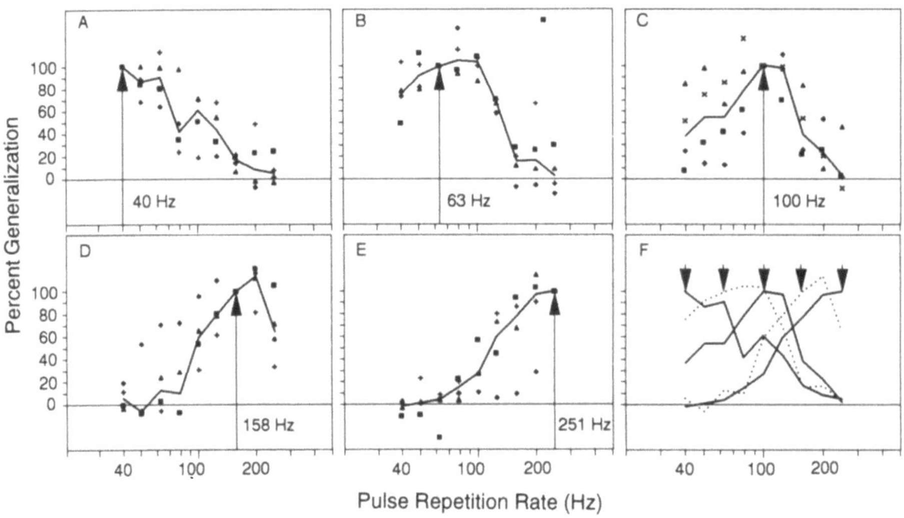

Figure 4. Group mean generalization gradients for five groups of animals initially conditioned at a different periodic pulse rate (vertical arrows). Panel F shows the five gradients superimposed (from Fay, 1994a).

These experiments demonstrate that the temporal pattern of a repeated acoustic pulse exerts a strong effect on perceptual quality for the goldfish, and that the stimulus generalization paradigm is useful for investigating complex aspects of sound perception in nonhuman animals.

8 Suggested Reading

Fay, R.R. (1970a). Auditory frequency discrimination in the goldfish. *Journal of Comparative and Physiological Psychology 73*, 175-180.

Fay, R.R. (1972). Perception of amplitude modulated auditory signals by the goldfish. *Journal of the Acoustical Society of America 52*, 660-666.

Fay, R.R. & Coombs, S. (1992). Psychometric functions for level discrimination in the goldfish (*Carassius auratus*): Psychophysics and neurophysiology. *Journal of the Acoustical Society of America 92*, 189-201.

Fay, R.R. (1988). *Hearing in Vertebrates: A Psychophysics Databook.* Winnetka, IL: Hill-Fay Associates.

Fay, R.R. (1992). Analytic listening by the goldfish. *Hearing Research 59*, 101-107.

9 References

Browman, H.I., & Hawryshyn, C.H. (1992). Thyroxine induces a precocial loss of ultraviolet photosensitivity in rainbow trout (*Oncorhynchus mykiss*, Teleostei). *Vision Research* 32, 2303-2312.

Coombs, S. & Fay, R.R. (1985). Adaptation effects on amplitude modulation detection: Behavioral and neurophysiological assessment in the goldfish auditory system. *Hearing Res*earch *19*, 57-71.

Coombs, S., & Fay, R.R. (1989). The temporal evolution of masking and frequency selectivity in the goldfish (*Carassius auratus*). *Journal of the Acoustical Society of America 86*, 925-933.

de Boer, E. (1976). On the "residue" and auditory pitch perception. In W.D. Keidel and W.D. Neff (Eds.) *Handbook of Sensory Physiology* vol V/3. Berlin: Springer-Verlag.

Fay, R.R. (1969a). Auditory sensitivity of the goldfish within the acoustic nearfield. U.S. Naval Submarine Medical Center, Submarine Base, Groton, Connecticut, Report No. 605, 1-11.

Fay, R.R. (1969b). Behavioral audiogram for the goldfish. *J. Aud. Res.*, *9*, 112-121.

Fay, R.R. (1970a). Auditory frequency discrimination in the goldfish. *Journal of Comparative and Physiological Psychology 73*, 175-180.

Fay, R.R. (1970b). Auditory frequency generalization in the goldfish. *J. Exp. Anal. Behav.*, *14*, 353-360.

Fay, R.R. (1972). Perception of amplitude modulated auditory signals by the goldfish. *Journal of the Acoustical Society of America 52*, 660-666.

Fay, R.R. (1974). The masking of tones by noise for the goldfish. *Journal of Comparative and Physiological Psychology 87*, 708-716.

Fay, R.R. (1980). Psychophysics and neurophysiology of temporal factors in hearing by the goldfish: Amplitude modulation detection. *Journal of Neurophysiology 44*, 312-332.

Fay, R.R. (1982). Neural mechanisms of an auditory temporal discrimination by the goldfish. *Journal of Comparative Physiology 147*, 201-216.

Fay, R.R. (1985). Sound Intensity Processing by the Goldfish. *Journal of the Acoustical Society of America 78*, 1296-1309.

Fay, R.R. (1988). *Hearing in Vertebrates: A Psychophysics Databook.* Winnetka, IL: Hill-Fay Associates.

Fay, R.R. (1989a). Frequency discrimination in the goldfish: Effects of roving intensity, sensation level, and the direction of frequency change. *Journal of the Acoustical Society of America 85*, 503-505.

Fay, R.R. (1989b). Intensity discrimination of pulsed tones by the goldfish. *Journal of the Acoustical Society of America 85*, 500-502.

Fay, R.R. (1992). Analytic listening by the goldfish. *Hearing Research 59*, 101-107.
Fay, R.R. (1994a). Perception of temporal acoustic patterns by the goldfish. *Hearing Research,* in press.
Fay, R.R. (1994b). The sense of hearing in fishes: Psychophysics and neurophysiology. *Sensory Systems* (A translation of the Russian language *Sensornyne Sistemy*), New York:Consultants Bureau, in press.
Fay, R.R. Ahroon, W.A. & Orawski, A.T. (1978). Auditory masking patterns in the goldfish: Psychophysical tuning curves. *Journal of Experimental Biology 74*, 83-100.
Fay, R.R. & Coombs, S. (1983). Neural mechanisms in sound detection and temporal summation. *Hearing Research 10*, 69-92.
Fay, R.R. & Coombs, S. (1988). Psychophysics and neurophysiology of frequency selectivity and masking in the goldfish. In H. Duifhuis, W. Horst, and H. Wit (Eds.) *Basic Issues in Hearing*, Groningen: Groningen University Press.
Fay, R.R. & Coombs, S. (1992). Psychometric functions for level discrimination in the goldfish (*Carassius auratus*): Psychophysics and neurophysiology. *Journal of the Acoustical Society of America 92*, 189-201.
Fay, R.R. & MacKinnon, J.R. (1969). A simplified technique for conditioning respiratory mouth movements in fish. *Behav. Res. Meth. & Instru., 1*, 123-124.
Fay, R.R. & Passow, B. (1982). Temporal discrimination in the goldfish. *Journal of the Acoustical Society of America 72*, 753-760.
Fay, R.R. & Patricoski, M. (1980). Sensory mechanisms for low frequency vibration detection in fishes. In R. Buskirk (Ed.), *Abnormal Animal Behavior Prior to Earthquakes, II.* U.S. Geological Survey Open File Report 80-453, 63-91.
Fay, R.R. & Patterson, R. (1994). Perception of sinusoid "ramps" and "damps" in the goldfish. *Journal of the Acoustical Society of America* 95, 2941 (abstract).
Fay, R.R., Yost, W.A. & Coombs, S. (1983). Repetition noise processing by a vertebrate auditory system. *Hearing Res*earch 12, 31-55.
Furukawa, T. & Ishii, Y. (1967). Neurophysiological studies on hearing in the goldfish. *Journal of Neurophysiol*ogy *30*, 1337-1403.
Macmillan, N.A. & Creelman, C.D. (1991). *Detection Theory: A User's Guide.* Cambridge: Cambridge University Press.
Otis, L., Cerf, L., & Thomas, G. (1957). Conditioned inhibition of respiration and heart rate in goldfish. *Science, 126*, 263-264.
Relkin, E.M. & Pelli, D.G. (1987). Probe tone thresholds in the auditory nerve measured by two-interval forced-choice procedures. *Journal of the Acoustical Society of America 82*, 1679-1691.
Watson, C.S., Rilling, M.E., & Bourbon, W.T. (1964). Receiver operating characteristics determined by a mechanical analog to the rating scale. *Journal of the Acoustical Society of America 36*, 283-288.

Methods in Comparative Psychoacoustics
G. M. Klump, R. J. Dooling, R. R. Fay & W. C. Stebbins (eds)

Investigations of Fish Hearing Ability Using an Automated Reward Method

Hong Young Yan

Thomas H. Morgan School of Biological Sciences, University of Kentucky, Lexington, KY 40506-0225, U. S. A.

Summary. A microcomputer-controlled positive reward operant conditioning paradigm was developed to measure auditory sensitivity of goldfish (*Carassius auratus*) and oscar (*Astronotus ocellatus*). Past studies of hearing in fish have proved challenging in part because of difficulties in precise measurement of the sound field and in part because their range of behavior seems so limited. Here a new positive reinforcement procedure is described which solves some of these problems. The audiogram obtained from this paradigm is in close agreement with thresholds obtained with instrumental avoidance and classical conditioning procedures involving aversive conditioning. The major advantages of the present paradigm are: (1) there is minimal stress to the fish, and (2) the versatility of the system allows easy modification for a variety of experimental needs.

1 Introduction

G. H. Parker and his student, H. B. Bigelow were two of the pioneers on fish hearing research at the turn of the 20th century. Parker (1902, 1903) demonstrated that *Fundulus heteroclitus* (Linnaeus) could respond to sounds produced by plucking a bass-viol string attached to the aquarium and cutting of the auditory nerves abolished most of the response. Bigelow (1904) later repeated Parker's procedures on goldfish and found similar results. He also found that by removing the semi-circular canals and utriculus, the fish were still able to respond to sound. The study also established that the sacculus-lagena portion (pars inferior) was probably responsible for sound perception in goldfish. Though stimulus control was crude by today's standards, quantitative work on the range of frequencies over which fish can hear was later carried out by von Frisch and his colleagues (see review by von Frisch, 1936). These work laid the foundation for more comprehensive psychophysical analyses (e.g., absolute thresholds and intensity difference limens) of fish hearing (Jacobs & Tavolga, 1967; Tavolga & Wodinsky, 1963; Tavolga, 1976).

The measure of hearing sensitivity is generally expressed in terms of the minimum detectable level, or threshold. This is a statistical value that is defined as the sound pressure level (SPL) at which the animal responds correctly in a specified proportion of signal presentations (e.g., 50%). A plot of thresholds over the range of audible frequencies is defined as an audiogram. Fay (1988) compiled a list of more than 50 behavioral audiograms for fish species including goldfish, carps, cods, catfish, damselfishes, minnows, knifefish, and sharks.

Behavioral audiograms of fish have generally been obtained with one of three different procedures. The simplest method involved training fish to approach a particular area on hearing a sound. A correct response was rewarded with food (Kritzler & Wood, 1961). With this method, the random location of the fish relative to the sound source during sound presentation, made it difficult to determine the exact SPL detected by the fish. In addition, the response of the fish was often ambiguous and erratic, especially near the threshold for the frequency being studied (Hawkins, 1981).

The second method consisted of instrumental avoidance conditioning typically using a shuttlebox (Horner, Longo, & Bitterman, 1961), in which fish were trained to cross a barrier in the tank on hearing a sound in order to avoid receiving an electric shock (Tavolga & Wodinsky, 1963). This method also suffered from some nonuniformity of the sound field, because of the free movement of the fish around the tank. In addition, training some fish like the cichlid *Tilapia macrocephala* (Tavolga, 1974) with electric shock took an unusually long time.

The third and the most widely used method was the classical conditioning of heartbeat or gill movement (Fay, 1969), in which the sound stimulus was paired with an electric shock. The fish were conditioned to associate the sound with the shock, which resulted in cardiac and ventilatory suppression whenever the acoustic stimulus was detected. In such studies, fish could be positioned at a specific point in the sound field where the SPL was known which is a tremendous advantage. This method has so far generated majority of the fish audiogram data (see Fay 1988 for detailed list). The procedure, however has minor drawbacks not only because the shock used to condition the fish might be stressful, but also because the animal had to be restrained. Restraining of fish may also lead to physiological stress (Donaldson, 1981). Furthermore, not every fish species could be conditioned with electric shock- including, for example, some sharks (Nelson, 1967), and a few cichlids (Tavolga, 1974; Allen and Fernald, 1985).

Operant conditioning procedures involving positive reinforcement have proved enormously successful in the investigation of visual processes in some fish (e.g. color vision in goldfish, Yager & Thorpe, 1970; spectral sensitivity in cichlids, Allen and Fernald, 1985) and in auditory studies in many other vertebrate species (see other studies, this volume). However, no such procedures have been applied to the study of acoustic sensitivity in fish until 1989 (Yan and Popper, 1991). Subsequently, the same method was used successfully in measuring audiograms of goldfish (Yan and Popper, 1991), oscar (*Astronotus ocellatus,* Yan and Popper, 1992) and two sunfishes, the bluegill (*Lepomis macrochirus)* and the ongear (*Lepomis megalotis*) (Yan, unpublished data). Intensity discrimination ability of oscar was also measured by a modified version of the procedure (Yan and Popper, 1993). My purpose in this chapter was to describe in details the automated operant conditioning protocol and hardware setup in which positive reinforcement was successfully used to measure fish acoustic sensitivity.

2 Method

2.1 Subjects

Both goldfish (*Carassius auratus*), about 8 cm, and oscar (*Astronotus ocellatus*), about 5.5 cm in standard length (measured from tip of the snout to the end of the vertebral column) were used as subjects. For goldfish each fish was maintained in individual 40-l glass tanks (50 x 30 x 26.6 cm). For oscar, fish were housed as a group in a 40-l tank. Both fish were fed daily with food pellets. The water was continuously aerated and filtered and 1/2 to 1/3 of the water was changed on weekly basis. The water temperature was maintained at about 26.5 (± 1) °C. The photoperiod of fish holding room was set at 16L8D cycle.

2.2 Apparatus

The test apparatus sat on a Plexiglass platform (32 cm x 20 cm) attached to a support made from wooden studs, which provided a structure that could be transferred easily from one tank to another (Figure 1).

An automatic feeder (solenoid-controlled) on the platform dispensed pellets of fish food on command from a one-shot timer (S52-53; Coulbourn Instruments (CI)) controlled by the computer. These floating food pellets (Hikari Gold for goldfish and Hikari Cichlid Staple for oscar; Kyorin Food) were dispensed through a plastic tube that contacted the surface of the water, so that a fixed feeding station (Figure 1, interface of T and water surface) was easily recognized by the fish. A 10-W, 28-volt ceiling light (C-light) was hung at the rear of the platform to illuminate the tank during the experiment when needed.

The response panel consisted of two steel rods 6 cm apart. These rods were attached to the underside of the platform. Each rod held a sealed clear plastic tube (diameter 1 cm, length 5 cm) that contained a miniature 10-W, 28-volt bulb (Sylvania Model 352110). A piece of pressure-sensitive Kynar piezo film (AMP, Inc.) was attached outside each tube. The pressure exerted by the fish "pecking" at the tube was transduced to an electrical impulse by this film.

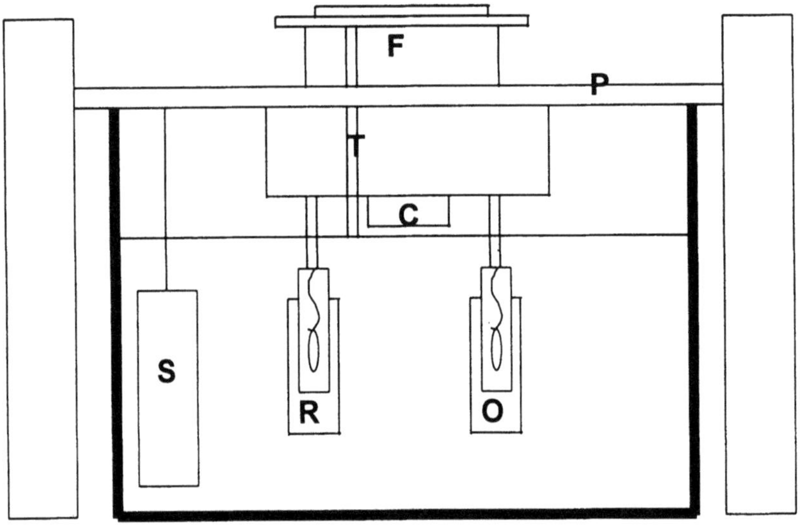

Figure 1. Schematic diagram of the test tank from fish-eye view, showing the apparatus. C, ceiling light; F, automatic feeder; O, observation paddle; P, plexiglas platform; R, report paddle; T, pellet delivery tube. Notice that both the O-light and the R-light are housed inside clear plastic waterproof tubes. See text for a detailed description of the apparatus.

One tube with piezo film was the observation paddle (O-paddle) and the other the report paddle (R-paddle). The piezo film only required a gentle touch or "peck" of the paddle by the fish in order to produce an above threshold analog response signal to a comparator with a schmitt-trigger circuit (Dual threshold detector, CI S21-10, Figure 2).

Acoustic signals were generated with a signal generator (CI, S81-06) connected to a digitally controlled attenuator (CI, S85-08) and an audio mixer-amplifier (CI, S82-24). The stimuli were presented through an underwater speaker (University Sound, Model UW-30) which was attached by wires to the underside of the Plexiglass platform. The signal generator and attenuator were calibrated with an LC-10 hydrophone (Celesco Co) and a wave analyzer (Hewlett-Packard 3581A). Sound pressure levels of various frequencies and attenuations were also measured and calibrated around the O-paddle. These data served as a reference when specific frequency and SPLs were used later in the acoustic sensitivity study.

A 24-bit parallel digital input/output board (Metrabyte, Model PIO12) was used as an interface between a microcomputer (IBM AT) and all electronic modules. The apparatus, training of animals, acoustic sensitivity and intensity discrimination studies were all controlled with programs written in Microsoft QuickBASIC .

Experiments were conducted in a soundproof IAC chamber (Model 402-A; 1.9 m x 1.8 m x 2.0 m, Industrial Acoustic Corp.) which helped to cut down on background noise and minimize masking of absolute thresholds.

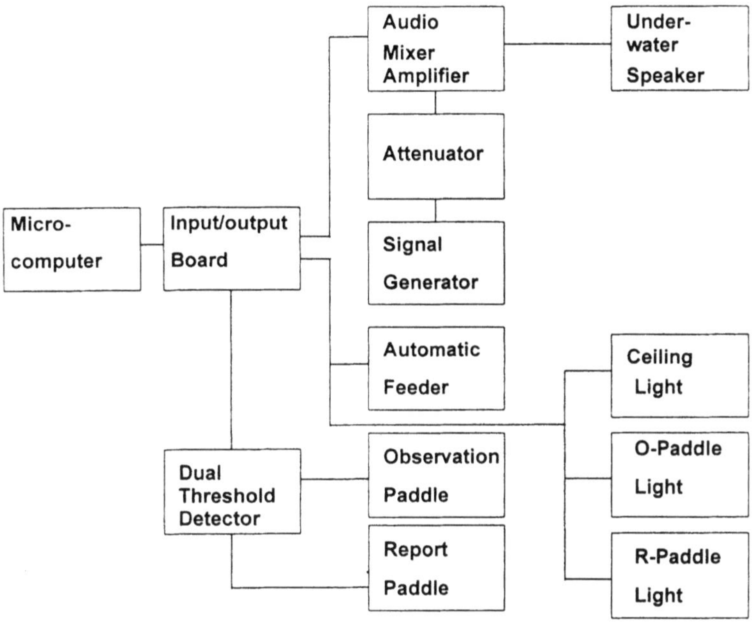

Figure 2. Block diagram of the automated electronic system and apparatus for behavioral acoustic study.

2.3 Training Procedure

a) Absolute Auditory Threshold Study

Autoshaping procedures were used to train both the goldfish and the oscar to make appropriate series of responses. The training procedure was divided into five stages. In stage I, food pellets were dropped into the feeding station at random intervals (5 to 300 s), so that the fish would recognize the location of the feeding site. In stage II, the O-paddle light was turned on and a medium size pellet (diameter 4mm) was fastened to the outside of the O-paddle with nearly transparent parafilm. When immersed in water, the sight and odor of the food pellet attracted a fish to peck on the O-paddle. However, the fastening of the pellet prevented it from immediately being taken by the fish. When pecked by the fish, the piezo film attached to the O-paddle generated an analog signal which was detected by comparator. The comparator then triggered the computer to activate the automatic feeder which dropped a food pellet. The fish thus learned to associate pecking the paddle with obtaining food rewards.

After the fish consistently pecked the O-paddle even after the attached pellet was deliberately removed it was possible to start stage III of training. In this stage no food pellet was fastened to the O-paddle. Instead, after the fish pecked the O-paddle its light was turned off and the R-paddle light was turned on. If the R-paddle was pecked within a defined time limit (no limit in the beginning, which was gradually reduced to 30 s), the R-paddle light was turned off and the fish was rewarded with food. However, if the R-paddle was not pecked within the time limit, or the R-paddle was pecked before the O-paddle, the ceiling light was turned off and the fish was subjected to complete darkness for either 90 s (goldfish) or 180 s (oscar). The blackout deprived the fish of feeding opportunities and served to train the fish to peck the paddle within the time limit. It also reduced the number of false positive responses made by the fish.

After a fish consistently pecked the R-paddle within 30 s of pecking the O-paddle, stage IV training was introduced. In this stage, the training was carried out with an almost identical protocol to that of stage III, except that the pecking of the O-paddle was followed by a 500 Hz tone (13 dB for goldfish, 33 dB for oscar; re: 1 μbar) instead of lighting of the R-paddle light. The fish had to peck the R-paddle when the sound was on. At the beginning, the sound duration was extended to either 90 s (for goldfish) or 180 s (for oscar) long, and it was gradually reduced to either 4 s (for goldfish) or 7 s (for oscar).

At stage V of the training, blank trials (no signal presented) were introduced in random order and in equal numbers with sound presentations. The pecking of the R-paddle during blank trials led to a blackout (lights temporarily extinguished). The purpose of the blank trial was to train the fish not to peck the R-paddle when sound was not heard, such as when the sound was below threshold. Training at stage V continued until the fish constantly attained an error rate of less than 10% over a session of 32 trials. The error rate was used to determine when the fish was ready for threshold testing and it was calculated by the following formula:

$$E\,(\%) = ((A + C)/(A + B + C + D)) \times 100 \qquad (1)$$

where A was the number of responses to the R-paddle when the blank was given (false alarm); B was the number of no responses to the R-paddle when blank was given (correct rejection); C was the number of no responses to the R-paddle when the sound was given (miss); D was the number of responses to R-paddle when sound was given (hit) (Yost and Nielsen, 1985).

b) Intensity Discrimination Study

The basic training procedures were the same as those for the audiogram study. However, some modifications were made to accommodate the needs of intensity discrimination studies and the oscar was the experimental species. In stage I, a background sound of 500 Hz at 25 dB (re 1 μbar, threshold value for oscar at 500 Hz, see Yan and Popper, 1992) was constantly presented through the underwater speaker. At stage II, pecking of the O-paddle increased the level of the background sound to 45 dB. At stage III, pecking the O-paddle changed the background sound to test sound and R-light was turned on. Stage IV training was almost identical to the previous stage except no R-light was turned on after O-paddle pecking. During the last stage (V) of training, background sound trials (analogous to "blanks" of audiogram study) were introduced in random order but in equal numbers with the test sound presentations. The purpose of the background sound trials was to train the fish not to peck the R-paddle when the fish could not detect a difference between the intensities (i.e., limits of discrimination limens) of two successive signals. The error rate criterion used here was the ame as that used in the audiogram study.

2.4 Testing

a) Acoustic Sensitivity Study

Fish were deprived of food for 16-23 h before a test. The test apparatus was placed onto to the test tank. For goldfish, thresholds were measured at 200, 500, 1000, 1500 and 2000 Hz; for oscar, thresholds were investigated from 200 to 1000 Hz in 100 Hz intervals, both using a modified method of constant stimuli (e.g., Yost and Nielsen, 1985). Four to 6 SPLs were used at each frequency. For each frequency tested, the initial SPL was always the maximum output from the speaker and this was subsequently attenuated either at 5 or 10 dB intervals. In each test run, 5 replicates of a chosen SPL and 5 blank trails were randomly presented to the fish. A minimum of 2 test runs were repeated for each fish at each SPL in order to calculate the response rate (R). A response rate was calculated by dividing the number of correct responses to the R-paddle by the total number of sound trials presented. The threshold was determined from a psychometric function relating percent correct response to SPL. To generate this function, a second-order polynomial was fitted to the percentage of correct responses R in relation to the sound-pressure level in dB (Richards, 1976):

$$\mathrm{SPL\ (dB)} = a + bR + cR^2 \qquad (2)$$

The threshold for a particular frequency was obtained by calculating the SPL that yielded a 50% correct response rate in the psychometric function. For each fish, final threshold estimations were based on a minimum of at least three polynomial psychometric functions at each frequency.

b) Intensity Discrimination Study

The intensity discrimination ability of oscars was investigated as function of sensation level (SL) and frequency. SL is defined as the sound pressure level above the threshold at a specific frequency. The fish was tested at 500 Hz at SLs of 10 dB, 20 dB, and 30 dB. Like the procedure used in the audiogram study, the intensity-difference limen for each SL was determined using a modified method of constant stimuli. Four to six ΔIs (ΔI=test sound-pressure level - background-sound pressure level) were used in each SL test. The threshold ΔI for a particular SL was obtained by the same curve-fitting method used in the absolute-threshold study. However, for each fish, threshold determinations for each SL at 500 Hz were repeated at least 4 times. Tests at an SL of 20 dB were repeated at different frequencies. Earlier studies showed that the oscar has a bandwidth of hearing that only extends from 200 Hz to 800 Hz (Yan and Popper, 1992). Therefore, only 200 Hz, 500 Hz, and 800 Hz were chosen and the intensity-difference limen was measured for each frequency. The protocol used here was identical to the previous test.

3 Results

3.1 Absolute Auditory Sensitivity Study

The threshold SPLs for five frequencies (200, 500, 1000, 1500, and 2000 Hz) obtained from the goldfish with the modified method of constant stimuli are shown in Figure 3.

All oscars showed no response to pure tone signals at 900 and 1000 Hz even to the maximum possible signal outputs possible in the setup (49 dB and 43 dB re: 1 μbar, respectively). The threshold SPLs for 7 frequencies tested are shown in Figure 3.

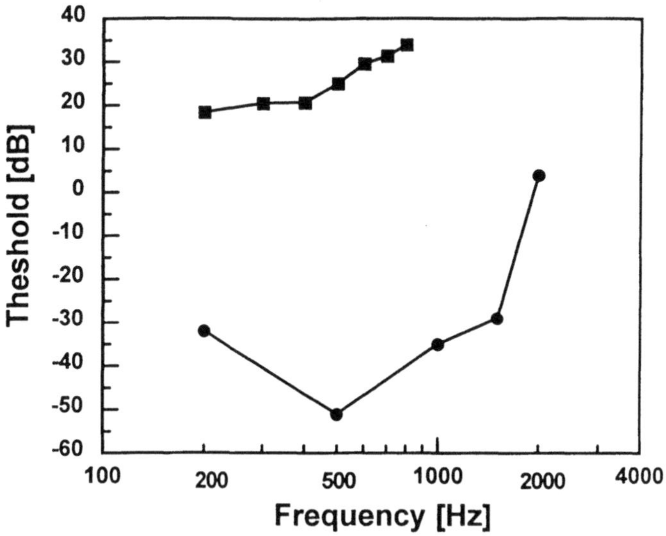

Figure 3. Audiograms of goldfish (*Carassius auratus*) (solid circle) and oscar (*Astronotus ocellatus*) (solid square) obtained from the same operant conditioning paradigm and test apparatus. Notice frequency axis is in common log scale. (re: dB/1 μbar)

3.2 Intensity Difference Limen Study

a) As a function of sensation level

The intensity discrimination ability for oscars at 500 Hz for 10 dB, 20 dB, and 30 dB SL were 8.9, 5.5 and 3.3 dB, respectively (Figure 4). The results showed that intensity discrimination ability of the oscar improved as the sensation level increased.

b) As a function of frequency

The intensity discrimination ability for oscar at 200 Hz, 500 Hz, and 800 Hz (all at 20 dB SL) were 4.5, 5.5, and 9.3 dB, respectively (Figure 5). The results indicated that oscars have better intensity discrimination ability at lower frequencies than at higher frequencies.

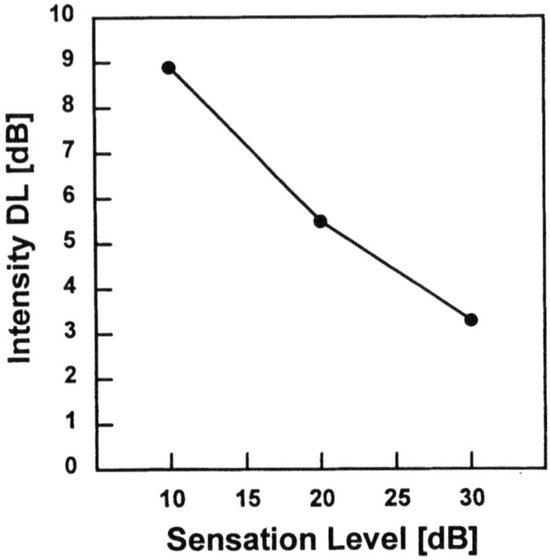

Figure 4. Mean intensity discrimination thresholds (ΔI) (intensity discrimination limen) as a function of sensation level (SL) in dB for oscar measured at 500 Hz (continuous tone). (re: dB/1 μbar)

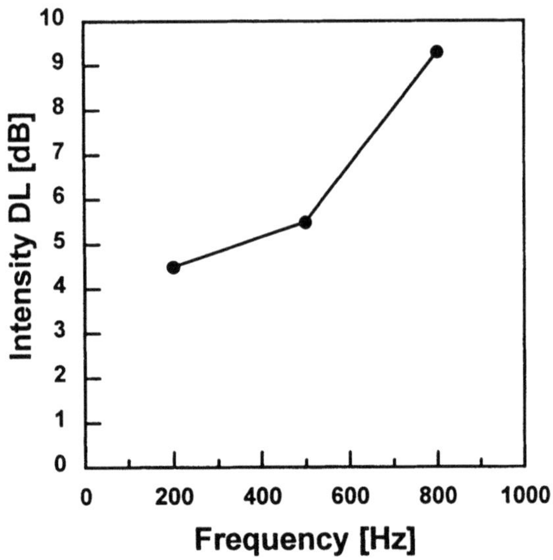

Figure 5. Mean intensity discrimination thresholds (ΔI) (intensity discrimination limen) as a function of frequency for oscar measured at +20 dB SL (continuous tone). (re: dB/1 μbar)

4 Discussion

The purpose of developing this automated operant conditioning paradigm was to provide a reliable measure of the hearing capacities of fish, without subjecting the fish to aversive shock conditioning. The audiogram of goldfish obtained here follows the general trend of audiograms of goldfish previously reported by several different workers who used either avoidance conditioning or classical conditioning methods (see the review by Popper & Fay, 1973; also see Fay, 1988). Despite completely different paradigms (positive reinforcement vs. classical conditioning), the present results agree with those from other investigators. This agreement both validates the effectiveness of the positive reinforcement paradigm employed here and strengthens confidence in existing data on hearing in goldfish.

The present paradigm avoids the problem of nonuniformity of the sound field encountered in the instrumental avoidance conditioning method for the fish always heard the sound at the same spot (O-paddle) for sound presentation. This spot could be thoroughly calibrated. In addition, the fish were allowed to swim freely in the tank, preventing physiological stress from restraint in the classical conditioning of cardiac and/or ventilatory suppression.

Classical conditioning of heartbeat or gill ventilation is often complete within 10 trials; it can be achieved 1 h after the start of training. Thus, several thresholds can be determined in a day (Hawkins, 1981). With the present reward reinforcement training method (stages I-V), on the average, 2-5 days are required to train a goldfish. It may take up to 14 days to train an oscar. The complexity of the paradigm, which involves two separate responses (O-paddle and R-paddle pecking) and control by the fish of the intertrial period in the beginning of training session, prolongs the training period. Observations indicated that the whole training process could be shortened by decreasing the intertrial interval. It was also observed that a long blackout period (i.e., 90 s for goldfish and 180 s for oscar) resulted in a significantly lower false alarm rate (< 10 %) than did a shorter blackout period (30 s for goldfish and 90 s for oscar).

Light cues (such as the O-paddle or R-paddle light as well as blackout) are used in the present experiment to facilitate training process. However, the light cue may not be equally effective for nocturnal species such as catfish or weakly electric fish. Bright light may have to be used instead of blackout to deter nocturnal species from making mistakes during training. The piezo film paddle setup may not work for fish species using different feeding modes, such as suction or scrapping. Necessary modifications are needed to tap into the feeding behavior of studied fish species in order to achieve the best results

Extensive observations on the response of the oscar to the experimental apparatus showed that an oscar housed alone would hide in the corners of the tank for an extended time (sometimes up to 10 days) before showing normal swimming activity again. This type of behavior presumably due to social isolation has also been reported in convict cichlids (*Cichlosoma nigrofasciatum*, Gallagher, Hertz, & Peeke, 1972). However, when 2 or 3 fish were housed in the same tank it took less than 3 h for them to resume normal activities. Therefore, training for each individual oscar fish was conducted so that the experimental animal could see two other fish but separated from the experimental animal by a transparent plexiglass divider. However, observation learning so far has not been observed in oscars receiving training under such arrangement. On the contrary, this type of training paradigm was not needed for goldfish (Yan and Popper, 1991). It is imperative that social behavior of the intended test fish species should be taken into account during the preliminary training sessions so necessary modifications can be made to this paradigm to ensure a successful operation.

Although the apparatus and paradigm presented here were originally intended for fish auditory sensitivity study, the system itself is very versatile and be easily modified to accommodate the needs for other studies of sensory systems in fish.

5 Acknowledgements

A. N. Popper and R. J. Dooling were instrumental in the development of this protocol. Financial supports were made possible from NIH (DC-00140; DC-01729); Office of Naval Research (N-00014-87-K-0604; N-00014-92-J-1114), and University of Kentucky (Vice Chancellor Office for Graduate Study and Research, College of Arts and Sciences and School of Biological Sciences), and Kentucky Water Resources Research Institute.

6 Suggested Reading

Clark, E. (1959). Instrumental conditioning of lemon sharks. *Science* 130, 217-218.

Couvillon, P. A., & Bitterman, M. E. (1985). Effect of experience with a preferred food on consummatory responding for a less preferred food in goldfish. *Animal Learning & Behavior*, 13, 433-438.

Fleishman, L. J., Zakon, H. H., & Lemon, W. C. (1992). Communication in the weakly electric fish *Sternopygus macrurus*. 2. Behavioral test of conspecific EOD detection ability. *Journal of Comparative Physiology A*, 170, 349-356.

Longo, N., & Bitterman, M. E. (1959). Improved apparatus for the study of learning in fish. *American Journal of Psychology*, 72, 612-620.

Okanoya, K., & Dooling, R. J. (1987). Hearing in passerine and psittacine birds: A comparative study of absolute and masked auditory thresholds. *Journal of Comparative Psychology*, 101, 7-15.

7 References

Allen, E. E., & Fernald, R. D. (1985). Spectral sensitivity of the African cichlid fish, *Haplochromis burtoni. Journal of Comparative Physiology A*, 157, 247-253.

Bigelow, H. B. (1904). The sense of hearing in the goldfish *Carassius auratus* L., *American Naturalists*, 38, 275-284.

Donaldson, E. M. (1981). The pituitary-interrenal axis as an indicator of stress in fish. In A. D. Pickering (Ed.), *Stress and fish* (pp. 11-47). London: Academic Press.

Fay, R. R. (1969). Behavioral audiogram for the goldfish. *Journal of Auditory Research*, 9, 112-121.

Fay, R. R. (1988). *Hearing in vertebrates: A psychophysics databook.* Winnetka, IL: Hill-Fay Associates.

Frisch, K. von (1936). Üeber den Gehörsinn der Fische. *Biological Review*, 11, 210-246.

Gallagher, T. E., Hertz, M. J., & Peeke, H. V. S. (1972). Habituation of aggression: the effects of visual social stimuli on behavior between adjacently territorial convict cichlids (*Cichlosoma nigrofasciatum*). *Behavioral Biology*, 7, 359-368.

Hawkins, A. D. (1981). The hearing abilities of fish. In W. N. Tavolga, A. N. Popper, & R. R. Fay (Eds.) *Hearing and sound communication in fishes* (pp. 109-137). New York: Springer.

Horner, J. L., Longo, N., & Bitterman, M. E. (1961). A shuttlebox for fish and a control circuit of general applicability. *American Journal of Psychology*, 74, 114-120.

Jacobs, D. W., & Tavolga, W. N. (1967). Acoustic intensity limens in the goldfish. *Animal Behaviour*, 15, 324-335.

Kritzler, H., & Wood, L. (1961). Provisional audiogram for the shark, *Carcharhinus leucas. Science*, 133, 1480-1482.

Nelson, D. R. (1967). Cardiac responses to sound in the lemon shark, *Negaprion brevirostris*. In P. W. Gilbert, R. F. Mathewson, & D. P. Rall (Eds.), *Sharks, skates, and rays* (pp. 533-544). Baltimore:Johns Hopkins University Press.

Parker, G. H. (1902). Hearing and allied senses in fishes. *Bulletin of the United States Fishery Commission*, 22, 45-64.

Parker, G. H. (1903). The sense of hearing in fishes. *American Naturalists*, 37, 185-204.

Popper, A. N., & Fay, R. R. (1973). Sound detection and processing by teleost fishes: A critical review. *Journal of the Acoustical Society of America*, 53, 1515-1529.

Richards, A. M. (1976). *Basic experimentation in psychoacoustics.* Baltimore: University Park Press.

Stebbins, W. C. (1970). *Animal psychophysics: the design and conduct of sensory experiments.* New York:Appleton-Century-Crofts.

Stebbins, W. C. (1983). *The acoustic sense of animals.* Cambridge, Massachusetts: Harvard University Press.

Tavolga, W. N. (1974). Signal/noise ratio and the critical band in fishes. *Journal of Acoustic Society of American*, 55, 1323-1333.

Tavolga, W. N. (1976). *Sound perception in fishes:Benchmark papers in animal behavior* (Vol. 7). Stroudsburg, PA: Dowden, Hutchinson & Ross.

Tavolga, W. N., & Wodinsky, J. (1963). Auditory capacities in fishes. Pure tone thresholds in nine species of marine teleosts. *Bulletin of the American Museum of Natural History*, 126, 177-240.

Yager, D., & Thrope, S. (1970). Investigations of goldfish color vision. In W. C. Stebbins (Ed.), *Animal psychophysics: the design and conduct of sensory experiments* (pp. 259-275). New York:Appleton-Century-Crofts.

Yan, H. Y., & Popper, A. N. (1991). An automated positive reward method for measuring acoustic sensitivity in fish. *Behavior Research Methods, Instruments, & Computers*, 23, 351-356.

Yan, H. Y., & Popper, A. N. (1992). Auditory sensitivity of the cichlid fish *Astronotus ocellatus* (Cuvier). *Journal of Comparative Physiology A*, 171, 105-109.

Yan, H. Y., & Popper, A. N. (1993). Acoustic intensity discrimination by the cichlid fish *Astronotus ocellatus* (Cuvier). *Journal of Comparative Physiology A*, 173, 347-351.

Yost, W. A., & Nielsen, D. W. (1985). Psychophysics: Discrimination procedures. In W. A. Yost & D. W. Nielsen (Eds.), *Fundamentals of hearing* (2nd ed., pp. 109-122). New York: Holt, Reinhart & Winston.

Integrative Perceptual Processes

Introductory Remarks

As research on bats and owls has shown, studies of the perception of complex, biologically-relevant sounds have successfully revealed specializations in both the peripheral and central auditory systems. Comparative studies of the perception of speech by humans and animals provide a clearer understanding of putative specializations for speech processing in humans. Studies of the perception of complex sounds offer an opportunity to explore the processes of the sensory system that permit the identification and spatial imaging of sound sources as they appear in the natural world.

As the chapters in this section demonstrate, complex sounds can be extremely difficult to define precisely and to manipulate effectively in psychoacoustic tests. This is so because such sounds are typically multidimensional and there is rarely a simple relation between a percept and the constellation of acoustic cues evoking that percept. Moreover, the behavioral paradigms for studying the detection and discrimination of simple sounds are not as useful in asking questions about the perception of complex sounds. Nonetheless, there is a large potential payoff in understanding the evolution of hearing and acoustic communication systems if biologically relevant signals are carefully chosen and used in well-designed experiments on complex sound perception.

The first chapter in this section asks whether there really is something special or species-specific in the processing of vocal communication sounds. Working with synthetic speech sounds, Sinnott describes several methodologies for the study of speech sound perception by monkeys and gerbils. Moody describes paradigms for studying the perception of species-specific vocalizations by macaques by looking for perceptual specializations and examining the critical acoustic features that support this perception. Using a different approach, Dooling and Okanoya describe a discrimination paradigm developed for use with birds in which response latencies to discriminate among human speech sounds or bird calls are analysed using multi-dimensional scaling and clustering techniques. Hulse, in his chapter, explores the subtleties of testing for perceptual invariance in starlings.

Finally, Stebbins ponders the uncertainties inherent in comparative psychoacoustic research. There may be limits to how much we can really know about the acoustic world of other species. Acoustic perceptual systems which have evolved by exploiting different ecological niches may always be, to some degree, unknowable by us. The challenge is in framing the correct questions and designing the appropriate paradigms to answer them.

G. M. Klump, R. J. Dooling, R. R. Fay & W. C. Stebbins (eds)

Methods to Assess the Processing of Speech Sounds by Animals

Joan M. Sinnott

Psychology Department, University of South Alabama,
Mobile, AL 36688, U.S.A.

Summary. Here are described some procedures designed to assess the capacities of animals to detect, discriminate, and identify human speech sounds. All procedures use a go/no-go response, which is taught to animals using operant conditioning techniques based on positive (food) reinforcement. All procedures are easily learned by animals, and allow for direct comparisons with human subjects, who can be tested with the same methods.

1 Introduction

The goal of our research is to determine to what extent animals can provide models for the process of human speech perception. Here we describe some procedures designed to compare the abilities of humans and animals to detect, discriminate, and identify speech sounds. In order to be as precise as possible in these comparisons, we have devised procedures that allow all subjects to be tested under, as far as possible, identical experimental conditions. We work with two Old World monkey species: the Asian Japanese macaque (Macaca fuscata) and the African Sykes' monkey (Cercopithecus albogularis). Monkeys, of course, are of inherent interest to study because of their relatively close phylogentic proximity to humans. We also have started to work with Mongolian gerbils (Meriones unguiculatus), which are small rodents with good hearing in the human speech range. The value of gerbil research is that it may illuminate aspects of speech perception that (presumably) emerged before the evolution of the primates. Thus these three-way comparisons with humans, monkeys and gerbils may ultimately lead to a more complete understanding of the evolution of speech perception in the mammalian order.

2 Speech Sound Detection

Detection refers to the ability to report the presence of a sound. The most fundamental capacity to explore in speech perception is the detection threshold for speech. We use a free-field test situation inside a double-walled IAC booth lined with sound-absorbing acoustic foam. Our basic go/no-go detection procedure is based on one first developed by Stebbins (1970). Animals are trained to make a reliable report response ("go") in the presence of a sound, and to withhold the response ("no-go") in the absence of the sound. We use a different response apparatus for monkeys and gerbils, but the procedures for each species are functionally equivalent, and can easily be modified to accommodate humans. Monkeys sit in a primate restraint chair and respond by contacting and releasing a metal disk mounted over a cuelight (see Figure 1). Gerbils are tested in a 20 cm x 20 cm x 20 cm wire mesh cage that contains a cuelight and a photobeam apparatus. Gerbils respond by jumping on and off a platform (see Figure 2).

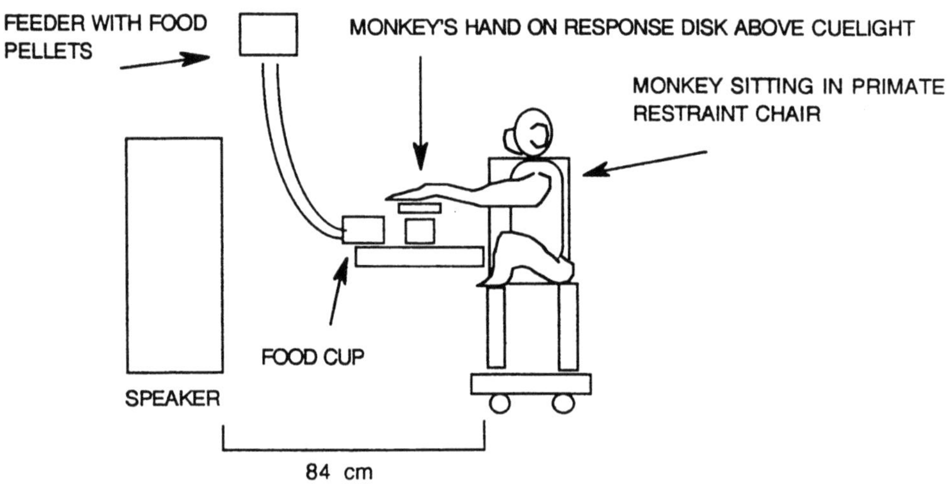

Figure 1. Schematized diagram of the monkey response apparatus. The monkey sits in a primate restraint chair inside an IAC booth. An automatic feeder delivers 190 mg banana flavored food pellets as rewards.

At the start of a trial, the cuelight flashes as a signal to contact the disk (monkeys), or to mount the platform (gerbils). This observing response insures that the animal presents trials only when ready to listen. Upon contact, the cuelight steadies and a variable hold begins ranging from 1600-6400 msec, so the animal cannot predict when the stimulus will occur on each trial. The hold terminates with two presentations of a vowel stimulus pulsed at a rate of 1 per 800 msec. The monkey must release the disk (or the gerbil must jump off the platform) during this 1600 msec response interval in order to score a hit. A hit is immediately rewarded with a food pellet dispensed from an automatic feeder, and then followed by 4-sec intertrial interval (ITI), which allows the animal time to eat the food pellet.

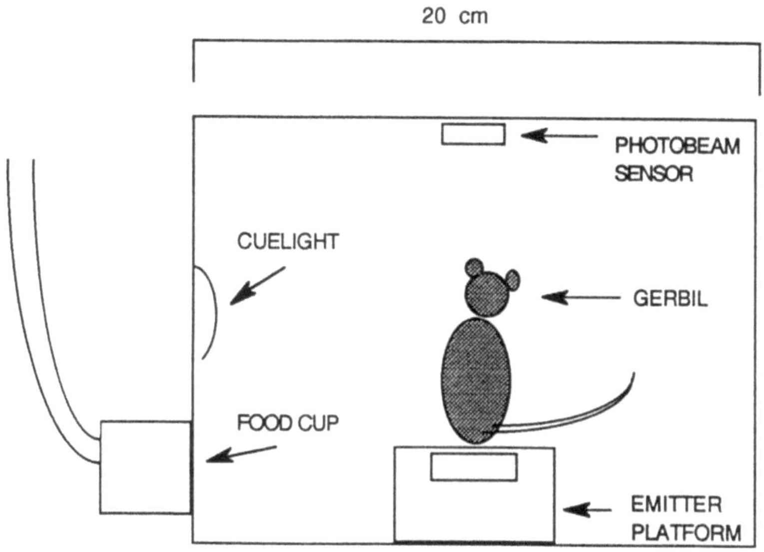

Figure 2. Schematized diagram of the gerbil response apparatus. The cage is mounted on a stand inside the IAC booth. A photobeam sensor is mounted on the ceiling of the cage, and the emitter is embedded in a small circular platform directly below the emitter. An automatic feeder delivers 20 mg bacon flavored food pellets as rewards.

Ideally, the subject will respond ("go") only when he hears a sound, and will not respond ("no-go") when he does not hear a sound. There are two kinds of errors the animal can make in the procedure. One error is to respond when there is no sound, referred to as a false alarm. High false alarm rates mean the animal is not a reliable observer and is guessing too much. To monitor false alarms, 25% of the trials are control trials, during which the attenuator is set to its maximum value. Responses to control trials are punished with 8-sec timeouts, during which the cuelight extinguishes and the animal cannot initiate trials. Correct rejections of control trials are reinforced by immediately following the control trial with a suprathreshold stimulus that is 100% detectable. A second error is to fail to respond during the allotted response interval when a sound occurs, referred to as a miss. Since subjects may in fact not be able to hear some of the low level sounds during threshold testing, a miss only results in a short 3-sec timeout as feedback.

Sound stimuli are the ten steady-state English vowels, synthesized on the Canadian Speech Research Environment (University of Western Ontario) and presented using a 12-bit D/A converter, manual and programmable attenuators, low-pass filter set at 5 kHz, amplifier, and loudspeaker. Stimuli are calibrated with a B&K microphone placed in the position of the subjects' heads during testing. Formant patterns of the vowel stimuli are shown in Figure 3.

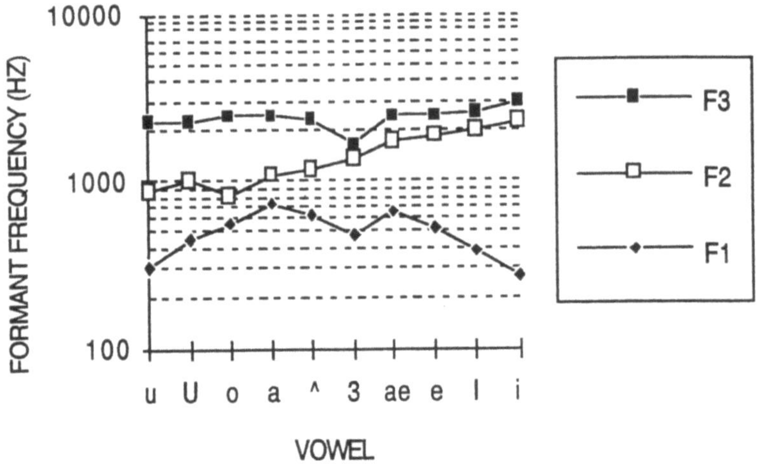

Figure 3. Formant patterns of the ten English vowels. F1 = the first formant; F2 = the second formant; F3 = the third formant. Vowels are 300 msec in duration, with a fundamental frequency beginning at 125 Hz that descends to 80 Hz at the end of the vowel (from Sinnott & Brown, 1992).

To measure thresholds, six vowel intensity levels using 10 dB step sizes are randomly selected on test trials according to the method of constant stimuli. The probability of detection

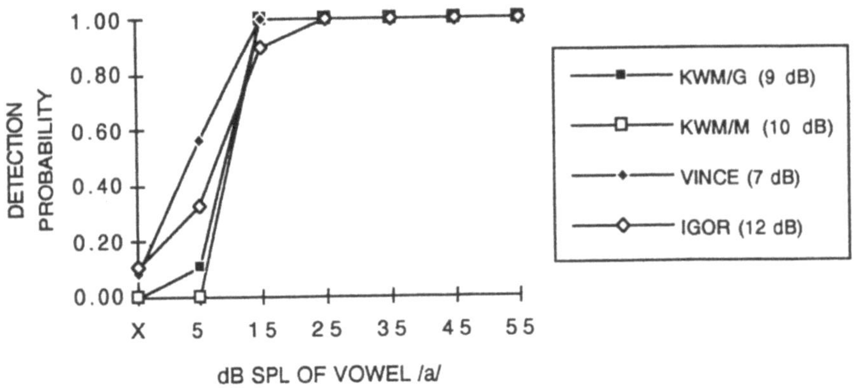

Figure 4. Psychometric functions for human KWM, Sykes monkey VINCE and gerbil IGOR detecting the vowel /a/. Probability of detection is plotted as a function of vowel intensity level. X = the control trial stimulus. Human KWM was tested using both the gerbil (G) and monkey (M) apparatus. Thresholds based on a d' of 2.0 are shown in parentheses for each subject (from Sinnott & Brown, 1992).

for each level is recorded in the form of a psychometric function (see Figure 4). The lowest level is set to be just at or slightly below a subject's threshold. We define threshold as that level on the function corresponding to a d' of 2.0, which equals the .5 level of detection for a .02-.03 false alarm rate on a yes-no procedure (Swets, 1964). Data are discarded from sessions with false alarm rates above .25.

Humans are tested using both the monkey and the gerbil aparatus. When using the monkey apparatus, they sit on a normal chair inside the booth, and use a contact key similar to that of the monkeys. When using the gerbil apparatus, they sit beside the gerbil cage and manipulate a small rod with a nickel-sized disk attached to the end that breaks the photobeam in the same way as the gerbil mounting the platform. We have not found any evidence of sensitivity differences for humans tested on the monkey versus gerbil apparatus.

Averaged vowel threshold functions for gerbils, Japanese macaques, Sykes monkeys and humans are shown in Figure 5. Human and monkey sensitivity is basically flat across vowels, but the gerbil function is U-shaped and appears to correlate with the vowel F1: Lowest gerbil thresholds, closer to those of primates, occur for vowels with higher frequency F1's, and higher thresholds occur for vowels with lower frequency F1s (see Figure 3). Since F1 is typically the most intense formant of the vowel, these results suggest that gerbils are less sensitive than primates to the lower frequency formants in the speech stimuli.

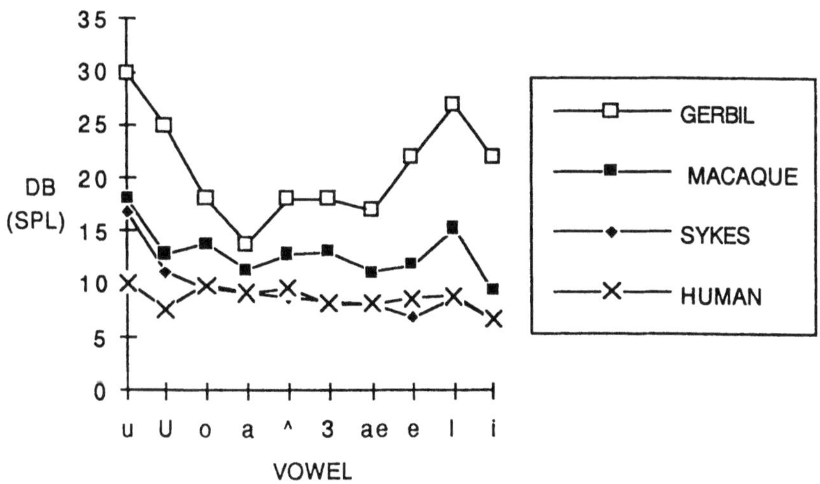

Figure 5. Averaged vowel thresholds for gerbils, monkeys and humans all tested in the same free-field acoustic test situation (from Sinnott & Brown, 1992).

3 Speech Sound Discrimination

Discrimination refers to the ability to tell the difference between two sounds. The detection procedure described above is easily modified to measure discrimination. We use a variant of the classic AX same-different discrimination procedure: The variable hold period contains a repeating background stimulus (A), pulsed at a rate of 1 per 800 msec, and the 1600 msec response interval contains two presentations of a different target stimulus (X). All other contingencies as to the reinforcement of hits, false alarms, misses, and correct rejections are in effect as in the detection procedure. Animals are easily transferred from detection to discrimination, by using a fading procedure that initially presents the background at a reduced intensity level, and then gradually increases it until it reaches the intensity level of the target.

For example, to assess vowel discrimination, one vowel is chosen as a repeating background, and the other nine vowels are presented as targets. Humans and monkeys discriminate all vowels from one another with almost 100% accuracy under these conditions, but gerbils do not discriminate well certain spectrally similar vowels, e.g. /u/ versus /U/, or /i/ versus /I/, indicating inferior sensory resolution for these stimuli compared to primates (see Figure 6).

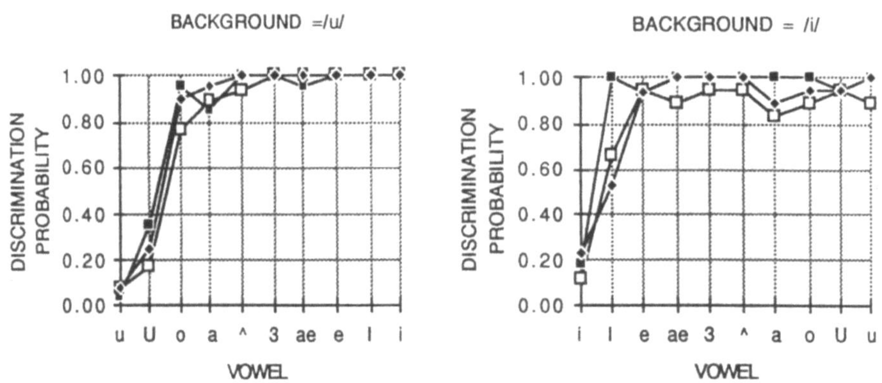

Figure 6. Psychometric functions for three gerbils discriminating nine target vowels from the repeating background /u/ (left) and /i/ (right), from Sinnott & Brown (1992).

Sensitivity differences between humans and monkeys may appear when using more complex speech sounds as stimuli, such as the multidimensional English liquid contrast /ra-la/. This contrast is rare among the languages of the world, and has attracted attention because native Japanese speakers learning English have much difficulty with it (Strange & Dittman, 1984). We synthesized an 8-stimulus continuum ranging from /ra/ to /la/ by varying two cues: a spectral cue in the third formant, and a temporal cue in the first formant (see Figure 7). Our procedure required subjects to hold to the /ra/ background and to respond to a change towards the /la/ target. Human subjects were all native speakers of American English, and monkey subjects were all Japansese macaques.

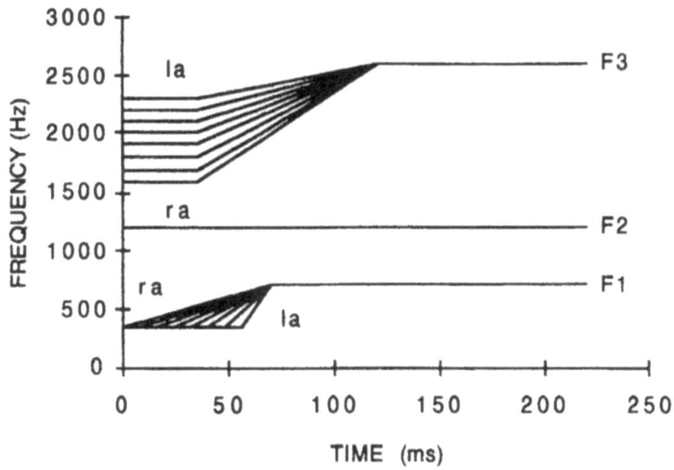

Figure 7. Formant patterns for a synthetic /ra-la/ continuum. /ra/ has a lower onset frequency for F3 and a more gradual transition in F1 than /la/ (from Sinnott & Barnett, 1992).

Two additional continua were also synthesized which varied only one of the cues, either the spectral or the temporal, and left the other constant at a middle value, in order to determine which cue was the more important of the two for humans and monkeys. A comparison of the psychometric functions for the full and partial cue continua is shown in Figure 8. Note that humans (left), find both cues more or less equally salient, but monkeys (right) appear more sensitive to the spectral than the temporal cue.

This discrimination procedure can also be used to assess phoneme boundary effects, or perception of a discontinuous phoneme boundary along a synthetic continuum that results in a discrimination peak. We presented 2-step stimulus pairs from the /ra-la/ continuum and measured sensitivity for each pair in terms of d'. Figure 9 shows that humans have a pronounced peak at pair 4-6, which coincides with their identified boundary, while monkeys are most sensitive to pair 1-3, which humans hear as two /ra/ stimuli. Of course, it is an open question as to whether or not this difference reflects a specialized phoneme processor in humans not available to monkeys, or rather general psychoacoustic capacities that reflect different sensitivity to the acoustic cues underlying this contrast, as shown in Figure 8.

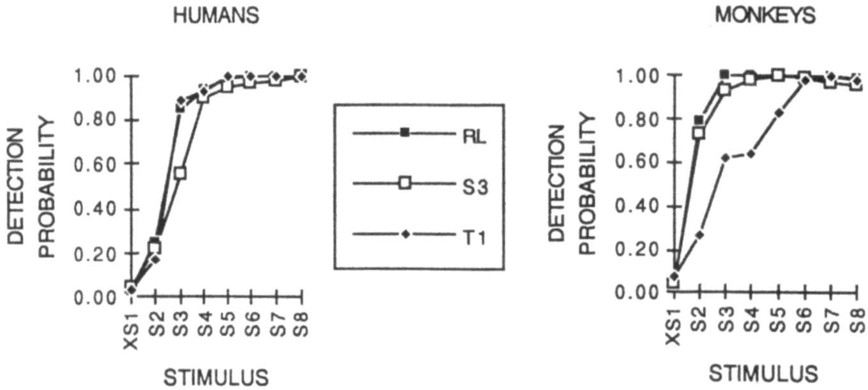

Figure 8. Averaged psychometric functions showing differential sensitivity of humans and monkeys to the normal and partial cue /ra-la/ continua. Continua: RL = both spectral and temporal cues vary; S3 = only spectral cue in F3 varies; T1 = only temporal cue in F1 varies. S1 = background /ra/; S8 = target /la/ (from Sinnott & Barnett, 1992).

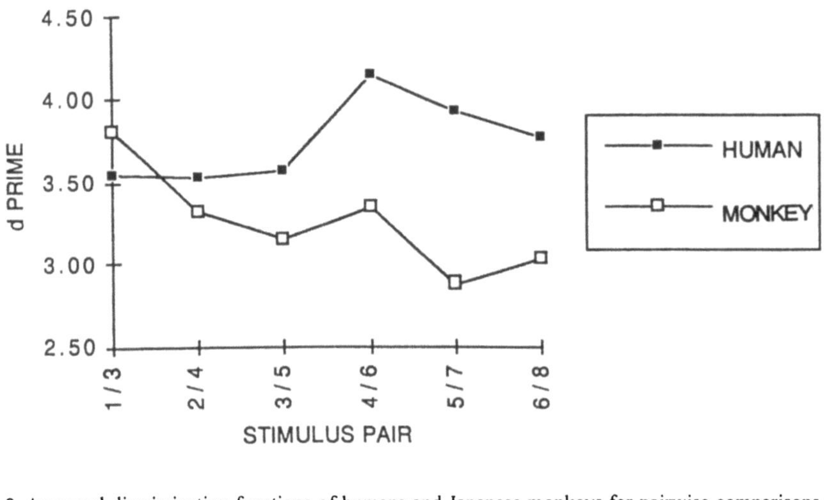

Figure 9. Averaged discrimination functions of humans and Japanese monkeys for pairwise comparisons along a synthetic /ra-la/ continuum (from Sinnott & Barnett, 1992).

4 Speech Sound Identification

Identification refers to the ability to classify sounds according to representations stored in memory. Human speech identification is typically assessed by presenting isolated speech sounds and having subjects to assign verbal labels to them. Identification procedures are used extensively to assess context effects in speech perception. Such an effect is the stop-glide phoneme boundary shift, by which the boundary along a /ba-wa/ continuum shifts to longer transition durations as the length of the vowel increases (Miller & Liberman, 1979). For example, a 40-msec transition followed by a short (100-msec) vowel sounds like a /wa/, but when followed by a long (300 msec) vowel it sounds like a /ba/. Thus stop-glide perception interacts with the whole syllable context in which it occurs. In order to compare this effect in humans and Japanese monkeys, we synthesized two synthetic continua ranging from /ba/ to /wa/, differentiated by having either a short or long vowel duration (see Figure 10).

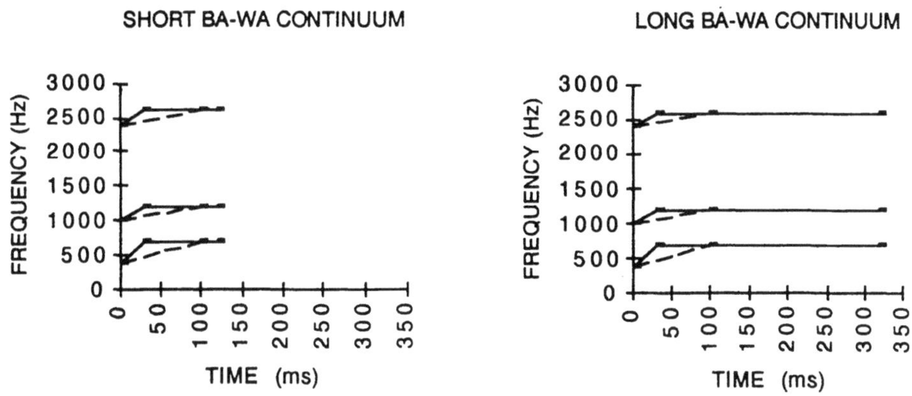

Figure 10. Formant patterns showing the ends points of a short (left) and a long (right) /ba-wa/ continuum. The 30-msec transition plotted with the solid line is heard as /ba/, and the 100-msec transition plotted with the dashed line is heard as /wa/. The short continuum has a 120-msec total syllable duration, the long continuum has a 320- msec syllable duration. (from Sinnott & Mosteller, in preparation)

We developed a go/no-go identification procedure as follows: At the start of the trial, a variable hold period begins ranging from 2-4 sec. Following this, two identical stimuli are pulsed at a rate of 1 per sec. Training stimuli are the end values from the continua consisting only of 30-msec /ba/ transitions and 100- msec /wa/ transitions. The "go" response is rewarded to /wa/ stimuli, and the "no-go" response is rewarded to /ba/ stimuli. If the subject makes an error on these training stimuli, a correction procedure goes into effect, by which the missed stimulus is repeated on subsequent trials until a correct response occurs. The correction procedure trials are not counted in the data analysis, they are simply to prevent the development of response biases towards one or the other response.

After a subject has learned to respond to the two training stimuli with at least 90% accuracy, test stimuli consisting of transitions between 30-100 msec are presented on about 25-30% of the trials to assess generalization. Test stimuli are randomly interspersed with the training stimuli, and for these the subject is rewarded for any response, either go or no-go. The identification functions in Figure 11 show how more /wa/ sounds are identified in the shorter continuum. Both humans and monkeys show this effect, which is consistent with the phoneme boundary shift to longer transtion durations with increased vowel duration.

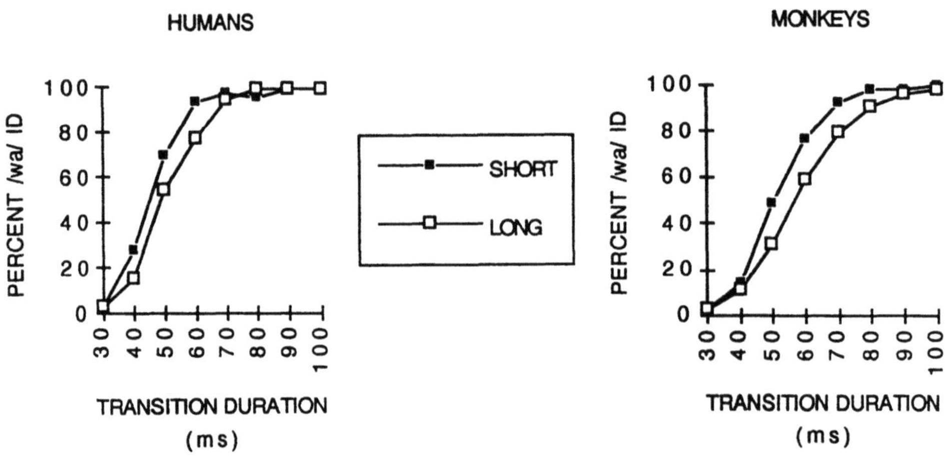

Figure 11. Averaged human and monkey identification functions along synthetic /ba-wa/ continua showing the shift of the phoneme boundary towards longer transition durations with increasing vowel duration (from Sinnott & Mosteller, in preparation).

5 Conclusion

The basic go/no-go procedure is an excellent one for quickly and efficiently training animals to detect, discriminate and identify complex human speech sounds. Experienced monkeys develop a learning set when working on all these procedures, which makes it easy to introduce them to new stimuli. Our experience has been that humans and monkeys attain similar degrees of stimulus control on all these procedures, e.g. with regard to false alarm rates. For gerbils, our experience is so far limited to detection and discrimination experiments, and here we tend to find slightly increased false alarm rates compared to primates, but this tendency is easily handled by using d' as a measure of sensitivity.

One minor problem with the go/no-go identification procedure is that many subjects, even humans, tend to develop a response bias towards the "go" response, which is most likely due to their past history on this procedure in detection and discrimination experiments, where rewards occur exclusively for the "go" response. However, this drawback is much outweighed by the advantages stemming from being able to transfer subjects back and forth among the various detection, discrimination and indentification procedures.

6 Acknowledgements

This work was supported by NIDCD grants K04 DC 00042 and R01 DC 00541 to JMS, R01 DC 00164 to Charles H. Brown, and the Carol P. Sinnott Memorial Fund of the University of South Alabama.

7 Suggested Reading

Dooling, R. & Brown, S. (1990). Speech perception by budgerigars: The voiced-voicless distinction. *Perception & Psychophysics*, 46, 65-71.

Dooling, R. Okanoya, K. & Brown, S. (1989). Speech perception by budgerigars: Spoken vowels. *Perception & Psychophysics*, 47, 568-574.

Hienz, R. & Brady, J. (1988). The acquisition of vowel discrimination by non-human primates. *Journal of the Acoustical Society of America*, 84, 186-194.

Kluender, K. (1991). Effects of first formant onset properties on voicing judgments result from processes not specific to humans. *Journal of the Acoustical Society of America*, 59, 83-96.

Kluender, K., Diehl, R. & Killeen, P. (1987). Japanese quail can learn phonetic categories. *Science*, 237, 1195-1197.

Kojima, S. & Kiritani, S. (1989). Vocal-auditory functions in the chimpanzee: Vowel perception. *International Journal of Primatology*, 10, 199-213.

Kuhl, P. (1986). Theoretical contributions of tests on animals to the special mechanisms debate in speech. *Experimental Biology*, 45, 233-265.

Kuhl, P. (1991). Human adults and human infants show a "perceptual magnet effect" for the prototypes of speech categories, monkeys do not. *Perception & Psychophysics*, 50, 93-107..

Sinnott, J. (1989). Detection and discrimination of synthetic English vowels by Old World monkeys and humans. *Journal of the Acoustical Society of America*, 86, 557-65.

Sinnott, J. & Adams, F. (1987). Differences in human and monkey sensitivity to acoustic cues underlying voicing contrasts. *Journal of the Acoustical Society of America*, 82, 1539-47.

Sinnott, J. & Kreiter, N. (1991). Differential sensitivity to vowel continua in Old World monkeys and humans. *Journal of the Acoustical Society of America*, 89, 2421-29.

8 References

Miller, J. M. & Liberman, A. (1979). Some effects of later-occurring information on the perception of stop consonant and semi-vowel. *Perception & Psychophysics*, 25, 457-465.

Sinnott, J. & Brown, C. (1992). Detection and discrimination of synthetic English vowels by Mongolian gerbils, African monkeys, and humans. Abstracts of the 15th Midwinter Research Meeting of the Association for Research in Otolaryngology, p.26.

Sinnott, J. & Barnett, D. (1992). Comparison of human and monkey sensitivity to the English liquid /ra-la/ contrast. *Journal of the Acoustical Society of America*, 92, pt. 2, 2340-41 (abstract only).

Stebbins, W. (1970). Studies of hearing and hearing loss in the monkeys. In W. Stebbins (Ed.), *Animal Psychophysics: The Design and Conduct of Sensory Experiments.* (pp. 41-66). New York: Appleton-Century-Crofts.

Strange, W. & Dittman, S. (1984). Effects of discrimination training on the perception of /r-l/ by Japanese adults learning English. *Perception & Psychophysics*, 36, 131-145.

Swets, J. (1964). *Signal detection and recognition by human observers.* Wiley, NY.

Methods in Comparative Psychoacoustics
G. M. Klump, R. J. Dooling, R. R. Fay & W. C. Stebbins (eds)

Classification and Categorization Procedures

David B. Moody

Kresge Hearing Research Institute
University of Michigan Medical School
Ann Arbor, Michigan 48109-0506 USA

Summary. In many instances, especially when dealing with biologically relevant stimuli such as communication signals, it is desirable to determine whether a subject treats a particular test stimulus as being equivalent to a set of other stimuli that may differ from the test stimulus along multiple dimensions. Specialized procedures such as those described in this chapter can be used to train subjects to report when a novel stimulus is perceived as a member of a particular stimulus category. These techniques can provide a better understanding of the ways in which real-life auditory stimuli are perceived and processed by the auditory system.

1 Introduction

Many of the chapters in this volume provide ample evidence that an array of procedures exist through which it is possible to determine the minimum stimulus or stimulus difference that an organism can detect. These procedures are designed to push a sensory system to its limits; that is, to ask what is the finest resolution possible. In the real world, however, sensory systems are seldom required to make such fine discriminations. While it is clearly advantageous for an animal to be able to detect the faint sound of another approaching animal at the earliest possible moment, such detections must be accompanied by the ability to classify the sound as that of a predator and not that of a potential mate. The critical decision of whether to approach the source of the sound or flee from it is based on the correct classification of the sound rather than on its detection. Similarly, it is clear from examination of acoustic communication systems, that a major task of the listener is to categorize signals into appropriate classes of "meaning" by attending to the relevant features of the signal while ignoring the irrelevant features. In human speech, for example, vast acoustic differences exist between the same word spoken by different individuals, yet most listeners are able to identify the word without interference from those acoustic differences.

Training animal subjects to perform the same task presents special challenges. Primarily, the issue is one of establishing differential responding to stimuli that differ in the characteristics that define a class, but not to stimuli that differ in characteristics that are irrelevant to that definition. In discussing the essence of concept formation, which for our purposes can be considered equivalent to classification and categorization, Keller and Schoenfeld (1950, p155) talked about "generalization *within* classes, and discrimination *between* classes ...". Our task in designing classification experiments is to teach our subjects to tell us when they are generalizing within a class and when they are discriminating between classes, but at the same time, we must try to avoid imposing on the subjects our own ideas of when to generalize and when to discriminate; that is, we must avoid the pitfall of training the animals to tell us about our classes rather than theirs.

2 A Simple Example

An example drawn from a simple experiment on frequency discrimination may help to clarify the issues. We want our subject to respond whenever we present a comparison stimulus that is a different frequency from a standard. To train this behavior, we deliver a reinforcer whenever our subject correctly reports a difference between standard and comparison stimulus frequency and perhaps we would also place the subject in a time-out whenever a response occurred when no stimulus difference was present. This procedure is fairly straightforward, and probably closely resembles those reported in other chapters in this volume. The problem comes because, as frequency varies, so does perceived loudness. As a result, our subject may not attend to frequency differences at all, but rather may listen only to differences in loudness, and may respond on that basis. Although, to our ears, frequency changes are relatively easy to detect, other animals require large frequency changes before they will report differences (e.g. Prosen, et al. 1990), particularly early in training, and if the animal is insensitive to frequency change, loudness may become a much more reliable cue. In this simple example, standard and comparison stimuli differed along two dimensions, frequency or pitch and loudness. Whenever such a situation exists, subjects will respond to whichever dimension provides the most readily accessible cues and although we report their thresholds as sensitivity to frequency change, they could easily be telling us the ease with which they can detect how much louder or softer a stimulus is than the standard.

Solutions to this particular problem include eliminating the loudness cues by presenting all stimuli at the same perceived loudness level or making the irrelevant dimension of loudness become an unreliable cue. The first alternative is probably not possible. We don't know enough about loudness perception in animals to arrive at equal-loudness values, and setting equal loudness based on our own perceptions ignores the fact that the subject's auditory system may be quite unlike ours. On the other hand, it is possible to arrange things so that, when the subject responds to a difference in loudness, he will usually be incorrect. To do so, we can deliberately vary the level of the standard stimulus over some range. Responses to level differences between standards will be errors, and no reinforcer will be delivered. Eventually, the lack of reinforcement for responses to level differences should cause such responses to disappear, and the only reliable cue remaining will be the difference in frequency. Hopefully, we have trained the subject to discriminate between frequencies and to generalize among intensities; in essence, we have trained a categorization.

On the other hand, it is often reasonable to assume that animal subjects will attend to any cue except the one the experimenter expects them to attend to, so it is necessary to consider whether alternative strategies might exist that would allow the animal to be reinforced for responses based on the "wrong" cue. In the frequency discrimination experiment, for example, only the level of the standard was varied. In such a procedure, comparison stimuli would be presented at the same level (although not necessarily the same loudness). It is conceivable that animals may learn to respond more frequently to stimuli presented at a particular level; that is, they may learn some absolute level discrimination rather than learning to discriminate between frequencies. The probability of such a strategy being adopted might be expected to be greater if the range over which the standard stimuli are varied is large relative to the range of loudness variation in the comparisons. In such a case, subjects may readily learn to identify stimuli having levels that are higher or lower than some mean value, and to ignore those stimuli while responding to stimuli closer to the mean. Indications of such behavior patterns might include a high rate of guessing when the level of the standard is near the mean, or a psychometric functions on the frequency dimension that are non-monotonic. A possible solution, of course, is to vary the level of both standard and comparison stimuli.

3 Using Complex Stimuli

It should be apparent from the above example that there are many issues involved in getting subjects to respond on the basis of the desired dimension. Now consider the case of complex stimuli that differ along a multitude of dimensions, such as natural communication signals. Animals emit a range of sounds that vary widely in frequency range, level, harmonic content, temporal pattern, and duration, and they exhibit different behavior patterns when they hear those sounds. The most frequent question when we bring those calls into the laboratory is to determine which acoustic differences in the calls are correlated with which behavior patterns. We are, in essence, asking how the structure of the call determines its function, or more specifically, which acoustic features are discriminated and which result in generalization. This question is fundamentally different from asking how much frequency difference must exist between two stimuli for them to be discriminated. In the frequency discrimination case, we know the critical dimension; in the complex stimulus case, we are trying to determine what the critical dimension is. Still, the strategies we can use to get the desired answer are similar in both cases; namely, to train the subject to respond only to stimuli that vary along that critical dimension.

Even without formal training in logic, it should be obvious that there is a circularity in the above statement. How can we determine which dimensions are relevant to subjects if we must first train the subjects to respond differentially to a particular dimension, and therefore to establish that dimension as the one that is relevant? In our experiments, we have broken this circularity by using differences in behavior that are correlated with different vocalizations in natural settings, and using those behaviors to define which vocalizations are functionally equivalent. These correlations can be either that the calls are most commonly emitted in certain social contexts, or that, when presented, the calls elicit particular behavior patterns. Stimuli that are correlated with one particular behavior are placed into one functional class, and stimuli correlated with a different behavior are placed into a second class. The basic procedure, then, is to train subjects to respond differentially between the two classes of stimuli. The stimuli within each class may exhibit a wide range of variation in acoustic structure, and initially, at least, it may not be apparent what acoustic features separate one class from another. By presenting a wide range of variation within each class, we hope to teach the animals to generalize to differences within the class, and to discriminate between classes. Once such behavior is established, the acoustic differences that control that behavior can be determined.

3.1 An Example: Japanese Monkey Vocalizations

Many of the early efforts to develop these classification procedures occurred during our investigations of the communication system of the Japanese macaque (*Macaca fuscata*). These animals had been extensively studied in the field by Green (1975) who observed what the animals were doing when certain vocalizations occurred, and showed that certain calls were more likely to occur in certain social contexts. In our experiments, we selected two classes of coo calls for study: the smooth early high and the smooth late high. Coos are tonal vocalizations consisting of several harmonics of a fundamental that is usually in the 500-Hz range. Smooth early high coos are emitted primarily by juvenile animals that are out of visual contact with the troop, and smooth late highs are emitted primarily by estrus females searching for mates. Green based his classification of the coos on the temporal position in the call at which the frequency modulation changed from rising to falling. Calls with this frequency inflection in the first two-thirds of the call were considered smooth early highs, and when the inflection was in the final third, the call was considered as a smooth late high. As with most naturally occurring sounds, however, there was a wide range of variation in other aspects of the call.

a) The Behavioral Procedure

As stimuli, we selected a number of examples of each type of call, and trained subjects to respond when examples of one type were presented and to withhold responses when the other type was presented. The behavioral procedure was modeled after one that had been successfully employed to measure difference thresholds for frequency and intensity. The subject was placed in a primate restraint chair, and tested in a soundproof room. The animal's head was restrained, and earphones were placed over the ears such that the stimuli could be presented separately to each ear. The response that was required from the subject was to initiate and maintain contact with a metal cylinder. For as long as contact was maintained, examples of the standard stimulus class were repetitively presented. A different standard stimulus was selected for each presentation so that the subject had to learn not to respond to differences between members of the standard class. After some randomly determined number of standard stimulus presentations, a stimulus was selected from the comparison class, and was presented in place of one of the standards. If the subject responded to that presentation by releasing contact with the cylinder, a reinforcer in the form of a banana-flavored food pellet was delivered. If the subject released the key at any other time, it was placed into a time-out from the experimental procedure. During the time-out, a cue light located in the center of the response cylinder was extinguished, and remained off until the subject refrained from making any contact with the cylinder for eight seconds. Since responses were only effective in producing stimuli, and ultimately food, when the cue light was on, turning it off served as a mild punisher that reduced inappropriate responses. In summary, if the animal responded when a comparison stimulus was presented, then it received a food reinforcer thereby strengthening responding to differences between classes; responses to a member of the standard class resulted in a time-out, thereby weakening the tendency to respond to differences within a class. Again, the idea is to get the animal to discriminate between classes and to generalize within classes.

One additional feature was included in these procedures in order to estimate the frequency with which subjects would respond when no comparison stimulus was presented. These "catch trials" were scheduled exactly the same as test trials, except that instead of presenting a comparison stimulus, another standard stimulus was presented. If subjects responded on a large number of these catch trials, it was assumed that the discrimination performance for that session was inadequate, and the data from the session were discarded. A more complete discussion of these procedures, along with consideration of the role and influence of the various parameters, is contained in Moody et al. (1976).

b) Initial Training

Although these final contingencies seem fairly straightforward, the optimum method for training subjects to perform on those contingencies is less so. We have tried two different approaches, both of which achieve the desired final behavior, and both of which have certain advantages and disadvantages. The primary difference between the two approaches is whether the training stimuli are introduced one at a time or all at once. In our initial experiments, we first trained animals using standard shaping procedures to maintain contact with the cylinder, and to release whenever any stimulus was presented. At first, only a single stimulus drawn from the comparison stimulus class was presented at the randomly scheduled intervals determined by the procedure. When the subjects were reliably maintaining contact until the comparison stimulus was presented and then releasing, a single repeating standard stimulus was introduced at an intensity level significantly below that of the comparison. When the subject learned to respond only to the comparison, then the level of the standard was gradually increased until, eventually, both standard and comparison were presented at equal levels.

It is important to note that subjects were exposed to only a single standard and a single comparison up to this point. Thus, the discriminations could be based on any of the many acoustic differences between the two stimuli. Once that discrimination was mastered, however, we then introduced a second standard and a second comparison, this time without any level difference. Now, for the first time, subjects had to begin to ignore some differences between stimuli, and attend to only those differences that distinguished one class from another. With only two stimuli from each class, however, the subjects might well employ the strategy of learning to identify individual stimuli and respond on an absolute basis, rather than responding on the basis of the general features that distinguish one class from another. As more stimuli are added to the standard and comparison sets, however, this strategy becomes less efficient, and the likelihood increases that behavior will be controlled by the acoustic features that distinguish one set from another.

One of the dependent variables that we found useful in these studies was the time required for a subject to progress through the training sequence to the point at which all of the examples of the standard and comparison classes were being presented. To increase the validity of that measure, it was necessary to establish rules that specified when a subject had achieved a criterion level of performance on one stimulus set so that the next pair of stimuli could be added to the set. The criterion used in the experiments being described required the subjects to respond to at least 90% overall correct responses to all comparison stimuli, at least 80% correct responses to each of the comparison stimuli, and no more than 10% responses to any standard stimulus. To some extent, such criteria must be tailored to specific experimental situations. For example, we now believe that the part of the criterion requiring no more than 10% responses to any standard stimulus may be too strict, and may result in subjects that are too conservative. We now use a 20% criterion. The important thing however is that, once a criterion is established, it should be strictly applied.

Although time to learn a task is gencrally not an ideal dependent variable, it proved useful in interpreting the results of these experiments (Beecher et al., 1979). Figure 1 presents the cumulative number of sessions required to meet criterion for each of the stimulus

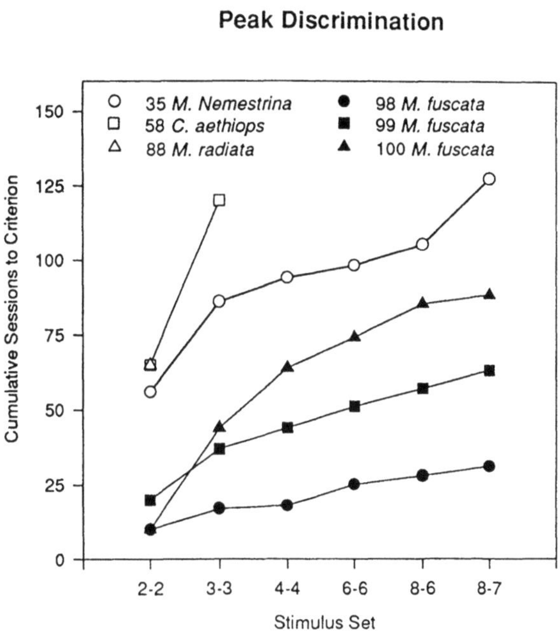

Figure 1. Cumulative number of sessions to reach criterion in the acquisition of a discrimination between smooth-early-high coos and smooth-late-high coos. The labels on the X axis indicate the numbers of stimuli present in the standard and comparison groups respectively, and serves as a measure of the complexity of the discrimination. Two of the control animals (58 and 88) failed to complete the original discrimination task. (From Beecher, et al., 1979; Reprinted by permission of S. Karger AG, Basel).

configurations for six subjects, three of which were Japanese macaques, and three of which were control species. This figure shows that the Japanese macaques learned the discrimination more rapidly than did the controls. In fact, only one of the control animals learned the final discrimination; the remaining two controls failed to reach criterion at some stage in training. These results are important in a number of ways. Most importantly, they demonstrate species differences in learning a discrimination based on species-specific communication sounds. Since the abilities of all of these species are remarkably similar on most measures of auditory system function, a difference such as shown by these experiments is an important indication that the stimulus groupings probably have some biological relevance to the test species. This assumption of biological relevance was reinforced when a second group of animals was trained on the same set of coo calls, except that the calls were grouped into high-pitch and low-pitch groups on the basis of starting frequency. In this second experiment, the Japanese macaques had significantly greater difficulty learning the discrimination than did the controls, suggesting that the high-pitch/low-pitch grouping of calls was inappropriate from the perspective of the communication system of that species. Another way of interpreting those results is that the success of the Japanese macaques on the smooth-early/smooth-late discrimination was not simply because they were "smarter" or more adept at forming discriminations in general. Although results such as these are useful in supporting the biological relevance of the groupings of stimuli selected for the training sets, they are probably the exception rather than the rule. Even in the absence of such results, other methods are available to determine the extent to which subjects have learned to behave appropriately on the classification task.

c) An Alternative Training Strategy

Earlier in this chapter, an alternative method for introducing training stimuli was mentioned; namely, beginning discrimination training with all stimuli present from the first training day. In this variation of the method, subjects are first trained to release whenever any stimulus is presented. In our experiments, we train this stage of behavior with pure-tone stimuli to avoid exposing the animal to any of the stimuli that will be used later in training. Once this behavior has been learned, all stimuli of both classes are introduced, with the standard stimuli presented at a greatly reduced sound level. Once the subject is reliably releasing to comparison stimulus presentations, the level of the standard stimuli is increased until, eventually, the standard stimuli are presented at a level equal to that of the comparisons. This means of introducing the training stimuli overcomes one of the problems of one-at-a-time stimulus introduction; namely, that a particular stimulus or stimulus pair may represent a particularly difficult discrimination for a subject. If that pair is introduced early in training, a particular subject may require an unusually large number of trials to reach the learning criterion. In the data shown earlier, for example, the control subjects had particular difficulty with the stimulus pairs that were introduced early in the discrimination. Those subjects later learned the task with all stimuli, but only after the order of stimulus introduction was varied. When all training stimuli are introduced simultaneously, subjects are able to learn the easier discriminations first, and then refine the rules for discriminating between stimulus classes using the more difficult stimuli. Time to acquire the discrimination should also be useful for detecting differences between control and experimental groups in the all-at-once procedure.

d) Testing New Stimuli: Probe Procedures

Once subjects have been trained by one of the two suggested methods to generalize to stimuli that are members of the same class and to discriminate between stimuli that are members of different classes, then that behavior can be used to study the strategies used by the animals to discriminate one class from another. In addition, but importantly, it is also possible to use the learned behavior to determine whether the animals will correctly classify stimuli to which they had not been previously exposed. This second possibility provides an important means of verifying the validity of both the initial training stimulus sets and the assumptions that were used in grouping stimuli into functional equivalence classes. Presumably, if the original selection of stimuli provided appropriate contrasts for the animals to learn the rules used to differentiate the classes, then other stimuli that represent those classes should be appropriately classified when they are first presented.

One means for testing this assumption is by using what we refer to as a *probe procedure*. In such a procedure, new stimuli (probes) are inserted into the testing set without classifying them as members of either of the two classes. In essence, we ask whether the probe stimulus is like the comparison stimuli, in which case the subject would release key contact when the stimulus was presented, or if it is like the standards, in which case the subject would continue to hold the key. The procedural difference between the outcomes of probe-stimulus trials and training trials has to do with what happens when a subject responds to the stimulus presentation. During normal training trials, reporting responses to comparison stimulus presentations lead to the delivery of a food reward, while responses to a standard stimulus presentation lead to punishment in the form of a brief time-out from the experiment. During probe-stimulus trials, reporting responses are neither rewarded nor punished; that is, they are treated in as neutral a manner as possible. From the subject's perspective, however, failure to receive a food reward for reporting responses to stimuli that are perceived as members of the comparison stimulus class may not be neutral at all, but rather may have negative consequences that result in a decreased tendency to respond to that stimulus in the future. As a result, the subject may begin to learn new and inappropriate rules for stimulus classification. To avoid this pitfall, the training procedure is modified so that not every response to comparison stimuli is rewarded with food. In our experiments, we typically reduce the frequency of reward during training sessions from 100% to 85% for correct responses. During probe testing sessions, reinforcement is switched to 100% for responses to all non-probe comparisons, and probe sessions are carried out relatively infrequently; usually not more than one probe session per week. With these modifications, subjects learn that not every correct response leads to a food reward, and therefore are less likely to be biased by the introduction of probe trials.

Once it has been demonstrated that the subjects are applying some generic rules for grouping stimuli into appropriate classes, then it is possible to begin to determine which acoustic features of the stimuli result in their inclusion in a particular class, and which features seem to be irrelevant for such classification. To carry out such an experiment, it is necessary to be able to manipulate certain acoustic features while at the same time avoiding changes in other characteristics of the stimuli. These manipulations usually involve using computer synthesis techniques to generate stimuli for which the acoustic parameters can be specified by the experimenter, but it is also possible to carry out simpler manipulations such as filtering and truncating to achieve some of the same effects. Details of the techniques for computer synthesis of animal communication signals are beyond the scope of this chapter. Software designed for synthesizing human speech sounds (e.g., Klatt, 1980) often has built-in constraints that prevent it from generating signals that resemble those produced by animals. For example, the fundamental frequency of Japanese monkey coo vocalizations is commonly around 500 Hz, well above that required for human speech synthesis. Often such limitations can be overcome by synthesizing stimuli having fundamentals half the desired frequency, but twice the desired length, and them playing them back twice as fast. Alternatively, specialized software is available that is suitable for many synthesis tasks.

Once techniques for stimulus synthesis have been developed, the first step is usually to generate stimuli that attempt to duplicate some of the training stimuli, and to test them using probe procedures to see if the subjects accept them as substitutes for natural calls. Given that the stimuli are treated as equivalents, then experimental manipulation can begin. For example, one of the first manipulations that May et al. (1988) carried out was to generate prototype smooth-early-high and smooth-late-high coos based on average values derived from the full set of training stimuli. Average values were determined for starting frequency, maximum frequency, number of harmonics, temporal position of the frequency inflections, and so on. Using these values, a typical exemplar was produced for each of the call classes and was then tested using probe testing procedures. May et al. (1988) found that their subjects readily accepted such stimuli as members of the appropriate classes, even though they did not attempt to duplicate particular individual natural calls.

Once prototype calls were generated, various parameters of the synthesis procedure could be manipulated, and the modified stimuli used in the probe procedure to determine whether the subjects would treat them as members of the class of comparisons. Among the manipulations that May et al. (1988) tried were 1) eliminating the frequency change while maintaining amplitude changes; 2) reducing the duration of the frequency modulation while maintaining the temporal position of the frequency inflection in the calls; 3) removal of the frequency inflection by placing the maximum frequency at the beginning or end of the call (linear sweeps only); 4) elimination of the fundamental while retaining the second and third harmonic; and 5) elimination of all harmonics except the fundamental. Although there was some variation between the subjects, condition 3 (linear sweeps only) produced the synthetic calls that were most readily accepted by the subjects as members of the training classes. Conditions 1 (amplitude only), 2 (shortened frequency modulation), and 5 (fundamental only) produced relatively little discrimination between classes; while condition 4 (second and third harmonics only) produced differing effects for different subjects. From this, May et al. (1988) concluded that a critical feature in the call was the direction, rising or falling, of the frequency sweep in the call having the longer duration.

4 Some Cautions and Concerns

Although these procedures have proven to be effective in allowing us to assess features of stimuli that are important to animals in their communication systems, two points about the procedures should be considered. The first point concerns interpretation of responding or lack thereof when probe stimuli are presented. The issue is whether reporting responses on such trials should be interpreted as "this new stimulus *isn't* like the standard class" or "this new stimulus *is* like the comparison class." The reverse interpretation is obviously possible when the subject fails to respond to probes. In essence, we have no way of allowing the subject to tell us that a new stimulus isn't a member of either of the training classes. In many cases, this problem can be circumvented by training different groups of animals and switching which stimulus groups are used as standards and comparisons, or by training different groups with different sets of stimuli and looking for consistency in the results of the probe trials.

The second point to consider is the extent to which the eventual outcomes of these experiments might be influenced by the selection of training stimuli. It is possible, for example, that training stimuli might be selected because they are "good" examples as defined by the experimenter; for example, stimuli might be included in the training set only if they contained a large amount of frequency modulation. In that case, subjects might learn a discrimination that did not allow them to correctly classify calls with smaller amounts of modulation. The best way to prevent such an occurrence is probably to ensure that the training stimuli reasonably represent the range of natural variation within the call classes, and also to train different groups with different samples from the training classes and observe the extent of agreement between groups in the data from the probe stimulus testing.

Although these procedures are a valuable and reliable means of assessing the perceptual processing of biologically relevant stimuli in animal subjects, they are also among the most fragile of the procedures available for understanding auditory function. All stages of training and testing must be carefully planned and carried out, and results must be interpreted with caution. When properly implemented, however, the techniques provide a window into the perceptual world of animals.

5 Acknowledgements

At various points in this chapter, I have used first person plural pronouns, certainly not intending the royal sense of the words, but rather to indicate that a large number of people have been involved in this research which was begun in the mid 1970's. At the extreme risk of omitting a critical member of the team, I would like to note the significant contributions of Bill Stebbins, Mike Beecher, Mike Petersen, Brad May, and Colleen Garbe from our laboratory. In addition Peter Marler, Steve Green, Steve Zoloth, and Marc Hauser have provided us not only with stimuli for these experiments, but also with their perspectives as field biologists on the interpretation of our results. The research and the preparation of this chapter were supported by Program Project Grant P01-DC00078 from NIDCD.

6 Suggested Reading

Moody, D.B., Beecher, M.D., & Stebbins, W.C. (1976) Behavioral methods in auditory research. In C.A. Smith, J.A. Vernon (Eds.) *Handbook of Auditory and Vestibular Research Methods* (pp. 439-497). Springfield, IL: Charles C. Thomas.

Moody, D.B., Stebbins, W.C., & May, B.J. (1990) Auditory perception of communication signals by Japanese monkeys. In W. C. Stebbins, M. A. Berkley (Eds.) *Comparative Perception, Volume II: Complex Signals* (pp. 311-343), New York: John Wiley.

May, B., Moody, D.B., & Stebbins, W.C. (1988) The significant features of Japanese macaque coo sounds. *Animal Behaviour*, *36*, 1432-1444.

7 References

Beecher, M.D., Petersen, M.R., Zoloth, S.R., Moody, D.B., & Stebbins, W.C. (1979) Perception of conspecific vocalizations by Japanese macaques. *Brain, Behavior and Evolution*, 16, 443-460.

Green, S. (1975) Variation of vocal pattern with social situation in the Japanese monkey (*Macaca fuscata*): a field study. In L. Rosenbloom (Ed.) *Primate Behavior (Vol. 4)*, (pp. 1-102), New York: Academic Press.

Keller, F.S. & Schoenfeld, W.N. (1950) *Principles of Psychology*. New York: Appleton-Century-Crofts.

Klatt, D.H. (1980) Software for a cascade/parallel formant synthesizer. *Journal of the Acoustical Society of America*, *67*, 971-995.

May, B., Moody, D.B., & Stebbins, W.C. (1988) The significant features of Japanese macaque coo sounds. *Animal Behaviour*, *36*, 1432-1444.

Moody, D.B., Beecher, M.D., & Stebbins, W.C. (1976) Behavioral methods in auditory research. In C.A. Smith, J.A. Vernon (Eds.) *Handbook of Auditory and Vestibular Research Methods* (pp. 439-497). Springfield, IL: Charles C. Thomas.

Prosen, C.A., Moody, D.B., Sommers, M.S., & Stebbins, W.C. (1990) Frequency discrimination in the monkey. *Journal of the Acoustical Society of America*, *88*, 2152-2158.

Methods in Comparative Psychoacoustics
G. M. Klump, R. J. Dooling, R. R. Fay & W. C. Stebbins (eds)

Psychophysical Methods for Assessing Perceptual Categories

Robert J. Dooling[1] and Kazuo Okanoya[2]

[1] *Department of Psychology, University of Maryland, College Park, MD 20742, USA*
[2] *Cognitive and Information Sciences, Chiba University, 1-33 Yayoi, Inage-ku, Chiba-city, 263 Japan*

Summary. This chapter describes two behavioral paradigms for generating correct response or response latency data from small birds suitable for the application of multidimensional scaling and clustering algorithms. These algorithms are powerful statistical procedures for inferring perceptual categories from discrimination data. We provide two examples of the applications of these procedures to understanding the perception of complex vocal signals.

1 Introduction

There is a limit to what we can learn about hearing when we restrict ourselves to the use of tones and noises. In fact, some of the most interesting questions in hearing come from the study of acoustic communication in animals and this often involves testing strategies aimed at inferring the animal's perceptual organization for complex communication signals. Perceptual studies using complex stimuli such as species-specific vocalizations, speech, or tone patterns, have often relied on some form of a classification task where learning rates for different classes of stimuli, or responses to intermediate and extreme forms of complex stimuli, are compared (see, for example, Burdick and Miller, 1975; Herrnstein, 1984; Kuhl, 1987; Sinnott, 1980).

2 Methods and Procedures

The following procedures represent a different approach relying instead on testing paradigms typically used for studying acoustic detection or discrimination. While it has become a relatively simple matter to ask an animal to detect a sound or discriminate between two different sounds, it is somewhat more challenging to ask the question how similar two easily distinguishable complex sounds are. The following procedures are similar to those used to study visual discrimination in pigeons (Blough, 1985; Schneider, 1978) and were developed precisely for the purpose of inferring perceptual categories for complex sounds in small birds. We do this by first training birds on one of two different tasks. The first task is often referred to as a Same/Different task. In this task, the bird is presented with a pair of sounds and trained to respond if the two sounds are different and withhold a response if they are the same. The second task is an Alternating Sound task. In this task the bird is required to detect a change in an on-going or repeating background of sound. Correct responses are rewarded with food, incorrect responses are punished with a blackout. In both procedures, the reinforcement contingencies are arranged in such a way that the birds respond faster the greater the perceptual difference between the stimuli (Dooling, Brown, Park, Soli, and Okanoya., 1987; Dooling, Park, Brown, Okanoya, and Soli, 1987; Okanoya and Dooling, 1988).

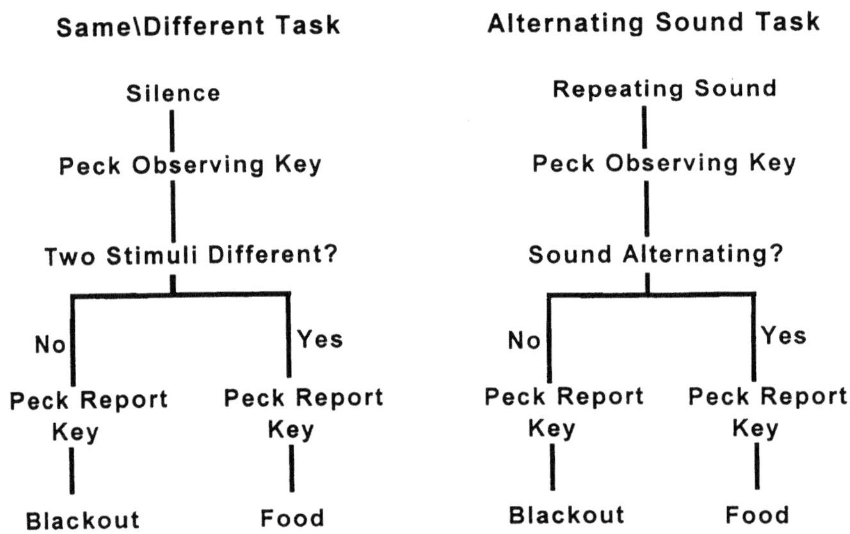

Figure 1. Schematic diagram of the trial events in the Same/Different task (left) and the Alternating Sound task (right).

We can demonstrate that there is a relation between stimulus similarity and reaction time or response latency using simple pure tones that differ along only a single acoustic dimension such as frequency, intensity, or duration (Dooling, Brown, et. al., 1987). Changes that are difficult to discriminate are responded to more slowly than changes that are easier to discriminate. We know from such experiments with simple pure tones that the speed with which birds detect a change in the repeating background can be taken as an index of stimulus similarity. Once we establish that response latency can be taken as a measure of stimulus similarity, we can use this dependent measure as an index of perceptual similarity in tests using more complex sounds such as bird calls or human speech sounds.

2.1 Apparatus

The apparatus for training and testing the birds is the same that is used for running detection and discrimination tests (Park, Okanoya, and Dooling, 1985). The birds are tested in wire cages mounted in sound-attenuated chambers. One wall of the wire test cage is modified by the addition of a custom-built response panel constructed of two sensitive microswitches with light emitting diodes (LEDs) attached (see Dooling and Okanoya, this volume). A bird can trip the microswitch by pecking at the LED. The right microswitch and LED serve as an observation key and the left microswitch and LED serve as a report key. Experimental events are controlled by a microcomputer using a digital I/O board operating optical relays. Acoustic stimuli are stored in digital form and output at a sampling rate of 20 kHz through D/A converters during a trial.

2.2 The Same/Different Task

In this procedure, a trial begins with the illumination of the observation LED. A response on the observation key results in the presentation of two stimuli separated by 300 ms. While the selection of an inter-stimulus-interval of 300 ms is arbitrary, we have found that inter-stimulus-intervals approaching 1000 ms or longer are difficult for the birds. A response on the report key within 2 s following the presentation of two different stimuli (hit) is rewarded with 2-s access to grain. On the other hand, a response on the report key during the 2-s interval following the presentation of two identical stimuli (false alarm) is punished with a 20-s timeout period during which the lights in the experimental chamber are extinguished. Regardless of the trial outcome (i.e. hit, false alarm, miss, correct rejection), each trial is followed by a 1-s inter-trial-interval and then by a new trial sequence starting with the illumination of the observation key.

A bird is tested in daily sessions until all possible pairwise combinations of the different stimuli are presented once with the added constraint that there be equal numbers of "Same" and "Different" trials. This constraint was met by repeating trials involving the pairing of each stimulus with itself. This results in a complete matrix of response latencies with repetitions in the cells of the diagonal. The total number of trials required to satisfy this criterion was

2N(N-1) where "N" is the number of stimuli in the test set. Once the bird completes a matrix of response latencies, the data are stored on disk, and the bird is tested again on all possible pairwise combinations of the different stimuli in the same manner. The order of stimulus presentations is always randomized across trials and each bird is typically tested until at least 5 complete response latency matrices are available for analysis.

2.3 The Alternating Sound Task

In this task, birds are trained to peck one key (observation key) repeatedly during the repetitive presentation of one sound (background) and to peck the other key (report key) when a new sound (target) is presented alternately with the background sound. A peck on the report key during this alternating stimulus pattern is defined as a correct response and is rewarded with a 2-s access to food. Typically, two pure tones (eg. 2 and 3 kHz) are used as training stimuli and both serve as a target and a background stimulus. Once the birds learn the task with pure tones, testing on a set of complex sounds begins.

The sequence of events during a trial actually consists of several distinct phases. In the initial phase or habituation phase, a peck on the observation key initiates an observation phase and begins a random time interval of 1-7 s. Following this variable interval, a second peck on the observation key initiates the report phase of the trial in which the target stimulus alternates with the background stimulus. A response on the report key within 2 s of the beginning of this alternating pattern was reinforced with a 2-s access to food. If the stimulus set consists of N sounds, 1 out of every N trials are sham trials in which the target stimulus was the same as the background stimulus. A response on the report key during a sham trial or during the waiting interval was punished with a timeout period (typically 5-10 sec) during which lights in the test chamber were extinguished but the repeating background sound continues. As above, regardless of the trial outcome (ie. hit, false alarm, miss, correct rejection), each trial is followed by a 1-s inter-trial-interval and then by a new trial sequence starting with the illumination of the observation LED.

In order to create symmetrical similarity matrices for multidimensional scaling or cluster analysis using this procedure, each sound in a set of complex sounds must serve both as a background and as a target. In other words, at the conclusion of testing the similarity of each sound in the test set to every other sound in the test set is assessed using either a percent correct measure or a response latency. Because all possible stimulus combinations are tested, it is useful to conceive of the set of test stimuli as forming a matrix (background by target) when considering the order of stimulus trials. When the animal begins testing, one row of background is randomly selected for testing from this matrix and every other sound in the row is tested as a target against this background. Each session proceeds a row at a time (i.e., same background stimulus) until all possible combinations within the row are tested some number of times (usually 10). Another row is then randomly selected and a new habituation phase begins with a new background sound. This procedure continues until all rows are tested thus ensuring that the same sound serves both as a background and as a target stimulus. Testing of a full matrix can extend over a number of sessions. Each cell in the full matrix then contains the average response latency from 10 trials of the same stimulus comparison.

3 Analysis of Results

Regardless of whether data are collected using is the Same/Different procedure or the Alternating Sound procedure, at the conclusion of testing, a matrix of response latencies is available which relates each complex acoustic stimulus to every other complex acoustic stimulus in the set. The upper and lower halves of the matrix are averaged to produce a single latency half-matrix and each cell in this half matrix then represents the average (or median depending on experimenter preference) response latency from 20 trials of a particular stimulus pair. This half-matrix can be log transformed, if necessary, to compensate for a positively skewed distribution of reaction times and then analyzed by a multidimensional scaling (MDS) procedure. MDS procedures compute distances between stimulus objects on the basis of error scores, confusion indices, response latencies, or other similarity measures (Kruskal and Wish, 1978; Shepard, 1980; Borg and Lingoes, 1987). MDS produces a spatial representation or "perceptual map" of complex stimuli where stimulus similarity (i.e. long response latencies) is represented by spatial proximity (i.e. stimuli are close together in multidimensional space). The process is iterative with the initial, random arrangement of all points being adjusted on subsequent iterations until the "best" (i.e. most variance accounted for) spatial arrangement for a given set of similarity matricies is found.

There are a number of scaling algorithms available. We typically use the SINDSCAL (Symmetric Individual Differences Scaling) algorithm and program to analyze the response latency matrices because it allows for the simultaneous analysis of a number of individual matricies (Arabie, Carroll, and Desarbo., 1987). SINDSCAL finds a common solution for a number of matrices (individuals). Each bird in these experiments provides a matrix though these procedures can also be used to compare the same bird across different conditions (e.g. following lesion or drug treatment). The variance in the similarity or proximity data (i.e. response latencies) accounted for by a spatial representation provides a measure of the goodness of fit. Subject weights provide a measure of how well the data for individual animals are accounted for by a common MDS solution. Similar to the varimax method of factor analysis, the spatial dimensions in MDS are orthogonal and ordered with the first dimension accounting for greater variance than the second dimension, and the second dimension accounting for greater variance than the third dimension, and so on. For practical reasons, data from our perceptual experiments are typically scaled in either two or three dimensions depending on the number of stimuli in the set and the increase in VAF in a three versus two dimensional solution (i.e. with ten or fewer stimuli the results were obtained from a two-dimensional solution).

3.1 Speech Perception - an Example of the Same/Different Procedure

The Same/Different procedure was used several years ago to address the question of whether budgerigars show evidence for perceptual categories for spoken vowels drawn from different phonetic categories in the face of normal acoustic variation across different talkers (Dooling and Brown, 1990). The stimuli in this experiment were the natural spoken vowels /a/, /i/, /ɜ/, and /u/ as occur in the words hot, heed, head, and hoot. Sustained vowels from each of these four categories were obtained from each of four male talkers. Each vowel token was edited to be 200 ms long with 5 ms rise and fall times. Three budgerigars were tested on these 16 vowels using the Same/Different procedure as described above. At the conclusion of the experiment, the response latency half-matricies from each bird were analyzed by MDS. The two-dimensional spatial representation by SINDSCAL of the results from the three birds is shown in Figure 2 (left).

This two dimensional solution by SINDSCAL accounts for a total of 55% of the variance in the birds' response latencies. The first and second dimensions accounting for 30% and 25% respectively. These results show that budgerigars hear the similarity among spoken vowels drawn from the same phonetic category in spite of the natural, and rather extensive, acoustic variation introduced by different talkers (represented by numbers). A similar result is obtained by performing a cluster analysis on the average response latency matrix from the three birds. Figure 2 (right) shows a dendogram from such an analysis and the grouping of stimuli by phonetic category rather than by talker is obvious.

Like MDS, cluster analysis also describes the structure of similarity data and the purpose of hierarchical cluster analysis is to group stimuli into subsets, each of which should correspond to a meaningful feature of the stimuli. Hierarchical cluster analysis groups or links stimuli by a predefined rule according to a Euclidean distance metric with more similar calls clustered at less-aggregated levels of a hierarchy. In the present experiment with speech sounds, the dashed lines used to highlight a stimulus cluster in MDS plots was determined by a cluster analysis. In general, cluster analysis and MDS provide complementary ways of defining a "cluster" and highlighting the different features of complex stimuli. Cluster analysis tends to produce a representation in discrete form while MDS yields continuous dimensions.

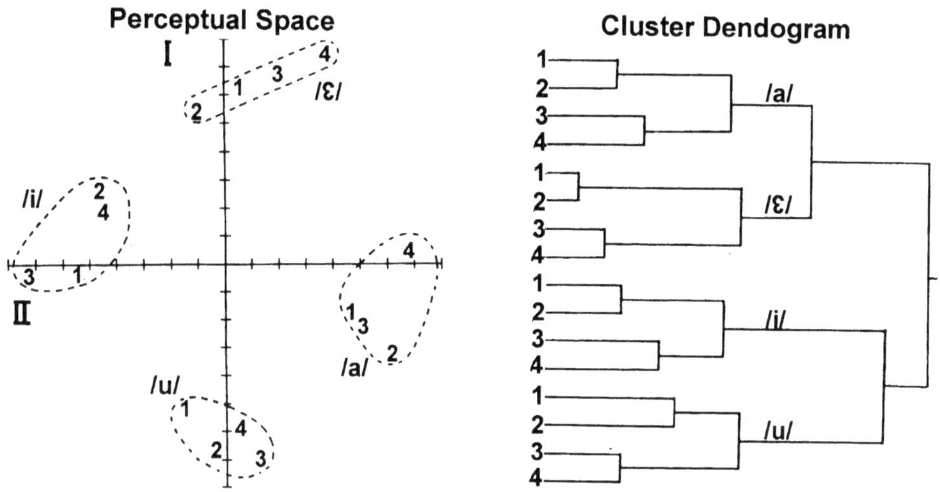

Figure 2. (Left) Two dimensional spatial map by SINDSCAL of the similarity among four vowels spoken by four male talkers. (Right). Cluster dendrogram from a hierarchical cluster analysis of the average response latency data from three budgerigars. In both figures, talkers are represented by numbers.

3.2 Perception of Natural and Synthetic Calls - an Example of the Alternating Sound Procedure.

In comparative psychoacoustic work, the last several decades have seen have a dramatic increase in the use of synthetic vocalizations, including human speech sounds, to assess specializations for the perception of species-specific vocal signals. One question that always must be addressed is the extent to which synthesis methods can produce an adequate model of a natural vocalization. In the case of bird calls, does a synthetic call sound like a natural call to a bird or does it sound like such a poor imitation that it would never be mistaken as a bird call? We addressed this issue with contact calls in budgerigars. We recorded a set of 5 natural contact calls (one from each of five birds) and then we prepared a set of synthetic analogs using a commercial software package called SIGNAL (Clark, Marler, and Beeman, 1987). The synthetic contact calls sounded very similar to their natural counterparts to human listeners but do they sound similar to the birds? Sonograms of these sounds are shown in Figure 3.

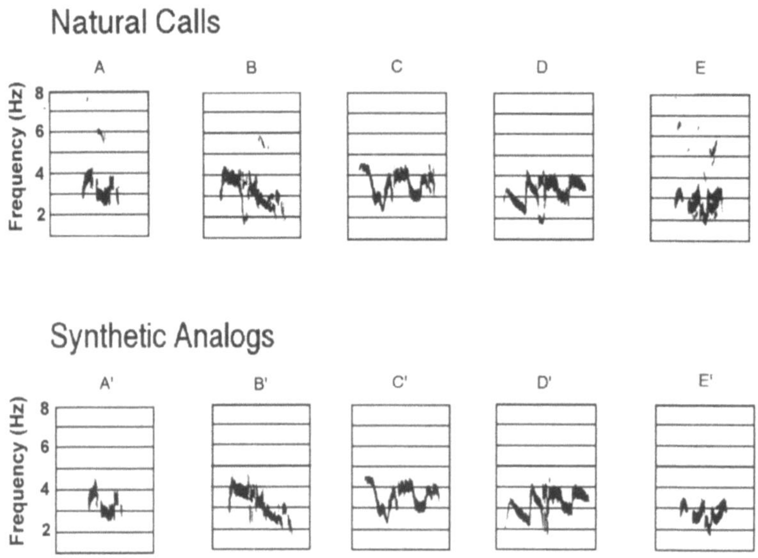

Figure 3. Sonograms of natural contact calls recorded from five different budgerigars and their synthetic analogs. The durations of these calls range from about 130 ms to 220 ms.

Three budgerigars were tested on these ten sounds using the Alternating Sound procedure described above. Each sound in the set of ten calls served as both a background and a target and birds were tested until each stimulus pair was tested 24 times. At the conclusion of testing an average response latency matrix was available for each bird. As in the experiment with speech sounds described above, these half-matrices were analyzed by SINDSCAL and the results plotted in multidimensional perceptual space. In this case, a three dimensional solution accounted for a significantly greater variance in the birds' response latencies than did a two dimensional solution so the results are plotted in three dimensions. The variance in responselatency accounted for by this three dimensional spatial map was 78% with the first, second, and third dimensions accounting for 28%, 28% and 22% respectively.

The three dimensional spatial map of these stimuli is shown in Figure 4. Calls are grouped in pairs in this space with each synthetic analog nearer to its natural counterpart than to other calls in the set. We can conclude that budgerigars hear a strong similarity between natural calls and their synthetic analogs. The demonstration that these synthetic vocalizations are perceived as similar to their natural counterparts by budgerigars allows one to design other psychoacoustic or physiological experiments using synthetic stimuli confident that such stimuli are biologically relevant to the animal.

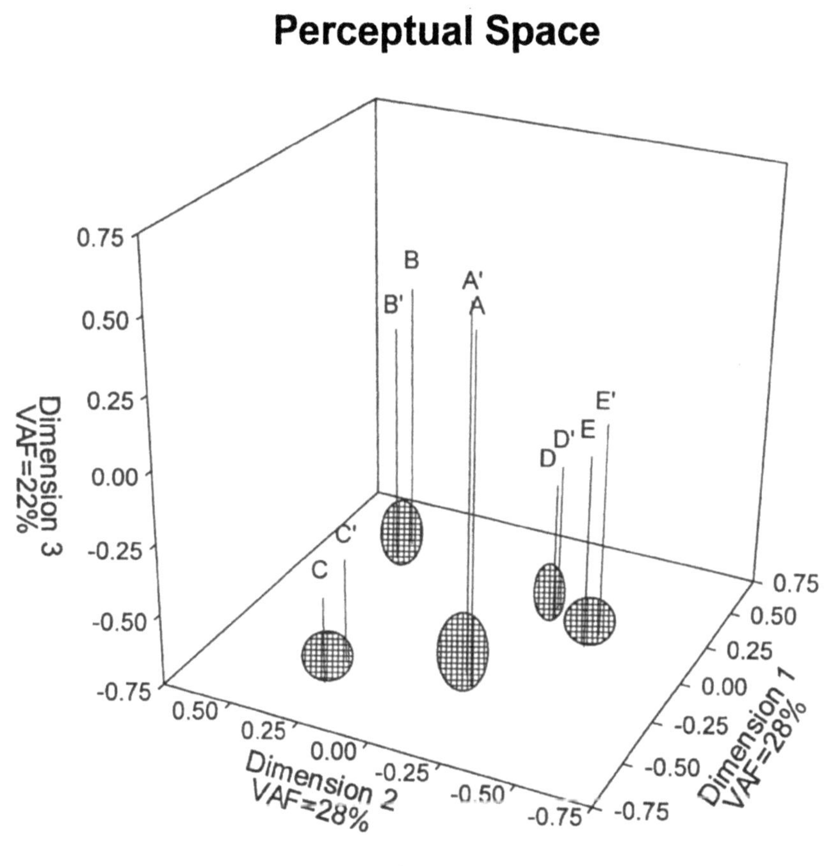

Figure 4. Three dimensional spatial map by SINDSCAL of the perceptual similarity among natural and synthetic contact calls as perceived by three budgerigars. Shaded areas represent stimulus pairs or clusters as determined by a hierarchical cluster analysis of the average response latency matrix for the three birds. Note that each synthetic call is nearer in multidimensional space to its natural analog than to other calls in the set.

3.3 Advantages and Disadvantages of these Procedures

We describe two procedures for collecting perceptual similarity data from animals in the form of response latencies in animals. We prefer to use response latency as the dependent measure (a continuous variable) rather than percent correct, though either will work (see, for instance, Blough, 1985). While both procedures produce similarity data, there are differences between the two procedures worth mentioning. It is generally more difficult (several months of daily test sessions) to train birds on the Same/Different procedure than it is to train them on the Alternating Background procedure (several weeks of daily test sessions). But, at least for budgerigars, once birds are trained in this procedure it generally requires fewer test trials per stimulus pair to obtain a matrix of response latencies suitable for MDS and cluster analysis than it does with the Alternating Sound procedure.

The use of individual difference models of scaling (eg. INDSCAL, SINDSCAL, etc.) offer the advantage that data from different subjects are analyzed together to produce a common perceptual map for all subjects. The assumption here is that all subjects share the same psychological space but that each subject attends differently to the different psychological dimensions. These subject differences are available in a subject weights space. Subject weights give the amount of variance in the response latencies of each subject that is accounted for by the spatial map of stimuli - providing a way of examining individual differences in perceptual space (Shepard, 1980). As one example, examination of subject weights in an experiment on calls perception by budgerigars showed that male and female budgerigars perceive vocalizations differently (Dooling, Brown, Park, and Okanoya, 1990.).

Another advantage of MDS is possible to gain some insight into the acoustic dimensions that animals may be using in discriminating among complex acoustic signals. By correlating stimulus characteristics with spatial location in multidimensional space, we were able to determine which stimulus characteristics budgerigars were using in discriminating among learned contact calls (Brown, Dooling, and O'Grady, 1988). A similar approach to the perception of vowels in budgerigars has shown that these birds rely on differences among formant frequencies in making phonetically relevant discriminations (Dooling and Brown, 1990; Dooling, 1982).

Finally, it is important to realize that the implementation of these procedures involves an equal number of presentations of all possible stimulus combinations and thus the animal is not trained or over-trained to any specific stimulus contrast. Moreover, differences in response latencies to different stimulus combinations are often evident on the animal's very first exposure to the stimulus combination. What this means that the stimulus groupings or categories that emerge from an MDS or cluster analysis of response latencies may represent more what an animal "does" do than what it can be "trained" to do. This represents an important advantage over other paradigms for assessing perceptual categories in animals. In some areas of comparative psychoacoustic research, such as the perception of speech sounds by animals (Kuhl, 1987), this advantage is important in deciding whether perceptual categories for speech sounds are "natural" categories or whether they arise through extensive experience and training with the stimuli.

4 Acknowledgements

We thank Susan Brown and Thomas Park for many helpful discussion in the development and application of these procedures and Michael Dent and Tracy Freeman for the preparation of figures and Michelle Chuen for important pilot experiments. This work has been supported by NIH grants NIH DC-00198 and MH-00982 to RJD.

5 Suggested Reading

Brown, S.D., Dooling, R.J., & O'Grady, K. (1988). Perceptual organization of acoustic stimuli by budgerigars (*Melopsittacus undulatus*): III. Contact calls. *Journal of Comparative Psychology*, *102*, 236-247.

Dooling, R. J., Brown, S.D., Park, T.J., & Okanoya, K. (1990). Natural perceptual categories for vocal signals in budgerigars (*Melopsittacus undulatus*). In M. Berkley and W. Stebbins (Eds.), *Comparative Perception, Vol. II: Complex Signals*. (pp. 345-374). New York: John Wiley & Sons.

Okanoya, K. & Dooling, R.J. (1988). Obtaining acoustic similarity measures from animals: A method for species comparisons. *Journal of the Acoustical Society of America*, *83*, 1690-1693.

Shepard, R.N. (1980). Multidimensional scaling, tree-fitting, and clustering. *Science*, *210*, 390-398.

6 References

Arabie, P., Carroll, J.D., & Desarbo, W.S. (1987). *Three-way Scaling and Clustering*. Beverly Hills, Sage Publications.

Blough, D.S. (1985). Discrimination of letters and random dot patterns by pigeons and humans. *Journal of Experimental Psychology: Animal Behavior Processes, 11*, 261-280.

Brown, S.D., Dooling, R.J., & O'Grady, K. (1988). Perceptual organization of acoustic stimuli by budgerigars (*Melopsittacus undulatus*): III. Contact calls. *Journal of Comparative Psychology, 102*, 236-247.

Burdick, C.K., & Miller, J.D. (1975). Speech perception by the chinchilla: Discrimination of sustained /a/ and /i/. *Journal of the Acoustical Society of America, 58*, 415-427.

Clark, C.W., Marler, P., & Beeman, K. (1987). Quantitative analysis of animal phonology: An application to swamp sparrow song. *Ethology, 76*, 1-1-115.

Dooling, R.J. (1992). Perception of speech sounds by birds. In Y. Cazals, L. Demany & K. Horner (Eds.). *The 9th International Symposium on Hearing: Auditory Physiology and Perception*. (pp. 407-413). Carcans, France.

Dooling, R.J., Brown, S.D., Park, T.J., Soli, S.D., & Okanoya, K. (1987). Perceptual organization of acoustic stimuli by budgerigars (*Melopsittacus undulatus*): I. Pure tones. *Journal of Comparative Psychology, 101*, 139-149.

Dooling, R. J., Brown, S.D., Park, T.J., & Okanoya, K. (1990). Natural perceptual categories for vocal signals in budgerigars (*Melopsittacus undulatus*). In M. Berkley & W. Stebbins (Eds.), *Comparative Perception, Vol. II: Complex Signals*. (pp. 345-374). New York: John Wiley & Sons.

Dooling, R.J., & Brown, S.D. (1990). Speech perception by budgerigars (Melopsittacus undulatus): Spoken vowels. *Perception and Psychophysics*, 47, 568-574.

Dooling, R.J., Park, T.J., Brown, S.D., Okanoya, K., & Soli, S.D. (1987). Perceptual organization of acoustic stimuli by budgerigars (*Melopsittacus undulatus*): II. Vocal signals. *Journal of Comparative Psychology, 101*, 367-381.

Herrnstein, R.J. (1984). Objects, categories, and discriminative stimuli. In H.L. Roitblat, T.G. Bever, & H.S. Terrace (Eds.), *Animal Cognition* (pp. 233-261). Hillsdale, NJ: Erlbaum.

Kruskal, J.R., & Wish, M. (1978). *Multidimensional Scaling*. Beverly Hills: Sage Publications.

Kuhl, P.K. (1987). The special-mechanisms debate in speech research: Categorization tests on animals and infants. In S. Harnad (Ed.), *Categorical Perception: The Groundwork of Cognition* (pp. 355-386), Cambridge: Cambridge University Press.

Park, T.J., Okanoya, K., & Dooling, R.J. (1985). Operant conditioning of small song birds for acoustic discrimination. *Journal of Ethology* (Japan), 3, 5-9.

Schneider, B. (1972). Multidimensional scaling of color differences in the pigeon. *Perception and Psychophysics*, 12, 373-378.

Sinnott, J.M. (1980). Species-specific coding in birdsong. *Journal of the Acoustical Society of America*, 68, 494-497.

Methods in Comparative Psychoacoustics
G. M. Klump, R. J. Dooling, R. R. Fay & W. C. Stebbins (eds)

The Discrimination-Transfer Procedure for Studying Auditory Perception and Perceptual Invariance in Animals

Stewart H. Hulse

Johns Hopkins University, Department of Psychology
Charles and 34th Streets, Baltimore, MD 21218 U. S. A.

Summary. This chapter describes methods for studying auditory perception of complex acoustic stimuli in animals. First, we describe techniques for establishing an initial baseline discrimination between two or more auditory patterns. Then, we discuss transfer procedures used to assay for the acoustic features that control the baseline discrimination. To illustrate the procedures, we use data from an experiment on the perception of auditory spectra by European starlings (*Sturnus vulgaris*). The procedures are adaptable, in principle, for use with a wide variety of species to study a broad array of problems in the comparative psychology of auditory perception.

1 Introduction

Suppose we train an animal to discriminate between two complex sound patterns. Given the discrimination, it is useful to ask how the animal solves the task. What acoustic features or perceptual processes does the animal use to distinguish between the patterns?

One way to answer that question is to change the patterns in some principled way and watch what the change does to the discrimination. If the change disturbs discrimination accuracy significantly, we then know that the change has tapped an acoustic feature that the animal was using. A series of such changes, or *transfer tests*, can be a very powerful tool to unravel the perceptual processes an animal uses in auditory discrimination. But what if the change between baseline and transfer sound patterns does not disturb the discrimination, and the animal continues to discriminate the changed patterns with high accuracy? We cannot safely conclude that the animal fails to notice the change. The baseline and transfer sounds might be readily discriminable as acoustic stimuli if the animal was asked directly to make the discrimination. What we can conclude, however, is that - from a *functional* point of view, at least - the animal perceives the sounds as equivalent. And that, too, can lead to useful information about auditory perception in animals. Sometimes it is useful to know not only when an animal perceives sounds and their acoustic features to be different, but also when an animal perceives the sounds to be functionally the same.

For example, we might imagine two male songbirds each singing the same set of song syllables. For some purposes - individual identification by a female, perhaps - it might be important for the song of the two birds to be readily discriminable. That might happen, say, on the basis of some acoustic feature of the syllables such as their overall pitch. But for other purposes - species identification by the female, for example - it might be important that the two sets of syllables be heard and treated as the same conspecific male-song syllables.

Consider another example. Suppose in the study of auditory perception by songbirds we were interested in their capacity to hear and organize acoustic patterns according to some abstracted rule, such as the rule that sequences of sounds rise (or fall) in pitch. We might train the birds to discrimination between sets of patterns that went up or down in pitch, and then transfer them to new patterns that also went up and down in pitch -- but were shifted upwards by an octave. If the birds had abstracted the relational rule distinguishing the baseline patterns, we might expect them to readily generalize the discrimination to the transfer patterns in the new pitch range. If they had not abstracted the rule, they might lose the discrimination with the transfer patterns. Thus, maintaining accurate discrimination between baseline and transfer sounds tells us something useful about auditory perception.

The foregoing leads to the idea of *perceptual equivalence* or *invariance* in auditory perception. Perceptual equivalence exists when an organism response to one auditory stimulus and some transformation of that stimulus in the same way, that is, as if the two stimuli were functionally equivalent. The idea is directly related to the concept of *congruence* in plane geometry. Two triangles, for example, are congruent if their sides differ in length, but if they share two identical angles (which determines that they also share the third angle). The two triangles will differ - and be readily discriminable - in *size*, but they will be identical in *shape*. Their common shape makes them *perceptually invariant* in our terms. Furthermore, they will be perceptually invariant over a very large number of discriminable changes in size.

The concept of perceptual invariance is obviously related to the concepts of stimulus generalization, stimulus classification, and stimulus categorization. Studies of perceptual invariance in auditory perception help us to understand how animals classify and structure their auditory world. Although the discrimination-transfer design can be used profitably to dissect acoustic features that control discriminations, it is the use of the design to study perceptual invariance that provides the primary focus of this article.

2 Discrimination Training

We turn then to a description of the discrimination-transfer procedures that we have used to study auditory discrimination and perceptual invariance. The techniques are based on familiar operant conditioning techniques that have been in use in many laboratories for some time. Details of procedure are available in the list of suggested readings at the end of this article.

2.1 Subjects

We have used the procedures with songbirds, chiefly European starlings (*Sturnus vulgaris*), but in principle, the procedures and experimental strategies are applicable to any species.

Mild food deprivation motivates the subjects for food rewards obtained in the experiments. The birds obtain much of their daily ration of food as rewards in the experiment in which they are tested each day. Daily weight checks assure that body weight remains stable, and supplemental feedings are given after the daily testing session if needed.

Typically, birds are maintained in the aviary under lights with a spectrum approximating that of daylight. The photoperiod is controlled by timers, and unless varied for experimental reasons, follows that prevailing in Baltimore, MD throughout the year.

2.2 Apparatus

Experiments are conducted inside sound-isolation chambers. We use Industrial Acoustics chambers (Bronx, NY, Model AC-3). Birds are transported from the aviary to the experimental chamber in test cages made from stainless steel weldwire. The test cage is attached to a stationary test panel suspended from the ceiling in the chamber. The test panel becomes the fourth wall of the test cage when a metal sliding door on the test cage is removed.

For starlings and other large birds, the test panel contains three horizontally-aligned response keys spaced 6 cm apart. The keys are translucent Plexiglas disks, 2 cm in diameter, attached to the levers of microswitches which are mounted on the rear of the test panel. The microswitches close when the bird pecks at the keys. The keys can be transilluminated by small lights located behind the disks. Gerbrands (Arlington, MA) pigeon feeders (Model 5610) are mounted behind openings below the right and left response keys.

For canaries and smaller birds, the test panel is similar, but the keys are constructed from light-emitting diodes mounted in the center of smaller plastic disks which are attached to microswitches (Park, Okanoya & Dooling, 1985). For smaller birds, we use a feeder mounted below the keys which consists of a tray 2 cm in diameter and 2 cm deep attached to a motor-driven lever. When food is to be delivered, the tray moves from outside into the cage for a pre-set time. A similar feeder is described by Njegovan, Hilhorst, Ferguson, and Weisman (1994) which should also work quite well. Keep the feeders clean for proper operation!

Sounds are delivered through speakers located either in the ceiling of the test chamber directly above the test cage, or above and behind the test panel and oriented toward the normal position of the bird in the test cage as it pecks at the keys.

Sounds are synthesized off line (see below) and stored on the hard disk of a PC computer for use in the experiments. A Data Translation (Marlborough, MA) Model DT 2812 D/A converter operating at a sampling rate of 20 Khz feeds the acoustic signal to a 10-Khz low-pass filter, and then to a Crown (Elkhart, IN) D-75 amplifier. The amplifier output drives the speakers.

The delivery of auditory stimuli, control of experimental contingencies, and data recording are managed with PC-type computers using software designed in the laboratory.

2.3 Stimuli

The auditory stimuli that are used in an experiment will, of course, depend on the experiment. If artificial sound patterns are called for, they are typically synthesized off line using a Macintosh computer and SoftSynth and Sound Designer II software packages (Digidesign, Inc., Menlo Park, CA). If natural stimuli such as samples of bird song, are to be used, they are first tape recorded or obtained on tape from appropriate sound libraries, then edited using the Sound Designer software package. Once constructed, all stimuli are digitized and stored on hard disk for on-line use in the experiments.

2.4 Operant Tasks

The procedures are all designed to establish first a discrimination between one or more relevant stimuli and, given the initial baseline discrimination, to test for perceptual invariance through transfer procedures. We consider these phases separately.

a) Initial Key Peck Training

Standard operant shaping procedures are used to establish key pecking. Once the bird has learned to peck the keys for food, the experimental procedures proper can begin. In typical experiments, one or the other of two procedures are used: a Go - No go procedure, or a two-alternative choice procedure. We consider each in turn.

b) The Go - No Go Procedure

Here, the bird is rewarded for pecking at the right or left key (arbitrarily selected for any experiment from the right or left keys on the test panel) in the presence of one acoustic stimulus set, say set A, and is not rewarded for pecking at the same key in the presence of the alternative acoustic stimulus set, set B. The discrimination between the two stimulus sets emerges as the bird learns to peck rapidly for food in the presence of stimulus A and to withhold pecking (because of no food) in the presence of stimulus B.

The bird is first trained to start a trial with a peck to the center of the 3 keys on the test panel - the trial initiation, or observation key. This turns on (with $p = .50$) an exemplar from Stimulus set A or B, and the bird listens to the auditory stimulus for a pre-set listening period of several seconds (depending on the experiment). During this time, pecks to the relevant side key have no consequences. Following the listening period, the sound may or may not continue (again, depending on the experiment), but now pecks to the side key have consequences. If Stimulus A (paired with food reward) appears, pecks to the side key are rewarded immediately if they occur within some present response period, say 4 s. If the bird fails to respond to Stimulus A (which rarely occurs), the procedure typically moves ahead to the next trial without further consequences. If Stimulus B appears, the bird must learn to withhold a response for the 4-s response period. If the bird instead pecks to Stimulus B, a mild punishment in the form of a "timeout" is administered: the lights in the chamber go out for some pre-set time (say 10 s) before the experiment can continue. If errors occur, they lead typically to *correction trials* in which the stimulus of a trial is repeated until the bird gets it right. Trials are separated by a short intertrial interval of 3 - 5 s.

Sometimes response opportunities are cued by using key lights. For example, the opportunity to begin a trial may be signalled by lighting the middle key on the test panel, and the response period may be signalled by lighting the appropriate side key. Birds are visual creatures, however, and evidence exists that the use of lights in this fashion may overshadow, bias, or otherwise distort the auditory discrimination of interest (Cynx, 1993). Accordingly, we recommend the use of key lights with appropriate caution.

The Go - No go discrimination appears in the form of changes in response latencies: rapid responses occur to Go stimuli, and slow responses occur to No go stimuli. Response latencies themselves can be used for subsequent statistical analysis. Alternatively, latencies can be translated to an arbitrary "percent correct" score by adopting an arbitrarily short latency as a cutoff for a correct Go response and an arbitrarily long latency (usually no response at all for the duration of the response period) as a cutoff for a correct No go response. Responses as fast or faster than the cutoff for Go responses are correct Go responses. Responses withheld for the duration of the response period are correct No go responses. Response failing to meet these cutoff criteria are errors. Of course, there can be different degrees of "correctness" depending on the stringency of the cutoff criteria, a potentially troublesome issue which suggests that the translation of latencies to percent correct should be used with care.

c) The Two-alternative Choice Procedure

This procedure is in most respects identical to the Go - No go procedure. The primary difference is that two response keys are used instead of one. The bird must learn to associate a peck to the left key, say, with Stimulus A, and a peck to the right key with Stimulus B in order to obtain food. The bird initiates a trial with a peck to the center observation key. Stimuli A or B then appear with $p = .50$ on a given trial, and play for an appropriate listening period. During the subsequent response period, correct key choices are rewarded with food, and errors produce "timeouts." Correction trials repeat stimuli until correct responses occur. The discrimination emerges as the bird learns to reserve responses to the appropriate keys for Stimuli A and B.

Of course, in principle, there is no reason to limit this procedure to two choices. In general, depending on the purpose of the experiment, there can be n choice keys associated with n stimulus sets. In fact, we have successfully used a 4-choice procedure.

2.5 Which Procedure to Use?

In principle, it should not matter which of the two procedures should be used because each reveals an animal's capacity to distinguish stimuli. As a practical matter, however, there are some further considerations that need to be kept in mind.

First of all, an experiment and its analysis may urge that the animal make a discrete choice for each stimulus set, i.e., that the two-alternative choice procedure should be used. This avoids problems like trying to decide under some test conditions with the Go - No go procedure, whether, from the subject's point of view, a middling response latency constitutes a correct or an incorrect choice. That can be important when one wants to know unequivocally the various stimulus factors that are controlling the discrimination. Other things equal, therefore, the two-alternative choice procedure seems the better method.

Unfortunately, as a practical matter, it is often more difficult to teach an animal an auditory discrimination using the two-alternative choice procedure. For reasons not fully understood (but certainly pondered - see Burdick, 1979), animals find it easier to demonstrate in their behavior what they know about stimuli if the Go - No go procedure is used. Nevertheless, the unambiguous response categorization that the two-alternative choice procedure offers may make an experiment with a lengthier period of initial discrimination training well worth the cost.

3 Analyzing the Discrimination: Transfer Tests

Once the baseline discrimination has been learned, the next step is to test for the things the animal has discovered about the relevant acoustic stimuli and their organization. That goal calls for transfer tests. The transfer tests modify the original baseline stimuli in some planned way so as to identify an acoustic feature or a perceptual principle the animal has been using to make the original discrimination. The presumption is that if the animal can generalize and maintain the discrimination with the new stimuli in the transfer test, then the new stimuli are being discriminated on the same basis as the old. On the other hand, if the animal loses the discrimination, then the transfer stimuli no longer contain some acoustic feature or organizational principle the animal was depending on to make the original discrimination. A series of such transfer tests can in principle, therefore, lead to a detailed analysis of the factors controlling the animal's auditory experience.

3.1 Transfer with a Complete Shift to New Stimuli

One way to do a transfer test following baseline acquisition of a discrimination task is to shift the animal all at once to trials which all contain the new stimuli. We have generally done this with the start of a daily test session; in principle the transfer could take place after the beginning of a test session - say half-way through - in order to obtain a sampling of initial baseline discrimination behavior on the day of the transfer itself.

Perception of the new transfer stimuli is assessed by comparing transfer performance with baseline discrimination just prior to the transfer test. Positive transfer appears to the extent that the animal responds immediately to the new stimuli just like the old. Negative transfer appears to the extent that the animal fails to discriminate the new stimuli.

A series of transfer tests can take place, each involving a different modification of the original baseline stimuli, to examine the many features that may control the baseline discrimination. Typically, each transfer test is preceded by a return to the baseline stimuli to recover the baseline discrimination, if necessary.

3.2 Transfer with Probe Stimuli

Another way to do a transfer test is to insert occasional probe trials with new stimuli among ongoing trials with the original baseline stimuli. Probe trial probabilities are set typically at $p = .10$ or less. This approach has the important advantage that the animal continues to respond on the baseline discrimination, and this permits a running comparison between performance on the old and new stimuli. The approach also maintains the overall stimulus context of the original discrimination - which may have effects on transfer performance under some conditions. For example, the animal's memory load for the original stimuli is low relative to the complete shift procedure outlined above because probe trials with the new stimuli are surrounded by frequent baseline trials with old stimuli.

3.3 To Reinforce or Not to Reinforce Transfer Trials?

Regardless of the transfer procedure, the introduction of transfer trials raises the question of whether or not responses on such trials should be differentially reinforced. If the animal is shifted completely to new stimuli in the transfer test, and if these are reinforced as in the baseline discrimination, one has to decide whether transfer performance reflects what the animal remembers about the old baseline stimuli or what the animal is learning about the new transfer stimuli. The problem is not important if the animal's behavior remains stable or changes immediately and unequivocally to some new level at the outset of the transfer test. But the problem is real if the animal's performance changes to some intermediate level at the outset of transfer and then moves gradually to some new level as transfer trials continue.

Alternatively, of course, transfer trials may not be differentially reinforced. We can either reinforce all or none of them regardless of what the animal does. But these procedures, too, are not without their consequences. If all trials are reinforced, the animal will learn to respond regardless of the transfer stimuli; if no trials are reinforced, responding will eventually extinguish.

One may use nondifferential reinforcement with the *probe-trial* procedure with relative safety because probe trials are infrequent, and it is less likely that the animal will learn that probe stimuli are always reinforced against the running background of reinforcement for the well-discriminated baseline stimuli. However, without special preparation, it is unwise to use nondifferential *nonreinforcement* with the probe trial procedure. Here, the probe trial stimuli - infrequent though they may be - may quickly stand out from the background baseline stimuli because they are not followed by food, and the animal may rapidly learn to distinguish them and stop responding to them altogethcr. A remedy for this problem is to prepare the animal for transfer probe trials by reducing reinforcement for correct trials at the end of baseline discrimination training - perhaps by 10 to 30 percent, depending on the frequency of probe trials to be used in the transfer tests. Reductions in reinforcement take place gradually over several baseline sessions. With that procedure, nonreinforced probe trials become indistinguishable from baseline trials in terms of their reinforcement consequences.

4 Methods at Work: An Experiment

An experiment by Braaten and Hulse (1991) tested European starlings' capacity for hearing spectra with a common harmonic structure as perceptually equivalent over changes in the fundamental frequency of the spectra. For humans, at least, sounds with a common harmonic structure sound highly similar; they have a constant *timbre*. That is one of the primary reasons musical instruments retain their characteristic quality over changes in the pitch of the sound they play. The question was whether starlings would perceive spectra similarly. If so, that would have some important implications for comparisons between the auditory perception of mammals and birds.

To start, 5 starlings were first trained in a series of discrimination problems to distinguish two spectral stimuli. The spectrum for the S+ stimulus was always the same from problem to problem and consisted of a sound with energy at the first harmonic (the fundamental) and fourth harmonic (coded F4). The spectrum for the S- stimulus took one of seven possible forms from problem to problem. The S- spectra always contained the fundamental and then included one or more of the second, third, and fourth harmonics (e.g., F2, F24, F234). The frequency of the fundamental was the same for the S+ and S- stimuli on any problem, but the fundamental frequency varied from problem to problem over a range of 10 values from 600 to 1400 Hz. The S- spectra and the different fundamentals were combined so as to produce a total of 35 different discrimination problems. Not all S- spectra were tested with each fundamental frequency, therefore, but all S- spectra were tested with fundamental frequencies of 700 Hz, 1000 Hz, and 1400 Hz. The main question, among some others we cannot address here, was whether the S+ stimulus would be heard as qualitatively the same from one problem to the next, regardless of the S- stimulus and the fundamental frequency. If so, discrimination of the S+/S- configuration should readily transfer from problem. And that would be a convincing demonstration of perceptual invariance for the S+ spectrum.

The Go - No go procedure was used as described above. After learning to peck the keys in the operant chamber, the starlings were trained to discriminate the positive, S+ stimulus (for which pecks were rewarded) from the negative, S- stimulus (for which pecks were not rewarded). The birds received just one day's practice on a given problem before moving on to the next problem, so the experiment was an especially stringent test for the capacity to learn a spectral discrimination and to perceive congruencies among exemplars of the spectra. Note, incidentally, that the experiment used a variation on the first design outlined above. Instead of just one transfer test for perceptual invariance, there was an initial discrimination problem followed, in a sense, by 34 such transfer tests!

The results showed that the starlings became very rapid spectrum discriminators, transferring above-chance performance from problem to problem on the basis of the common S+ spectrum almost immediately. By the time they were working on the last 14 of the 35 discrimination problems, performance was well above chance of 50 percent correct within the first 20 trials on a problem (response latencies were translated to percent correct according to the procedures outlined above). Apparently, perceptual invariance for the S+ spectrum prevailed.

Figure 1 shows detailed data on this point. The figure presents percent correct for the three fundamental frequencies that were used with all seven S- spectra. Clearly, the common S+ spectrum facilitated good discrimination regardless of the fundamental frequency used with the spectra. It is interesting to note, however, that the birds had relatively more difficulty if the S- spectrum lacked the second harmonic. For one or more of the three fundamentals, performance was poorer if the second harmonic was missing (although performance remained reliably above chance). Subsequent experimentation (described in the original paper) supported this fact.

Note, incidentally, that the experiment wasn't designed to show how well starlings can do in the limit with spectral discriminations because the birds only had a single practice session to do

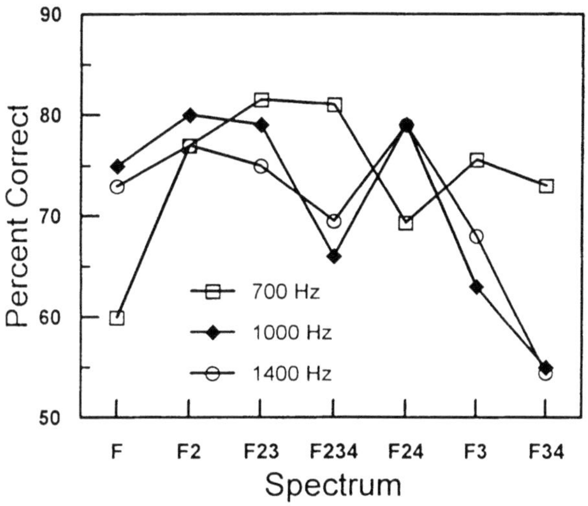

Figure 1. Mean percent correct on the S+ vs. S- spectral discrimination for 7 S- stimuli played with fundamental frequencies of 700, 1000, and 1400 Hz. The spectra are coded by whether they contained the fundamental (F), and the second, third, and fourth harmonic (2, 3, 4).

as best they could on any given spectral discrimination. A later test in which a starling was given continuous discrimination practice on a pool of six S+ and six S- spectral exemplars showed that the bird reached an asymptote of 97% correct by the 10th session.

To sum up, starlings show immediate transfer across spectral discriminations with a common S+ spectrum - in spite of changes in fundamental frequency associated with the spectrum. We can safely conclude that starlings showed perceptual invariance for this spectral structure regardless of the S- spectrum and for all fundamental frequencies tested.

5 Acknowledgements

Preparation of this article was supported by National Science Foundation Research Grant BNS 8911046. I thank Cynthia Gray, Scott Shackleton and Timothy Gentner for their helpful comments about the manuscript. E-mail address for the author is hulse@jhunix.hcf.jhu.edu.

6 Suggested Reading

Hulse, S. H., & Cynx, J. (1985). Relative pitch perception is constrained by absolute pitch in songbirds (*Mimus*, *Molothrus*, and *Sturnus*). *Journal of Comparative Psychology, 99*, 176-196.

Cynx, J., Hulse, S. H., & Polyzois, S. (1986). A psychophysical measurement of pitch discrimination loss resulting from a frequency range constraint in European Starlings (*Sturnus vulgaris*). *Journal of Experimental Psychology: Animal Behavior Processes, 12*, 394-40.

Bernard, D. J., & Hulse, S. H. (1992). Transfer of serial stimulus relations by European starlings (*Sturnus vulgaris*): Loudness. *Journal of Experimental Psychology: Animal Behavior Processes, 18*, 323-334.

Hulse, S. H., & Kline, C. (1993). The perception of time relations in auditory tempo discrimination. *Animal Learning and Behavior, 21*, 281-288.

Hulse, S. H., Takeuchi, A. H., & Braaten, R. F. (1993). Perceptual invariances in the comparative psychology of music. *Music Perception, 10*, 151-184.

7 References

Braaten, R. F., & Hulse, S. H. (1991). A songbird, the European starling (*Sturnus vulgaris*), shows perceptual constancy for acoustic spectral structure. *Journal of Comparative Psychology, 105*, 222-231.

Burdick, C. K. (1979). The effect of behavioral paradigm on auditory discrimination learning: A literature review. *Journal of Auditory Research, 19*, 59-82.

Cynx, J. (1993). Auditory frequency generalization and a failure to find octave generalization in a songbird, the European starling (*Sturnus vulgaris*). *Journal of Comparative Psychology, 107*, 140-146.

Njegovan, M., Hilhorst, B., Ferguson, S., & Weisman, R. (1994). A motor-driven feeder for operant training in song birds. *Behavior Research Methods, Instruments, and Computers, 26*, 26-27.

Park, T., Okanoya, K., & Dooling, R. J. (1985). Operant conditioning of small birds for acoustic discrimination. *Journal of Ethology, 3*, 5-9.

Methods in Comparative Psychoacoustics
G. M. Klump, R. J. Dooling, R. R. Fay & W. C. Stebbins (eds)

Uncertainty in the Study of Comparative Perception: A Methodological Challenge

William C. Stebbins

Kresge Hearing Research Institute, Departments of Psychology and Otolaryngology, University of Michigan 1301 E. Ann Street, Ann Arbor, MI 48109-0506, USA

Summary. The senses of animals continually engage a complex and ever changing world. That they are remarkably adapted to this level of complexity is evident, and the lives of their ancestors have depended on it. In their exploration of animal senses scientists have employed the right mix of behavioral methods and highly developed technological instrumentation. We now know and understand much about the sensory acuity of animals and the limits of their resolution of the environment. Unfortunately, we stop short of being able to grasp a rich variety of complex perceptual phenomena that include the sensory illusions, aftereffects, certain forms of perceptual categorization, sensory matching, and perceptual judgment. Our methods are unable to yield precise and exact measurements of these phenomena, and in fact, these methods may interfere with what they purport to measure. For this reason our results are uncertain and even indeterminate. The chapter describes our efforts to break this impasse; they have not been entirely successful.

The methodological advances in the study of animal perception in the last quarter century have been substantial. Examples abound in the contributions to this book. A veritable comparative psychology of sensory perception has incorporated accurate measurements of sensory capacity and even sensory judgment, together with well established relations between sensory acuity and receptor physiology (Dooling and Hulse, 1989; Stebbins, 1970b; Berkley and Stebbins, 1990; Stebbins and Berkley, 1990). While these are significant accomplishments, some formidable challenges remain to the experimenter's ingenuity that may be illustrated in the following questions: Do animals experience sensory illusions or aftereffects? Are they able to make judgments about relative loudness or brightness as humans seem to be able to do? Can they and do they classify and categorize stimuli (an important hallmark of human cognition) and yet discriminate within these categories if the environmental context leads them to do so? The proper asking of these questions is what prompted the uncertainty referred to in the title. Although experiments have been done to answer some of these questions, it is not at all clear that these experiments have always been designed in the appropriate way. In some instances, the experimenters may be leading their animal subjects; in other instances, they may be simply confusing them.

The idea that merely attempting to measure certain phenomena can alter those phenomena in unpredictable ways, thus rendering the results or solution indeterminate, was introduced by the physicist Werner Heisenberg (Heisenberg, 1927). The effect became better known as Heisenberg's "uncertainity principle". Previously animal and human psychophysicists have used a variety of words like judgment, subjective, corrected for the observer, evaluative, categorical, and illusory when describing a class of phenomena that is still poorly understood and resistant to our standard methods for studying sensory perception. Recently, physicists claim to have found a way out of the uncertainity impasse in quantum mechanics, which is the subject to which Heisenberg attached his famous principle (Taubes, 1994). It is unlikely that psychologists and psychophysicsts can benefit directly from this new discovery, but the principles by which the process is undertaken in quantum mechanics may have some generality, as, of course, did the original principle. Certainly, the concept of a measurement taken interfering with the thing measured has a clear application to the analysis of certain perceptual phenomena.

In contrast to sensory threshold measures that usually have a secure and well recognized relationship to the peripheral physiology, these other phenomena often seem more top-down, not always tightly related to what goes on in the receptor but frequently dependent on experience and, likely, on some form of more central involvement. Our own earlier research (Moody et al., 1990) and that of the Heffners (1990) that demonstrate cerebral laterality for complex biologically relevant signals such as species calls in nonhuman primates is but one example. These calls seem to be lateralized to the left temporal lobe of the brain in these animals.

Such perceptual phenomena are no less of a challenge for human psychophysics. The only difference is that human psychophysicists either fail to engage the issues or ignore them as Stevens (1951) did, and he incurred the skepticism of his colleagues for doing so. The visual physiologist Brindley, in his seminal bookon retinal physiology, struggled with the problem but could offer no solution and suggested that if one could not tie the observations to the relevant physiology then, until one could, these observations really should be considered unmentionable (Brindley, 1970). He referred to these psychological observations as Class B as opposed to the usual sensory threshold phenomena which he labeled Class A, which probably tells us more about Brindley's experimental prejudices than anything else. In this regard Brindley considered S.S. Stevens quite beyond the pale - "The extreme liberal opinion has been forcefully expressed by Stevens" (Brindley, 1970 p. 134) - an interesting use of the word "liberal". Brindley struggled to establish what he called psycho-physical linking hypotheses but found the task daunting and claimed little success for himself or anyone else ("Like other theorists, I have not yet succeeded", Brindley, 1970 p. 138). It is of note that he should consider this a theoretical issue, and for the record, some twenty-five years later, he still has not succeeded. Later, the visual physiologists Barlow and Mollon (1982) considered that "Although such judgments can often be made reliably and repeatedly, there is a serious difficulty in interpreting them, for there is a linguistic or verbal element involved in asking people to judge a quality such as brightness in the presence of another quality such as colour that is also variable. This verbal element tends to be even stronger if we ask for a more complex judgment, such as expressing brightness by rating it on a 10-point scale" (Barlow and Mollon, 1982 p. 117). Barlow did not expand on why he found the verbal element difficult to deal with. One wonders what Brindley or Barlow and Mollon would do in raising these questions with animal subjects.

Animal psychophysicists have often sought to determine some form of sensory threshold or discrimination between stimuli. In these more traditional experiments, animal subjects are instructed by the manner in which the reinforcement contingency is established in the design of the experiment. If, for example, the investigator is trying to estimate a tonal threshold, a discrimination protocol is used, and the reinforcer is given contingent upon a response to the tone but not for a response in its absence. The animal is being trained to respond to tones and withhold responses in silence. The conditions are specified in advance of the experiment, and the reasonable assumption is made that the animal will not differentiate between a tone it cannot hear and no tone at all. Upon this simple premise rests much of threshold animal psychophysics. Likewise, for stimulus difference, the subject will respond in the same manner to stimuli that are physically equal as to those that are physically indistinguishable to it.

But what if the situation is illusory? In the Muller-Lyer figure, for example, physically equal is not physically indistinguishable, at least not for human subjects. If the illusion is experienced, the line with the inward pointing arrowheads appears longer than the line with the outward pointing arrowheads when the lengths of the two lines between the arrowheads are, in fact, physically equal. At the perceptual point of equality, the line with the outer pointing arrowheads is longer than the line with the inner pointing arrowheads, and that difference in line length between the two lines is referred to as the extent of the illusion. It is not known if other animals experience the illusion, or, even if they do, whether they perceive it to the same extent that we do. Therefore, the standard discrimination protocol described above cannot be used because it is unclear how to structure the reinforcement contingency. The investigator finds herself in the somewhat problematic position of needing to know the answer before she begins the experiment. Put another way, she cannot, before doing the experiment, state a relationship between behavior and some stimulus value as correct or incorrect - that is, reinforceable or not. But what about using her knowledge of the human perceiver to structure the experiment? That would be a serious error of judgment, and she would end up teaching the animal subject to behave like a human observer.

The illusions like the Muller-Lyer figure are just one set of a rather large class of quite varied perceptual phenomena that includes aftereffects, stimulus matching, categorization, and classification, among others. The set is defined simply as those phenomena about which the subject cannot be questioned directly - that is, the reinforcement contingency cannot be specified in advance of the experiment. Rather than come at the problem head on the investigator has to deceive the animal by coming at it obliquely or indirectly or use some form of factor analytic procedure like multidimensional scaling in the design of the experiment and in the treatment of the data. The phenomena include the judgments that animals make about sensory events, usually well within the limits imposed by their sensory systems as defined by threshold measures. Humans and other animals are seldom operating out at these limits but, far more often, are well within them in their daily interaction with the environment. For this reason they are important processes and worth knowing more about. Many of us have attempted a variety of strategies for getting at the problem (see Stebbins, 1970a,b; Stebbins and Berkley, 1990; Dooling and Hulse, 1989). Some appear to work, but all seem to me to be flawed in one respect or another. In some instances we are almost unavoidably distorting the results by the method we choose, or we are tolerating excessive variability. It is incredible that we have, for the most part, ignored the problem for so long. It is entirely possible that its solution will remain indeterminate and our results uncertain.

Berkley and his colleagues have recently reported on some interesting research on visual aftereffects in monkeys that demonstrates several facets of the problem (Berkley et al., 1992; Pauly et al., 1993). Human subjects perceive the orientation of a contour (such as a grating) to be shifted several degrees from its true orientation if they have been preadapted with a tilted contour of differing orientation from the one they are asked to judge (see Figure 1). Scan the tilted grating at the top of Figure 1 for 15-30 seconds but maintain your gaze within the circle. After adapting, transfer your gaze to one of the gratings below. If you choose the central target, which is true vertical, it should appear tilted to the left. If you choose the right target, which is tilted 2 degrees to the right, it should appear vertical. If you choose the left target, which is tilted 2 degrees to the left, it should appear even more tilted.

Berkley and his colleagues (Berkley et al, 1992, Pauly et al, 1993) found that the monkeys reported the same illusion, although its extent was less than that perceived by the human subjects, as shown in Figure 2. They suspected that the reinforcement procedure was at fault rather than that monkeys differ from humans in their perception of this illusory aftereffect. When the test stimulus was actually vertical, the monkeys were reinforced half of the time regardless of their response (left or right). For the other test stimuli, which were tilted in one direction or the other, subjects were reinforced on half of the trials for "correct" responses. Thus, they were never paid off for making an error. Pauly et al (1993) then proceeded in a subsequent experiment to test humans under comparable conditions using a cash reward and found that the illusion was reduced by 25 to 100 percent. It is an elegant demonstration of the

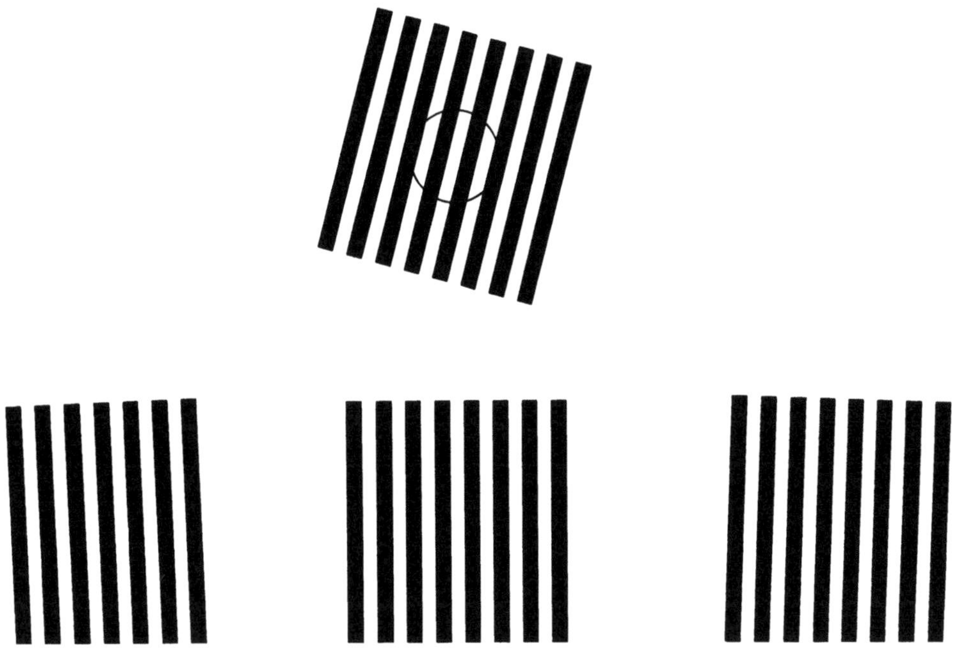

Figure 1. Adaptation grating (top) and test gratings (below) for measurement of the tilt aftereffect (after Berkley et al., 1992).

effects of reinforcement but does not tell us the extent of the illusion for the monkeys nor perhaps for their human counterparts. What is this interaction between reinforcement and the visual system - that is, between the response criterion, i.e. response-dependent mechanisms subject to manipulation via reward and or instructions, on the one hand, and visual perception, on the other? Scott and Milligan (1970) found themselves in a similar predicament in demonstrating the visual motion aftereffect in monkeys with the rotating Archimedes spiral. In fact, they suggested "a principle of psychophysical indeterminacy" (Scott and Milligan, 1970 p. 343) and asked appropriately "Does it (the measurement of the aftereffect) represent an impassable boundary which will always resist the trespass of objective behavioral science?" (Scott and Milligan, 1970 p. 358)

Another interesting methodological approach to the problem is the use of the classical generalization gradient. By training an animal with reinforcement to one stimulus and then

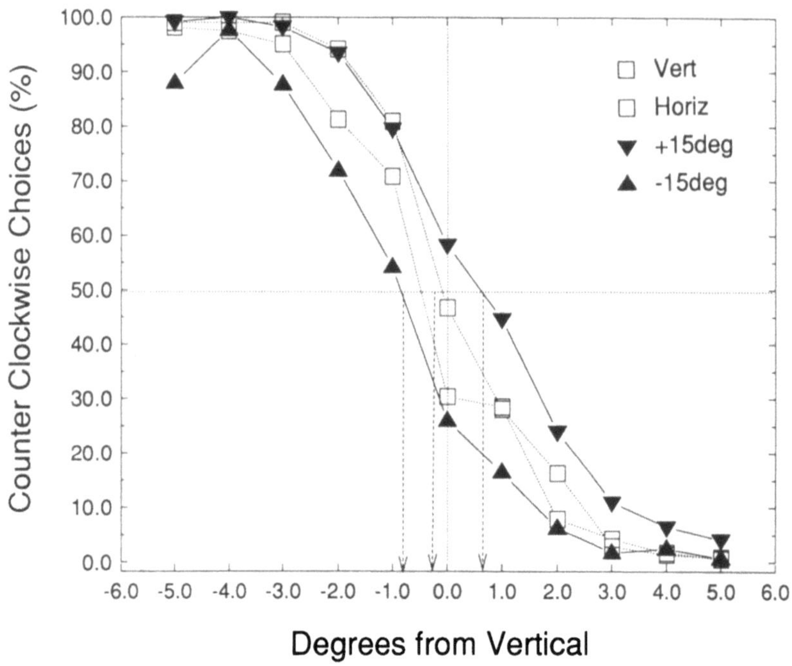

Figure 2. Tilt aftereffect in two monkeys (*M. mulatta*). Perceived tilt to the left (counterclockwise choices) is measured as a function of the angle (degrees from vertical) of the adapting stimulus (after Berkley et al., 1992).

testing it to an array of stimuli on the same dimension as the original stimulus and in the absence of reinforcement, one may be able to gauge the differential salience of these stimuli for the animal by some measure of the magnitude or rate of its response to them without the kind of reinforcement problem that Berkley and his colleagues and Scott and Milligan confronted. Presumably the animal is not being guided or led under these conditions. The Malotts (1970) attempted to use the generalization procedure with the Muller-Lyer illusion in pigeons and were able to demonstrate, at least partially, the operation of an illusion of the sort that humans experience on viewing these arrowheads. The problem with this approach seems to be in the variability encountered both within and between subjects. It is difficult to be precise, for example, in measuring the illusion's extent. Fay (1992) refers, I think, to the same thing when he talks about "the resolution observed in these generalization experiments is probably limited by the methods used rather than the animals' capabilities" (Fay, 1992 p. 105). But how can we know that if we are uncertain in our measurement of those capabilities.

May, Moody, and colleagues (May et al., 1988; Moody et al., 1990) have described a variant of the generalization procedure that we have referred to as the generalization probe, in attempting to measure the perception of species calls by animal subjects. For example, in trying to determine whether animals categorize along a continuum of different call types, we began by training a discrimination between the stimuli (calls) at the two ends of the continuum, reinforcing responses to the call at one end of the continuum and not to the call at the other end. Reinforcement was partial (75-85 percent of the time) because, subsequently and on infrequent occasions we presented intermediate call types without reinforcement. The partial reinforcement schedule presumably prepares the animal for this eventuality. We are trying not to lead the subject by this procedure. but could it be, that by not reinforcing the intermediate calls, we are inadvertently linking them with that call at the end of the continuum that was not reinforced during original training? There is an obvious control that could be run here, but the fact remains that in making the decision to reinforce or not reinforce, we may be influencing the outcome of the experiment in ways that we as yet fail to understand. Herrnstein (1985) faces this same dilemma in his experiments in which pigeons categorize photographs as does Raslear (1985, 1989) in his research with rats on loudness and temporal bisection. "What an animal can do may differ from what it does on any occasion. The distinction is a corollary of the environment's control over psychological processes, as illustrated by the power of reinforcement. An animal that fails to perform in a particular way may indeed lack the necessary capacity or it may just be that the particular test has failed to activate it" (Herrnstein, 1985 p. 141). In doing these experiments we are defining the category by the way in which we set up the reinforcement contingencies. Are we showing that these are natural categories that the animals possess prior to the experiment, or are we revealing that the animals are capable of learning the categories?

Yet another class of related experiments involves sensory scaling and includes such phenomena as brightness bisection and loudness judgment in other animals. Can the "verbal element" that Barlow found so difficult to deal with be avoided? Herrnstein and Van Sommers (1962), in an elegant tour de force, reinforced pigeons for responding on a key at five different specified rates to five distinct light intensities on an intensive continuum. When the behavior was stable, four additional intensities that bisected the interval between successive pairs of the training stimuli at the geometric mean of the interval were introduced without reinforcement. In the test, subjects responded to these intermediate intensities with rates that approximated the geometric means of the rates at each of the two bracketing training stimuli (see Figure 3). It was on the basis of these data that the authors argued, in spite of some variability, that the pigeon's brightness scale was at least not incompatible with a power function, as Stevens had shown earlier for humans (Stevens, 1951). Unfortunately such an experiment demonstrates very clearly the trap we fall into when we overly restrict the subjects in the design of the experiment by the instructions that we give. In all fairness to Herrnstein and Van Sommers, they also recognized the problem.

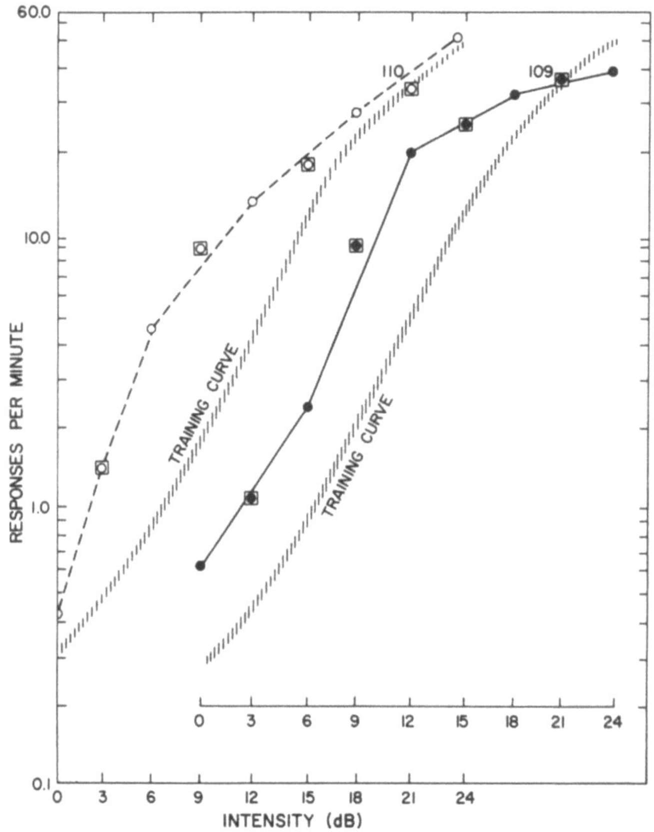

Figure 3. Rate of key pecking as a function of the luminance of the stimulus for two pigeons (Nos. 109 and 110). The training curve shows the prescribed rate of responding at the training stimuli (0, 6, 12, 18, and 24 dB). Points enclosed in squares give rates obtained with test stimuli (3, 9, 15, and 21 dB), in whose presence responding was never reinforced (after Herrnstein and VanSommers, 1962).

Our own earlier research brings up yet another example. but the problem with it is of a different kind altogether. It was our objective to attempt to derive loudness functions from a nonhuman animal. With human subjects, we exploit their familiarity with the language in general and the term "loud" in particular when we ask them to tell us which tones sound equally loud to them (Barlow's "verbal element"). With monkeys, we resorted to an indirect method that used their reaction time to acoustic stimuli as the dependent variable and made the assumption that equal reaction times to different acoustic frequencies represented measures of equal sensory effect. Thus, identical reaction times to pure tones differing in both level and frequency could be said to represent equal loudness to the listener. The data (from Pfingst et al., 1975) were taken from reaction time-intensity functions, and equal loudness functions were thereby constructed (Figure 4). We also contrived a loudness scale by simply inverting the reaction time-intensity function to show the growth of loudness with stimulus level (Stebbins, 1966). In these experiments we were not leading the animal, and our assumptions were defensible. The problem lies in the reaction time data, which are quite variable at lower stimulus levels. This is a problem in other sensory modalities as well (see Moody's (1969) earlier work on brightness in rats, for example).

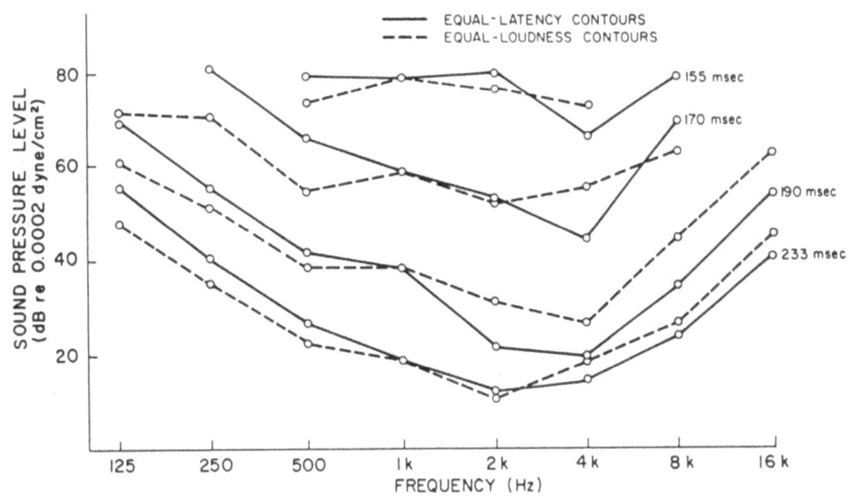

Figure 4. Equal latency contours for two monkeys (*M. mulatta*). Contours were derived from reaction time - stimulus intensity functions (after Pfingst et al., 1975).

Another very different strategy for dealing with the problem comes from the perspective of experimental design and data analysis. It is a type of factor analytic methodology called multidimensional scaling (MDS) and has been taken from human psychophysics (Shephard, 1980). Dooling (1990) and Moody (1993) in animal hearing and Blough (1985) in animal vision have used it in a variety of experimental paradigms. The difficulties with MDS seem more subtle. When an animal is reinforced for responding to stimuli that are different, the experimenter justifies this by pointing out that, in fact, they are physically dissimilar based on some measurable physical difference between them. But suppose that the animal either doesn't hear it that way, or, before being trained, has a different perceptual take on it. Perhaps what the investigator has called different, the animal, prior to the experiment saw or heard as the same or sufficiently similar; however, if the difference is within its sensory capabilities, it could still be trained to respond to that difference. By training an animal to respond to any difference, the experimenter may be breaking down whatever generalization might have occurred. In effect he may be measuring what the animal can do rather than what it does do under any given set of circumstances, or its abilities rather than its proclivities as Prosen et al. (1990) have described it.

Finally, I leave you with what I see as an unsolved problem or set of problems in animal psychophysics. It is, in a sense, an impasse. This is in no way meant to denigrate the research that has been done (some of it our own) and the progress that has been made, but to call attention to what most of us already know but about which we seldom speak. I have described a class of phenomena about which we have learned a great deal with a diverse battery of procedures, but the precise nature and the exact properties of these phenomena have yet to be revealed and continue to elude us.

Acknowledgement

My thanks to my good friend and colleague Mark Berkley for his helpful comments on an earlier draft of this chapter.

References

Barlow, H.B. & Mollon, J.D. (Eds.) (1982). *The Senses*. Cambridge, U.K.: Cambridge University Press.

Berkley, M.A., Vogels, R., Sary, G. & Orban, G.A. (1992). Visual aftereffects in the monkey: The judgement of vertical after adaptation to oblique contours. *Society for Neuroscience Abstracts*, *18*, 587.5.

Berkley, M.A. & Stebbins, W.C. (Eds.) (1990). *Comparative Perception: Basic Mechanisms, Vol. 1*. New York: John Wiley.

Blough, D. S. (1985). Discrimination of letters and random dot patterns by pigeons and humans. *Journal of Experimental Psychology: Animal Behavior Processes*, *11*, 261-280.

Brindley, G. S. (1970). *Physiology of the Retina and Visual Pathways*, 2nd ed., London: Edward Arnold.

Dooling, R.J. & Hulse, S.H. (Eds.) (1989). *The Comparative Psychology of Audition: Perceiving Complex Sounds*. Hillsdale, N.J.: Erlbaum.

Dooling, R. J., Brown, S.D., Park, T.J. & Okanoya, K. (1990). *Natural Perceptual Categories for Vocal Signals in Budgerigars (Melopsittacus undulatus) in Comparative Perception: Vol. 2, Complex Signals*, W.C. Stebbins and M. A. Berkley, Eds. New York: John Wiley.

Fay, R.R. (1992). Analytic listening by the goldfish. *Hearing Research*, *59*, 101-107.

Heffner, H.E. & Heffner, R.S. (1990). *Role of Primary Auditory Cortex in Hearing in Comparative Perception: Vol. 2, Complex Signals*, Stebbins, W.C. and M.A. Berkley, Eds. New York: John Wiley.

Heisenberg, W. (1927). Uber den anschaulichen Inhalt der Quantentheoretischen Kinematik und Mechanik. *Zeitschrift fur Physik*, *43*, 179-198.

Herrnstein, R.J. (1985). *Riddles of Natural Categorization in Animal Intelligence*, L. Weiskrantz, Ed. Oxford: Clarendon.

Herrnstein, R. J. & Van Sommers, P. (1962). Method for sensory scaling in animals. *Science*, *135*, 40-41.

Malott, R.W. & Malott, K.M. (1970). *Perception and Stimulus Generalization in Animal Psychophysics: The Design and Conduct of Sensory Experiments*, W.C. Stebbins, Ed. New York: Appleton-Century-Crofts.

May, B., Moody, D.B. & Stebbins, W.C. (1988). The significant features of Japanese macaque coo sounds. *Animal Behavior*, *36*, 1432-1444.

Moody, D. B. (1969). Equal brightness functions for supra-threshold stimuli in the pigmented rat: A behavioral determination. *Vision Research*, *9*, 1381-1389.

Moody, D.B. (1993). Discrimination of acoustic communication signals by macaque monkeys: evidence for relative spectral amplitude as a cue. *Abstracts of the 16th Midwinter Research Meeting*, Association for Research in Otolaryngology, 66.

Moody, D.B., Stebbins, W.C. & May, B.J. (1990). *Auditory Perception of Communication Signals by Japanese Monkeys in Comparative Perception: Vol. 2, Complex Signals*, W.C. Stebbins and M.A. Berkley, Eds. New York: John Wiley.

Pauly, R.S., Berkley, M.A. & Vogels, R. (1993). Feedback and reward effects on the tilt aftereffect (TAE). *Society for Neuroscience Abstracts*, *19*, 738.5.

Pfingst, B.E., Heinz, R., Kim, J. & Miller, J.M. (1975). Reaction time procedure for measurement of hearing. I. Suprathreshold functions. *Journal of the Acoustical Society of America*, *57*, 421-430.

Prosen, C.A., Moody, D.B. & Stebbins, W.C. (1990). A changing focus of animal perception - from abilities to proclivities. *International Journal of Comparative Psychology*, *4*, 5-23.

Raslear, T.G. (1985). Perceptual bias and response bias in temporal bisection. *Perception and Psychophysics*, *38* (*3*), 261-268.

Raslear, T.G. (1989). Discriminability, loudness, and masking in the rat (*Rattus Norvegicus*): A confirmation and extension. *Journal of Comparative Psychology*, *103* (*3*), 289-296.

Scott, T.R. & Milligan, W.L. (1970). *The Psychophysical Study of Visual Motion Aftereffect Rate in Monkeys in Animal Psychophysics: The Design and Conduct of Sensory Experiments*, W.C. Stebbins, Ed. New York: Appleton-Century-Crofts.

Shepard, R.N. (1980). Multidimensional scaling, tree fitting, and clustering. *Science*, *210*, 390-398.

Stebbins, W.C. (1966). Auditory reaction time and the derivation of equal loudness contours for the monkey. *Journal of the Experimental Analysis of Behavior*, *9*, 135-142.

Stebbins, W.C. (1970a). *Principles of Animal Psychophysics in Animal Psychophysics: The Design and Conduct of Sensory Experiments*. W.C. Stebbins, Ed. New York: Appleton-Century-Crofts.

Stebbins, W.C. (Ed.) (1970b). *Animal Psychophysics: The Design and Conduct of Sensory Experiments*. New York: Appleton-Century-Crofts.

Stebbins, W.C. (1990). *Perception in Animal Behavior in Comparative Perception: Vol. 1, Basic Mechanisms*, M.A. Berkley and W.C. Stebbins, Eds. New York: John Wiley.

Stebbins, W.C. & Berkley, M.A. (Eds.) (1990). *Comparative Perception: Complex Signals, Vol. 2*. New York: John Wiley.

Stevens, S.S. (1951). *Mathematics, Measurement, and Psychophysics in Handbook of Experimental Psychology*. S.S. Stevens, Ed. London: Chapman and Hall.

Taubes, G. (1994). Heisenberg's heirs exploit loopholes in his law. *Science*, *263*, 1376-1377.

Subject index

MIX
Papier aus verantwortungsvollen Quellen
Paper from responsible sources
FSC® C105338

If you have any concerns about our products,
you can contact us on
ProductSafety@springernature.com

In case Publisher is established outside the EU,
the EU authorized representative is:
Springer Nature Customer Service Center GmbH
Europaplatz 3, 69115 Heidelberg, Germany

Printed by Libri Plureos GmbH
in Hamburg, Germany